Masanao Aoki

State Space Modeling of Time Series

Second, Revised and Enlarged Edition

With 13 Figures

Springer-Verlag
Berlin Heidelberg New York
London Paris Tokyo
Hong Kong Barcelona

Prof. Masanao Aoki
Department of Computer Science
University of California
4731 Boelter Hall
Los Angeles, CA 90024-1596
USA

ISBN 3-540-52870-9 2. Auflage brosch. Ausgabe Springer-Verlag Berlin Heidelberg New York
ISBN 0-387-52870-9 2nd Edition soft cover Springer-Verlag New York Berlin Heidelberg

ISBN 3-540-52869-5 2. Auflage geb. Ausgabe Springer-Verlag Berlin Heidelberg New York
ISBN 3-387-52869-5 2nd Edition hard cover Springer-Verlag New York Berlin Heidelberg

ISBN 3-540-17257-2 1. Auflage brosch. Ausgabe Springer-Verlag Berlin Heidelberg New York
ISBN 0-387-17257-2 1st Edition soft cover Springer-Verlag New York Berlin Heidelberg

ISBN 3-540-17256-4 1. Auflage geb. Ausgabe Springer-Verlag Berlin Heidelberg New York
ISBN 0-387-17256-4 1st Edition hard cover Springer-Verlag New York Berlin Heidelberg

Printing: Zechnersche Buchdruckerei GmbH & Co. KG., 6720 Speyer
Bookbinding: G. Schäffer GmbH u. Co. KG., Grünstadt
2142/7130-543210

Preface to the Second Edition

Since the appearance of the first edition many computational experiences have been accumulated by myself, some of my friends, and students in applying the modeling algorithm announced there. These support the claim made in the first edition that the algorithm is robust and computationally efficient. Lack of coverage of statistical properties of the proposed estimators, however, was a source of complaints and criticisms.

In this second edition, we include description of asymptotic properties of the estimators proposed in the first edition, and those of the instrumental-variable estimators, newly introduced in the second edition,which use the state vectors as the vectors of instruments. Normally state vectors are not directly available to be used as instruments. The two decompositions of the Hankel matrix--one by the singular value decomposition, and the other as the product of an observability matrix and a matrix, called Ω matrix in this book and exhibiting the same structure as a reachability matrix--allows us to use them as instruments. In particular in some classes of vector autoregressive (VAR) processes, we show that the estimators of system matrices originally announced in Aoki [1983] are asymptotically as efficient as the improved estimators using the state vector as instruments. We also include in this new edition results on the closed form solutions of the Riccati equation. Here again, VAR processes are special. For them the Riccati equation solution step can be bypassed. In addition to these new results, this second edition includes more extensive descriptions on modeling integrated and nearly-integrated processes.

The chapters devoted to preparatory material have been completely reorganized in this second edition. After two brief introductory chapters, Part I of the book introduces in four chapters material preparatory to Part II, which is the main body of this book. In Chapter 3 we note two ways of decomposing spectral density functions in which two of the system matrices of the state space model naturally appear. One is the matrix, B, needed to express impulse response matrices, and the other is the matrix, M, which relates auto-covariance matrices of the data series to its state space model. In this second edition the relationships of the two key matrices, M and B, are much more clearly drawn than in the first edition. Chapter 4 shows how to put autoregressive moving average models into observable state space forms by undoing a sequence of lag operations used to express ARMA models. This chapter is more streamlined than in the first edition. Chapter 5 introduces Lyapunov equations for the system Gramians, the observability, the reachability, and what we call Ω Gramians. Chapter 6 is one of the key chapters of Part I. Two alternative decomposition schemes of a Hankel matrix are used to introduce the notion of balanced model representation. These alternative decompositions allow us later to characterize the state vector of a forward innovation state space model as asymptotically the most efficient vector of instruments.

Part II (Chapter 7-10) describes innovation models in balanced representation, introduces two methods for model parameter estimation, and presents some results on sensitivity and error analysis. Chapter 7 introduces backward innovation models explicitly in this second edition in addition to the forward innovation models of the first edition, because the former provides a useful interpretation to one of the matrices we later estimate, the matrix M. This chapter shows how the Riccati equation naturally arises as the equation for the state vector covariance matrix of the innovation model. One important finding of this chapter, new in this second edition, is the closed form expression for the solution of the Riccati equation for autoregressive processes. In fact, we can bypass the Riccati equation for these processes. This chapter also includes several identification exercises which are important in policy or decision analytic applications of estimated models. Chapter 8 is a short chapter which considers several alternative ways of introducing state vectors to provide approximate summaries of information contained in the data vectors, and relates the singular values to the canonical correlation coefficients used by Akaike in his modeling algorithm.

Chapters 9 and 10, completely rewritten and much expanded in this second edition, are the main contribution of this book. Chapter 9 presents two algorithms for estimating system matrices in the balanced forward innovation state space model for time series. One type is the same as that presented originally in Aoki [1983] and carried over in Aoki [1987], now called stochastic realization estimator. The other, new in this edition, is an instrumental-variable estimator where the state vector is used as a vector of instruments. This chapter also presents a number of small analytical examples using exact covariance matrices to separate out the statistical errors from other sources of errors in the proposed modeling algorithm. Chapter 10 presents the asymptotic error covariance matrices of the two types of estimators introduced in Chapter 9. Although the instrumental-variable estimator is asymptotically more efficient than the stochastic realization estimator in general, it is shown that for a certain subclass of vector autoregressive processes, the original stochastic realization estimator is as efficient asymptotically as the instrumental- variable estimator.
Part III, consisting of an entirely new single chapter, Chapter 11, is devoted to topics important in modeling time series with unit or near-unit roots. Part IV, also consisting of a single chapter, presents several numerical examples based on macroeconomic time series. Part V comprises some background and supporting material in appendices, and discusses time series models generated by intertemporal optimization and by rational expectations models. Even though Appendices 11 and 12 of this part do not directly deal with model constructions and estimations from data, models of these two chapters are of interest to applied econometricians and macroeconomists.

As usual, a number of people helped me in revising the first edition. A. M. Havenner has as usual been a constant source of inspiration by urging me to new results. Monte Carlo experiments were conducted on the University of Minnesota supercomputer. I thank C. Sims for making my visit to the Institute for Empirical Macroeconomics possible and making the supercomputer available. I also thank him for explaining his views on the identification questions. G. Hall and D. Zha provided able resaerch assistance while I was visiting the Institute. Susanna Reyn and Tin Do typed several versions of rough drafts and I thank them for their help.

I have given a number of talks on topics related to the subject matters of this book at various conferences, workshops and seminars during the last few years. For their invitations, helpful comments, and hospitality, I wish to thank J. Balchen of the University of Trondheim; O. Bjerkhold of the Central Statistical Bureau, Oslo; F. Carlucci, G. Gandolfo, and M. L. Petit of the University of Rome; R. Craine of the University of California, Berkeley; C. W. J. Granger of UC San Diego; W. Wecker of UC Davis; D. Hendry of the University of Oxford; J. de Gooijer of the University of Amsterdam; C. Gourieroux of INSEE, Paris; S. Johansen of the University of Copenhagen; N. Kunimoto and H. Yoshikawa of the University of Tokyo; M. Hatanaka of the University of Osaka; P. Caines of Magill University; P. Perron of Princeton University; S. Mittnik of the State University of New York, Stony Brook; J. Taylor of Stanford University; J.-M. Dufour and E. Ghysels of the University of Montreal; A. Lindquist of the Royal Institute of Technology, Stockholm; L. Ljung of the University of Linköping; A. Maravall, formerly of Bank of Spain, now of the European University Institute, Florence; D. Pena of ETS of Madrid, and D. Sartore of the University of Padova. Comments by M. Deistler of the Technical University, Vienna, are also gratefully acknowledged.

Preface to the First Edition

Many books are now available on data analysis. Some show how to fit data to linear regression or other equations. Others discuss how to separate data into various clusters. Such books are directed primarily at discovering static relations among data.

Books on time series, i.e., those which explicitly deal with time-indexed data, are also numerous, reflecting many advances made in the last four decades or so. Probably the single most influential book is that by Box and Jenkins [1970], which showed how to fit autoregressive moving average models to (differenced) time series data. No book, however, incorporates the advances in the construction and approximation of dynamic systems that have occurred in system theory since the late 1970s. This seems to be particularly true for books on time series of economic origin. Because of fundamental differences in the nature of data generation mechanisms, it is not surprising that litte communication has taken place between econometricians and systems theorists - two important groups of people engaged in time series modeling.

I believe that modeling of (economic) time series (some of the objectives of which are mentioned a little later) can benefit from recent advances in system theory. I have already compiled my lectures delivered at several universities as *Notes on Economic Time Series Analysis* to facilitate communication between the two groups, or more precisely, to bring some new methods for model construction to the attention of econometricians and macroeconomists, (Aoki [1983]). This book builds on *Notes* by expanding theoretical materials and improving the exposition as well as adding more computational examples based on further classroom testing of *Notes*.

Model building is the essence of time series analysis. Models are then used (i) to describe the data behavior succinctly, (ii) to explain behavior of the time series in terms of exogenous variables, (iii) to forecast, and even (iv) to control, e.g., to synthesize macroeconomic policies under rational expectations. Thus, one major objective of this book is to describe and evaluate several model construction methods, yielding markovian or state space models of time series not easily found elsewhere. Two justifications may be offered for treating several time series as a single vector-valued time series. First, such a treatment allows us to understand the dynamic relations between the series. For example, leading indicators exploit dynamic relations that are assumed to exist either explicitly or implicitly between various time series. Second, more accurate forecasting is provided by using joint information contained in vector-valued time series.

A second objective is to assess computational, structural and other properties of models thus constructed, and to compare them with those of more traditional models. For example, our construction method leads to stable Markovian models, while autoregressive models constructed using the traditional methods may not.

How is a state space model constructed?, i.e., how is the dimension of the state vector determined and the model parameters estimated? How good is the model's predictive capability? These are some of the questions that need to be answered in proposing any time series model construction method. This book addresses these questions in Part II. Briefly, the covariance matrices between past data and future realizations of time series are used to build a matrix called the Hankel matrix. Information needed for constructing models is extracted from the Hankel matrix. For example, its numerically determined rank will be the dimension of the state model. Thus the model dimension is determined by the data, after balancing several sources of error for such model construction. The covariance matrix of the model forecasting error vector is determined by solving a certain matrix Riccati equation. This matrix is also the covariance matrix of the innovation process which drives the model in generating model forecasts. In these model construction steps, a particular model representation, here referred to as balanced, is used extensively. This mode of model representation facilitates error analysis, such as assessing the error of using a lower dimensional model than that indicated by the rank of the Hankel matrix.

The well-known Akaike's canonical correlation method for model construction is similar to the one used in this book. There are some important differences, however. Akaike uses the normalized Hankel matrix to extract canonical vectors, while the method used in this book does not normalize the Hankel matrix. Consequently, Akaike's choice of state vector to summarize information contained in the data also differs from the one in this book. What is more his models are generally not balanced.

This book falls naturally into four parts: After a brief explanatory chapter, Part I introduces, in three chapters, material preparatory to Part II, which is the main body of this book. The notion of state is basic to our representation of time series dynamics by Markovian models; hence it is introduced early in Chapter 2. Chapter 3 deals with preparing raw data for analysis as explained in this book, and time domain representation of time series is summarized. Theoretical equivalence of this newer mode of representation of time series by state space or Markovian models to the more traditional one is established in Chapter 4.

Part II describes innovation models in balanced representation, describes a method for model construction, and presents some results on sensitivity and error analysis in Chapters 5 through 10. Part III, consisting of a single chapter, comprises several numerical examples based on macroeconomic time series. Other, smaller examples are found throughout the book. Part IV comprises some background material in appendices, and discusses time series models generated by intertemporal optimization and by rational expectations. Even though Appendices 17 and 18 of this part do not directly deal with model constructions from data, models of these two appendices are of interest to macroeconomists.

Chapter 5 is one of the key chapters of the book because it develops properties of Hankel matrices and balanced innovation models, and for easy reference collects in one place many of the theoretical constructs used in this book. Chapter 6 discusses how to predict time series when their models have a special structure. The important notion of innovation is introduced, and a method for calculating orthogonal projections to produce innovation sequences is explained.

The most general way to generate innovation sequences is provided by Kalman filters as discussed in Chapter 7. Kalman filters can even deal with nonstationary time series which traditional methods have difficulty in handling.

Chapters 8 through 10 embody the new contributions of this book. Chapter 8 considers several alternative ways of introducing state vectors to provide approximate summaries of information contained in the data vectors. For example, Akaike's canonical correlation method is shown to lead to a slightly different state vector specification from ours. These alternative methods for introducing state vectors are shown to be optimal with respect to different optimality criteria. Chapter 9 describes how to compute the system matrices of the innovation models and the innovation vector covariance matrices from the Hankel matrices. Structural sensitivity and error analysis of approximate models are conducted in Chapter 10. Chapter 11 contains several numerical examples drawn from some macroeconomic time series of OECD countries.

The research activities reported here have been partially supported by a grant from the National Science Foundation, program for economics, for which the author is very grateful. Many people helped me in preparing this book. I would like to mention Dr. Genshiro Kitagawa, and Professor Hirotugu Akaike for making TIMSAC programs available to the author. Without the initial help of Professor Arthur Havenner and Mr. Bjorn Ellertsson with computation, the book would never have been started. Professor Havenner has also read most of the manuscript and made many useful comments. Dr. Jorma Rissanen made available to the author some references of which he was unaware. Mr. Brian Young served as a capable research assistant for some of the computations reported in this book, and proofread some of the manuscript.

Contents

1 Introduction

Study of time series has a history much older than modern system theory. Probability theorists, statisticians and econometricians have all contributed to our understanding of time series over the past several decades, as is evidenced by the publication of numerous scholarly books. One may wonder what system theory can add to this well established field and doubt if any new perspective or insight can be gained by this relative newcomer to the field. The history of science shows us, however, that a given problem can fruitfully be examined by different disciplines, partly because when it is viewed from new perspectives, implications of alternative assumptions are explored by researchers with different backgrounds or interests, and partly because new techniques developed elsewhere are imported to explore areas left untouched by the discipline in which the problem originated. Although a latecomer to the field of time series analysis, system theory has introduced a set of viewpoints, concepts, and tools sufficiently different from the traditional ones, and these have proved effective in dealing with vector-valued time-indexed data.

I believe that the new modeling procedure and insights thus obtained into vector-valued time series are of potential interest to econometricians, economists, statisticians, and to researchers in biology, medicine, and social sciences in general. I hope that this book will help them overcome the language barriers and make these results and new tools accessible to all those interested in time series, so that they may benefit from system theory in their own research activities.

What are the new perspectives and results referred to? First, time series are represented differently. Loosely put, traditional time series analysis is primarily directed toward scalar-valued data, and usually represents time series or their differenced version by (scalar) autoregressive, moving average, or autoregressive-moving average models. Modern control and system theory treat several variables simultaneously as vector-valued variables, and have introduced an alternative mode of representing dynamic phenomena, called state space or Markovian representation, by defining internal or state space variables as useful auxiliary variables. This book uses this alternative mode of representing vector-valued time-indexed data which directly connects exogenous variables or innovations of endogenous variables at several time points with endogenous variables also at serveral time points. As shown in this book, these two ways of representing dynamic phenomena are theoretically equivalent. However, their numerical and statistical properties are different. State space representation is decidedly superior in several ways, as is detailed in this book. At any rate, having two alternative ways of dealing with vector-valued time series is definitely worthwhile.

Prompted by this different view of dynamic phenomena, or different means of describing such phenomena, and by the necessity of paying greater attention to the interrelation of vector components, system theory has introduced theoretical notions that are nontrivial only for vector-valued time series, such as reachability, observability, and minimal realization which are not found in the traditional, i.e., scalar-value-oriented time series literature. These notions turn out to be essential in many considerations of modeling of time series by "minimal" dimensional Markovian representations and in examining the robustness of various algorithms for identification. For example, the problem of common factors in the AR portion and *MA* portion of ARMA models is exactly that of minimal realization of given time series by Markovian models.

Because the mathematical tools available for time series analysis basically require at least covariance-stationarity--i.e., first and second moments of the time series must exist for all time and they must be invariant to shifts of the time origin--data used to be routinely transformed to fit this requirement of covariance stationarity by removing deterministic functions of time from the data or by taking differences. A more recent approach to nonstationarity is to fit explicitly ARMA models in which the AR parts contain unit or near-unit roots which are shared by several components of time series. In state space representations such common trend representation is easy to implement, using the procedure of this book. Error-correction models used by econometricians turn out to be a "dual" representation of this common trend representation.

This book focuses on a few aspects of the analysis and modeling of time series that are primarily system theoretic in origin or in orientation in order to provide some new perspectives and analytical techniques. I have chosen Hankel matrices as a unifying theme in the treatment of prediction, (approximate) representation of data by (lower-order) state space models, and examination of identification and identifiability conditions from system viewpoint. Singular value decomposition of certain Hankel matrices followed by suitable scaling produces so-called balanced state space models of vector-valued time series, which are the main topic of this book. There are some important links between the estimators based on the system theory and the instruemntal-variable estimators extensively used in econometrics. I have included some materials on these links, with the hope that they may make the method more easily accessible to those trained in econometrics. I leave many important statistical considerations untouched, because there are many excellent books on statistical aspects of time series modeling.

Unlike the computationally-demanding maximum likelihood method, the model-building procedure of this book is very easy to carry out computationally. It is quick, it produces "nested" estimates of system matrices which are consistent even when the model size (the dimension of the state vector) is misspecified, it is guaranteed to produce asymptotically stable models when certain technical conditions are met, and it produces models with good predictive performance, especially when only a few dynamic factors are involved in generating data. Because of the ease of model building by this method, the method may be used in other ways, for example, as a way of generating initial guesses to be further improved by the maximum likelihood method.

2 The Notion of State

System theory has shown that the behavior of dynamic systems can be conveniently and succinctly described by introducing the notions of state space and state vectors. Some time series behavior may be so complex that it can't be described by a finite number of parameters, i.e., its description may go beyond the framework of finite dimensional dynamic models. When a finite dimensional state space model does not suffice to capture time series behavior, we attempt to approximate the series by another of lesser complexity which admits a finite dimensional characterization. In the transform domain description, spectral density functions of finite dimensional dynamics are rational functions of frequencies. They are used to approximate irrational spectral density functions of infinite dimensional dynamics.

This book examines only those time series models with associated finite dimensional dynamic structures, i.e., models with rational spectral densities and rational transfer functions. These models require only finite-dimensional state vectors to represent their dynamics. Economic time series may not possess rational spectral densities. If so, their approximate descriptions by finite-dimensional dynamics are sought in this book.

Although the notion of state vector is essential in our contruction of innovation models for time series, this concept does not enjoy widespread use in time series modeling. Models in state space representation are found less often in the literature than those in the popular ARMA type representation. (For example, Harvey [1981] seems to have written the only book on time series which mentions state space. He devotes only a chapter to this topic, however.) This chapter provides a brief introduction to the concept of the state vector in time series modeling in particular and in dynamic systems in general.

Time series may be viewed as being generated by systems which transform information in present and past exogenous signals into future observations. States are summaries of the information contained in the exogenous signal that is transmitted by the dynamics generating the time series. The state vector of a deterministic dynamic system is a minimal collection of information necessary to determine uniquely the future evolution of the dynamic system, given future time paths of all relevant exogenous variables affecting the system, including decision or control variables. For example, if $\{z_t\}$ is generated by a (vector) difference equation, $z_{t+1} = f(z_t, x_t)$, where $\{x_t\}$ is a sequence of exogenous variables, the vector z_t is a state vector of the system, since all future z's are well defined given z_t and the time path for the exogenous variables. This equation is an example of state (space) representation of $\{z_t\}$.

Suppose next that dynamic equations also depend on some predetermined variables such as lagged endogenous and exogenous variables. An example is given by

$$z_{t+1} = f(z_t, z_{t-1}, x_t, x_{t-1}, t), \quad t=1, 2, \cdots$$

where z_1 is also given, together with z_0. The time variable t in $f(\cdot)$ indicates time-varying dynamics. For this system knowledge of z_t and of x_s, $s \geq t$ is not sufficient to determine the future values of the z's. The values of z_{t-1} and x_{t-1} must also be known; the vector z_t is not a state vector of this dynamic system. If more information is added, and a vector $s_t{}' = (z_t{}', y_t{}', w_t{}')$ is defined, where $y_t = z_{t-1}$ and $w_t = x_{t-1}$, then the dynamic equation is expressible as

$$s_{t+1} = F(s_t, x_t, t)$$

where the function F is defined by a vector formed by stacking $f(z_t, y_t, w_t, x_t, t)$ above z_t and x_t. Now the vector s_t qualifies as a state vector for this dynamic system and $F(\cdot, \cdot, \cdot)$ is its state (space) equation. Chapter 4 describes several systematic methods for expressing linear dynamic models in state space form. The above brief discussion shows that nonlinear dynamic systems can also be converted into state space form.

When stochastic processes are involved, one must properly reinterpret the phrase "uniquely determine" used in describing the notion of state. With stochastic processes, specifying probability laws that govern time evolution of dynamic processes is the best one can do in describing the future time paths of the processes. Only in special cases can probability laws be uniquely specified by a few sufficient statistics. In those special cases, however, sufficient statistics can serve as a finite-dimensional summary of the stochastic processes. Otherwise, "state vectors" would be of infinite dimensions, because probability distributions themselves are the state vectors. Even when no sufficient statistics are available, it is useful to have approximate, finite-dimensional summaries of the past realizations of time series. Aoki [1989, Chapter 9] presents some examples for using approximate sufficient statistics in decision making. Similar approximations are also possible for forecasting time series. Such representation of time series by finite-dimensional state-space models would summarize information contained in past data approximately, and would predict future realizations with some error but adequately over a short time span. Later this book describes three types of such approximate summaries of information contained in time series data sets.

State vectors in some guise have actually been used by economists from time to time. Errors-in-variables models, (dynamic) index models, and (dynamic) factor models are all found in the economics literature. See Geweke [1977], Nerlove et al. [1979], or Litterman and Sargent [1979] for some examples. Time series models in state space representation may be thought of as unobserved component models or generalized dynamic factor models. Nerlove et al. discuss unobserved component models including errors-in-variables models [Chapter IV, 1979]. Although they do not explicitly mention state space, they do come quite close to the idea of state space. Changes over time in unobserved variables--i.e., state vectors--are governed by dynamic or state transition equations which are stochastic because of the presence of additive disturbances.

Akaike [1975, 1976] introduced state vectors as vectors of canonical variates between the data and future observations in building state space models for time series. We describe his method in Chapter 8. His construction may be regarded as a generalization of the similar notion for two vectors in factor models. It is a genuine generalization because, unlike static factor models, dynamic factors can't be a mixture of past and unobserved future realization of the data vectors. The direction of time is important.

The notion of common factors in dynamic models has been made more precise by Lindquist and Picci [1979] in the Hilbert space context. In particular they adapted the notion of the minimal splitting subspaces to introduce state vectors in a Markovian representation. Consider two Hilbert spaces, one spanned by the vectors y_s , $s \geq t$ and the other by y_s , $s \leq t$. Denote them by $\mathcal{H}^+$ and $\mathcal{H}^-$. A subspace is called a splitting subspace if $\mathcal{H}^+$ and $\mathcal{H}^-$ are conditionally orthogonal given this subspace--i.e., if the residuals of the orthogonal projection of $\mathcal{H}^+$ onto X and those of $\mathcal{H}^-$ onto X are orthogonal. Such a subspace with the smallest dimension is a minimal splitting subspace. Stated differently, a minimal Markovian splitting subspace X is a subspace of smallest dimension containing all the information from the past needed to predict the future; i.e., the orthogonal projection of any element of $\mathcal{H}^+$ onto the subspace spanned by $\mathcal{H}^-$ and X is the same as that onto the subspace X alone. This subspace is the space of state vectors we discuss in this book. This formulation turns out exactly the dynamic counterpart of common factors in factor models. In other words, the state vectors of state space models belong to a minimal splitting space and serve as common factors which link the past to the future evolution of the series. The number of common factors is equal to the dimension of the state vector. This book provides an explicit coordinate in which a state vector is introduced for a minimal state space innovation representation of a time series model.

The next two examples illustrate this aspect as well as indicating a way of representing random growth in time series by state space models. Chapters 3 and 4 provide more information on the nature of state vectors and how they are related to arrays of lagged endogenous variables in the more familiar ARMA or ARIMA representations of time series models.

Random Trend (Harvey [1984]). Consider a scalar variable y_t which is known or hypothesized to have a random trend and a residual term specified by

$$y_t = \mu_t + u_t$$

where μ_t evolves with time according to

$$\mu_t = \rho\mu_{t-1} + \beta_{t-1} + \eta_t$$

and where β is the growth rate of the trend with random disturbances

$$\beta_t = \gamma\beta_{t-1} + \zeta_t \ .$$

The term μ_t can be thought of as a random trend because it reduces to a constant plus a linear term in time when the additive noises are absent. The dynamics of this model are given by

$$x_t = Ax_{t-1} + n_t$$

where

$$A = \begin{bmatrix} 1 & \rho \\ 0 & \gamma \end{bmatrix}, \quad x_t = \begin{bmatrix} \mu_t \\ \beta_t \end{bmatrix}, \quad \text{and } n_t = \begin{bmatrix} \eta_t \\ \zeta_t \end{bmatrix}.$$

Since neither μ_t nor β_t is directly observable, the state vector x_t is not directly observed. It is observed only through the expression

$$y_t = [1\ 0]\, x_t + u_t ,$$

which is the observation equation of this state space model. The notion of random trends or growth rates is further developed in Chapters 3, 4, and 11.

A Regression Model. A regression model

$$y_t = \sum_{i=1}^{m} \theta_i x_i(t) + \eta_t ,$$

in which the regression coefficients are not constant but change according to

$$\theta_i(t+1) = \alpha_i \theta_i(t) + \xi_t , \quad i=1, \ldots, m ,$$

can be put into state space representation by defining a state vector $z_t = (\theta_1(t), \cdots, \theta_m(t))'$. The state transition equation becomes

$$z_{t+1} = A\, z_t + w_t ,$$

where

$$A = diag\ (\alpha_1, \ldots, \alpha_m), \text{ and } w_t = (\xi_1(t), \ldots, \xi_m(t))',$$

and the observation equation is specified by

$$y_t = x_t'\, z_t + \eta_t .$$

Alternatively, starting from the same regression equation, suppose that $x_i(t)$ is observed only through noise

$$z_i(t) = x_i(t) + e_i(t) ,$$

and that $x_i(t)$ evolves with time according to

$$x_i(t+1) = \phi_i x_i(t) + \zeta_i(t) .$$

Then, the regression equation can be put as

$$z_{t+1} = \Phi z_t + w_t ,$$

and

$$y_t = (\theta_1, ..., \theta_m) z_t + v_t ,$$

where $v_t = \eta_t - \sum_i \theta_i e_i(t)$, and $\Phi = diag\ (\phi_1, \ldots, \phi_m)$.

Example. Seasonal components can likewise be included in state space models. For example, monthly seasonal adjustment components s_{t-i} satisfy the relation

$$\sum_{i=0}^{11} s_{t-i} = \eta_t$$

where η_t represents some weakly stationary uncorrelated exogenous random disturbance to account for deviations from a theoretical relation. These components may be represented in the state space form as

$$\theta_t = \Theta \theta_{t-1} + h\ \eta_t$$

where

$$\theta'_t = [\ s_t\ s_{t-1}\ \cdots\ s_{t-10}\ ],\ \text{and}\ h' = [1\ 0\ 0\ \cdots\ 0],$$

and the matrix Θ has -1 throughout the first row, the remainder being the 10-dimensional identity matrix. This vector θ_t can be added as another direct sum component to form the complete state vector. This procedure is described more fully in Kitagawa and Gersch [1984]. Kitagawa has also developed a computer program, TIMSAC-decomposition, to take a scalar time series and extract a trend of specified order and seasonal components. Models of this type are more general than the one used by Hodrick and Prescott [1981].

3 Data Generating Processes

Before we attempt to build models for some data series, we need to describe the scope of coverage, i.e., the class of data generating processes for which we design model-building algorithms. First, we assume that data are generated by linear processes. Second, throughout most of the book, we further assume that the processes are at least covariance (or weakly) stationary. We assume, that is, that the mean of the process is a constant not varying with time, and that the covariance matrices are functions of the differences of the two time instants and not of these two instants themselves. Later, when we discuss series with apparent trends, we drop this constant-mean assumption. Third, the spectral density matrix of the processes are rational--i.e., the elements of the matrix are ratios of finite order polynomials. This last assumption means that we can use a finite number of parameters to describe the process, such as finite-order *ARMA* processes or finite-dimensional state space models which are discussed in Chapter 4.

Given the scope of the processes just described, several alternative schemes of parametrization of the class of the processes are available to us. Traditional ones are in terms of autoregressive (AR), moving-average (MA) or mixed autoregressive moving-average (ARMA) processes, or their extensions to differenced series or to series with exogenous variables (variously abbreviated as ARIMA, ARX or ARMAX, where the letter X denotes the presence of some exogenous processes in the model). See Hannan and Deistler [1989] or Ljung [1987] for these and other related models. Among these, vector-valued AR models (VAR) are used most extensively, because they are particularly easy to estimate. These are (vector-valued) discrete-time stochastic processes generated by finite-order (constant-coefficient) difference equations for vector-valued processes, sometimes containing explicit input or decision variables in addition to noise processes. Alternative schemes of parametrization for the same class of processes are available in terms of state space models. In the state space approach to time series modeling, time series are described with the help of auxiliary variables, called state variables. They are linearly related to the observed processes, possibly through noises. No matter how complicated the processes, state space models contain state variables at most at single lag. This representation of dynamics by the single lag leads to a uniform treatment of time series and facilitates constructions of algorithms for processing data.

Theoretically state space models are equivalent to the traditional parametrization of time series by ARMA type models, because generic models in one representation can be transformed into the generic ones in the other. This is the topic of the next two chapters. When judged on other grounds--such as numerical stability, sensitivity with respect to small specification errors, statistical properties of parameter estimators, or simply ease of dealing with vector-valued or nonstationary time series represented in these alternative modes--they are not equivalent. Furthermore, to avail oneself of theoretical results and computational algorithms that have been developed over the years in systems literature, time series must be put in this newer state space representation. Kalman filters, for been developed for dynamic models in state space form. It would be cumbersome to write Kalman filters for vector-valued ARMA models. Calculations of dynamic multiplier or impulse response analysis, which is one of the reasons for building models in the first place, is also easier to perform in state space representation.

3.1 Statistical Data Descriptions

Weakly stationary mean zero data streams are described by their joint probability distribution functions. In most cases the joint distribution functions are not known, and instead, data sets are statistically characterized by their (sample) first and second moments--i.e., by their means and auto-covariance matrices at all lags. Alternatively their spectral density matrix, which is the generating function for the covariances, may be used in their characterization. We denote the theoretical covariance matrices of a mean zero process by

$$\Lambda_k = E(y_t y_{t-k}') , \quad k = 0, 1, \cdots .$$

The covariance matrices with negative indices are the transpose of the corresponding ones with positive indices because

$$\Lambda_{-k} = E(y_t y_{t+k}') = E(y_{t+k} y_t')' . \tag{1}$$

In applications, these theoretical matrices must be estimated by sample ones. See Appendix 5 on the way to calculate them. Statistical properties of the estimated covariance matrices are discussed, for example, in Anderson [1971, Chapter 8] and in Porat and Friedlander [1985]. See Chapter 10 of this book also. Arato [1961] and Anderson [1963] have shown that only AR processes have sufficient statistics which are smaller in number than the number of data points. Covariance matrices are part of the sufficient statistics for AR processes. We return to this in Chapter 7 in discussing the likelihood functions for these processes.

By our assumptions on the data-generating processes, the covariance matrices are not all independent, but rather admit parametrization by a finite set of parameters. Eq.(3) below presents one such example in state space parametrization. In the traditional $AR(p)$ parametrization, the so-called Yule-Walker equation constrains the set of covariance matrices. Suppose that $y_t = \sum_1^p \phi_i y_{t-i} + u_t$. Multiplying from the right by the transpose of y_s, $s = 1, 2, \cdots$ and taking expectations, the covariance matrices must satisfy

$$\Lambda_0 = \sum_j \phi_j \Lambda_{-j} + \Delta , \quad \text{and} \quad \Lambda_i = \sum_j \phi_j \Lambda_{i-j} , \quad j=1,\ldots,$$

As we discuss in detail in Chapters 4 and 5, data-generating processes can be represented by a (vector-valued) first-order difference equation for a mean zero vector z_t, called the state vector, which is driven by another mean zero vector e_t, called the innovation vector and represented by the equation

$$z_{t+1} = Az_t + Be_t \ , \text{and} \ \ y_t = Cz_t + e_t \ , \tag{2}$$

for some matrices A, B, and C. The state and innovation vectors are mutually uncorrelated weakly stationary processes, and e's are serially uncorrelated as well. In particular y_t is uncorrelated with e_{t+s} for $s = 1, 2, \cdots$. Therefore, we can represent the covariance matrices as

$$\Lambda_k = CA^{k-1} E(z_{t+1}y_t') = CA^{k-1}M \ , \tag{3}$$

by introducing an important matrix M by

$$M = E(z_{t+1}y_t') = A\Pi C' + B\Delta \ , \tag{4}$$

where we denote $cov(z_t)$ and $cov(e_t)$ by Π and Δ respectively. They are constant matrices, because the processes are weakly stationary. They satisfy the next two algebraic equations,

$$\Pi = A\Pi A' + B\Delta B' \ , \tag{5}$$

and

$$\Lambda_0 = C\Pi C' + \Delta \ . \tag{6}$$

When the noise covariance matrix Δ is eliminated from (5) using (4) and (6), the resulting equation is known as the Riccati equation for the covariance matrix of the state vector, Π.

3.2 Spectral Factorization

Much of stochastic realization theory in the systems litrature is based on the model represented as (2). The method for estimating the matrices in (2) using a sequence of sample covariance matrices of the time series is explained in Chapter 9.

Define the generating function of the theoretical covariances by a formal power series

$$S_{y,y}(q) = \sum_{-\infty}^{\infty} \Lambda_k q^{-k} \ , \tag{7}$$

which is also called the spectral density matrix of the y-process, when $e^{i\omega}$ is substituted for q. The variable q is merely a place marker in this generating function. For example, given $S_{y,y}(q)$, to retrieve Λ_3, say, we merely look for the coefficient matrix multiplying q^{-3}, directly or for example by evaluating the contour integral in the complex plane

$$\Lambda_3 = \frac{1}{2\pi i} \int S_{y,y}(q) q^2 dq \ ,$$

where q is treated as a complex variable for the analytic function $S_{y,y}(q)$. Appendix 10 collects more facts on the spectrum.

Noting (1), the terms in (7) can be grouped into the sum of two factors

$$S_{y,y}(q) = T(q) + T'(q^{-1}), \tag{8}$$

where

$$T(q) = \Lambda_0/2 + \sum_1^{\infty} \Lambda_k q^{-k} = \Lambda_0/2 + C(qI - A)^{-1}M$$

after (3) is used to substitute out the Λ_k's. Here we use the formal expansion of $(I - X)^{-1}$ into an infinite series $I + X + X^2 + \cdots$.

On the other hand, we know from the general relation that the spectral density matrix of the output process $\{y_t\}$ is related to that of the input process $\{e_t\}$, which is just Δ, by the transfer function of the dynamics as

$$S_{y,y}(q) = W(q)\Delta W'(q^{-1}), \tag{9}$$

where $W(q)$ is the transfer matrix of the state space model (2)--i.e.,

$$W(q) = I + C(qI - A)^{-1}B\ . \tag{10}$$

To see directly that (8) and (9) are equal, substitute Δ out by $\Lambda_0 - C\Pi C'$, $B\Delta$ by $M - A\Pi C'$, and $B\Delta B'$ by $\Pi - A\Pi A'$, using (6), (4), and (5) after forming the products in (9). Then the individual terms on the right-hand side of (9) can be grouped into

$$\Lambda_0 + C(qI - A)^{-1}M + [C(q^{-1}I - A)^{-1}M]' + C(qI - A)^{-1}X(q^{-1}I - A')^{-1}C'\ ,$$

in which the expression X can be easily proved to vanish, thus establishing the equality of the two alternative expressions given in (8) and (9).

The expression of (9) is known as the spectral factorization theorem. By construction the transfer matrix is real, rational, and stable (i.e., has poles inside the unit disc), and has no zeros outside the unit disc. This last fact is known as the minimum phase property of transfer matrices, a topic to be discussed further at the end of this chapter. More formally, we have

Theorem. Let $S(q)$ be a real, rational, full-rank covariance-generating matrix. Then, it can be factored as $S(q) = W(q)\Delta W'(q^{-1})$, where $W(q)$ is real, rational, stable and of minimal phase, and Δ is symmetric and positive definite.

Appendix 10 summarizes the main steps of this derivation.

Consequently, the inverse $W^{-1}(q)$ is analytic in $|q|>1$, and hence has the Taylor series expansion $W^{-1}(q) = \sum_0^{\infty} c_k q^{-k}$ in $|q| > 1$. It is the z-transform of a stable causal (one-sided) dynamic system. It is called the "whitening filter", because it produces white noise from $\{y_t\}$ by $W^{-1}(q)y_t$. Actually, the expression given in (10) is the transfer function of the model, given in (1) of Chapter 4, which we later construct for the time series.

Note that the matrix M defined in (4) appears in the sum decomposition of the spectral density matrix and the matrix B (2) appears in the product decomposition of the same. The matrix M is needed to express the covariance matrices compactly as shown in (3), and from (2) the impulse response matrices are expressed compactly as

$$H_k = CA^{k-1}B, \quad k=1, 2, \cdots, \tag{11}$$

where the matrix Hs appear in the expansion of the transfer function representation of (2) as

$$y_t = [I + LC(I - AL)^{-1}B]e_t$$

$$= [I + LH_1 + L^2H_2 + \cdots]e_t .$$

In this sense the matrices M and B play a similar role in representing the covariance matrices and the impulse response matrices. The similarity of these two matrices do not end here. Later in Chapter 5 we show the parallel roles of these two matrices in the decompostion of two Hankel matrices related to the same data processes.

3.3 Decomposition of Time Series

Dynamic Modes

Time series usually contain both slowly-changing and rapidly-changing components which may be visible by casual inspection of the time series plots against time. According to Kendall [1976], general time series are often viewed as a mixture of four components: a trend, a seasonal component, fluctuations about the trend of greater or lesser regularity, and residual, irregular, or random effects. What is slow and what is fast are of course relative to the time span of reference, and we can't give any definitive guide as to how to classify various frequency components in time series. A trend is a long-term smooth and monotonic movement relative to the time span of the data set. Over a longer time span a trend may turn out to be a (short) segment of an oscillatory or cyclical component. For example, a steady increase in temperature in the morning may be a slow rise in temperature if time span is a matter of minutes or several hours. The same change in temperature constitutes a rapid fluctuation if yearly changes or temperature changes over several centuries are of primary interest.

With respect to a given time span or frame of reference, it is usually possible for us to decide which components may be regarded as slow changes and which part of rapid variations. Trends arising from random walk components, which require separate handling, are discussed in Chapter 11. It is convenient to have a scheme for selectively modeling components of time series of specified frequency ranges, such as slow or fast components. Slowly-varying components are sometimes called (random) "trend" components, whereas those that change more rapidly are called cyclical variations or residuals. Time series may not possess any well-defined frequency components, because they may be randomly changing with time. Still, the notions of fast and slow frequency components may be of practical use.

We will use the notion of dynamic modes of linear time-invariant deterministic dynamic systems and classify modes into fast and slow ones and model them separately as submodels which are aggregations of the total model. Dynamic modes decompose behavior of time-invariant linear dynamics into component responses. If some specific models for trends or shorter-run dynamic models are suggested by theory, then such a model-based decomposition is preferable to a purely phenomenological one such as the one by dynamic modes which is entirely data-based. It is an exception rather than a rule, however, that a well specified dynamic model is available to perform the decompostion in economic time series. An example is Yoshikawa [1990] which proposes a theoretical model for longer-run movements in exchange rates. This is an exception in economic time serie modeling. Each mode corresponds to an eigenvalue of a matrix which determines how a current state is moved with time. Consider

$$x_{t+1} = Ax_t$$

where A is a constant matrix. For simplicity, suppose that the matrix A has linearly independent eigenvectors

$$Au_i = \lambda_i u_i \quad , \quad i = 1 \cdots n \quad , \quad u_i' u_i = 1$$

where $n = dim\, x_t$.
Then using u_i , $i = 1 \cdots n$ as a basis vector, x_t is expressible as

$$x_t = \Sigma \xi_i(t) u_i$$

where $\xi_1(t) \cdots \xi_n(t)$ are the coordinates. This is called a modal representation, because each $\xi_i(t)$ represents a distinct behavior pattern due to a single eigenvalue, called modal response.

$$\begin{aligned} \xi_i(t+1) &= \lambda_i \xi_i(t) \\ &= \lambda_i^t \xi_i(0) \quad , \quad i = 1 \cdots n \ . \end{aligned}$$

The initial vector x_0 contains the i-th mode if $\xi_i(0)$ is not zero, which is determined by

$$\xi_i(0) = v_i' x_0 \, ,$$

where

$$[v_1 \cdots v_n] = [u_1 \cdots u_n]^{-1} \ .$$

By writing powers of the matrix A^t as $A^t = \Sigma \lambda_i^t u_i v_i'$, the expression $x_t = A^t x_0 = \Sigma \lambda_i^t \xi_i(0) u_i$ shows that x_t is composed of the sum of individual modal responses. If $\xi_j(0)$ is zero for some j, then that mode is absent from $x(t)$ for all

t. Suppose that the data vector is related to x_t by $y_t = Cx_t + u_t$. Since the solution of $x_{t+1} = Ax_t + Bu_t$ is $x_t = A^t x_0 + \sum_{s=0}^{t-1} A^{t-1-s} Bu_s$, we can decompose y_t into dynamic modes even when there is an exogenous signal, because

$$A^{t-1-s} Bu_s = \Sigma \lambda_i^{t-1-s} (v_i' Be_s) u_i \quad .$$

For example, if the largest eigenvalue is real and if we want to separate out the effects of this eigenvalue from the rest, we can write

$$y_t = g_1(t) Cu_1 + \sum_{i=2}^{n} g_i(t) Cu_i \ ,$$

where $g_i(t) = \lambda_i^t \xi_i(0) + \sum_{s=0}^{t-1-s} \lambda_i^{t-1-s} v_i' Be_s$, $i = 1, \cdots n$.

Two Aggregation Schemes

A convenient way to separate modes into slow and fast ones is to classify them by the magnitudes of the eigenvalues. The larger the magnitude, the more persistent is the response.

Pick a positive number $\rho < 1$ and group eigenvalues into two mutually-exclusive categories

$$C_1 = \{\lambda_i : |\lambda_i| \geq \rho\}$$

and

$$C_2 = \{\lambda_i : |\lambda_i| < \rho\} \quad .$$

Relative to the time span (which dictates a choice of ρ value), those eigenvalues in C_1 represent slow modes and those in C_2 fast modes. Let k be the number of eigenvalues in C_1 counting multiplicities. An extreme case of the decomposition obtains when ρ is chosen to be 1. Then C_1 contains only unit roots, i.e., eigenvalues of unit magnitude such as those representing random walks, or pure cyclical components representing seasonal components.

Collect the basis column vectors of the right invariant subspace of A associated with the eigenvalues in C into an $n \times k$ matrix P normalized by $P'P = I_k$. We have

$$AP = P\Lambda$$

where

$$\Lambda : k \times k \quad .$$

Next collect the basis vector for the left invariant subspace of the eigenvalues in C_2 as $Q' : (n-k) \times n$

$$Q'A = \Gamma Q'$$

$$Q'Q = I_{n-k} \quad .$$

By construction the matrices Λ and Γ have no eigenvalues in common. Therefore the right and left invariant subspaces are orthogonal (see Householder [1964, p. 169] for example)

$$Q'P = 0 \quad .$$

Select a new coordinate with the basis $[P, Q]$, i.e.,

$$x_t = P\xi_t + Q\eta_t \ ,$$

where ξ_t and η_t are k and $(n-k)$- dimensional (sub) state vectors.

Then we can construct two smaller models for each of the two classes of eigenvalues. Advance time by one in the above representation

$$x_{t+1} = P\xi_{t+1} + Q\eta_{t+1}$$

$$= A(P\xi_t + Q\eta_t) + Be_t \quad ,$$

and multiply it from the left by P' to obtain a model for slow modes

$$\xi_{t+1} = P'AP\xi_t + P'AQ\eta_t + P'Be_t$$

$$= \Lambda\xi_t + P'AQ\eta_t + P'Be_t \quad .$$

The model for fast modes is derived by multiplying the same equation from the left by Q

$$\eta_{t+1} = Q'AP\xi_t + Q'AQ\eta_t + Q'Be_t$$

$$= \Gamma y_t + Q'Be_t \quad .$$

In this representation fast dynamic modes affect the slow ones

$$\begin{bmatrix} \xi_{t+1} \\ \eta_{t+1} \end{bmatrix} = \begin{bmatrix} \Lambda & P'AQ \\ 0 & \Gamma \end{bmatrix} \begin{bmatrix} \xi_t \\ \eta_t \end{bmatrix} + \begin{bmatrix} P'B \\ Q'B \end{bmatrix} e_t \quad . \tag{12}$$

This model reduces to the common trend representation of Stock and Watson (1988) if we take p to be one, because C_1 contains then only unit roots. This representation is a common slow-mode representation, since in the observation equation

$$y_t = Cx_t + e_t$$

$$= CP\xi_t + CQ\eta_t + e_t \ .$$

Here, the $p \times k$ matrix CP shows how ξ_t is shared by the components of η_t.

By reversing the association of the invariant subspaces, i.e, the left invariant subspace with the eigenvalues in C_1 and the right invariant subspace with C_2

$$S'A = \Theta S' \ , \ S'S = I_k$$

$$AR = R\Omega \ , \ R'R = I_{n-k}$$

and using $[S, R]$ as a basis

$$x_t = S\tau_t + R\zeta_t \ ,$$

another representation of the same dynamics is obtained. Here the orthogonality condition is $R'S = 0$ and the joint dynamic model has the representation

$$\begin{bmatrix} \tau_{t+1} \\ \zeta_{t+1} \end{bmatrix} = \begin{bmatrix} \Theta & 0 \\ R'AS & \Omega \end{bmatrix} \begin{bmatrix} \tau_t \\ \zeta_t \end{bmatrix} + \begin{bmatrix} S'B \\ R'B \end{bmatrix} e_t \ . \tag{13}$$

In this representation the slow modes act on the fast modes. By interpreting $\tau = 0$ as the long-run equilibrium relation, this representation is an error-correction model in the sense of Sargan [1964] because deviation from the equilibrium affects short-run (adjustment) dynamics.

We next present two simple examples which may be considered to be model-based decompositions in which some a priori considerations impose particular forms of trend or seasonal variational components of time series.

Example. Later in Section 11.4 we apply these aggregation schemes to vector-valued data-generating processes. As a simple example consider a VAR(2) process

$$y_t = A_1 y_{t-1} + A_2 y_{t-2} + u_t \, .$$

In Chapter 4 we describe a general method for converting such models into state space form. Here it is simple to see that if we set x_{1t} to $y_t - u_t$, and x_{2t} to $A_2 y_{t-1}$, then we arrive at a state space model

$$x_t = Ax_{t-1} + Bu_t, \text{ and } y_t = [I \ \ 0]x_t + u_t \ ,$$

where the dynamic matrix A is given by

$$A = \begin{bmatrix} A_1 & I \\ A_2 & 0 \end{bmatrix} .$$

To link this representation with that used by Johansen [1988] in his discussions on the cointegration vectors, assume that $A_1 + A_2$ is equal to the identity matrix plus $\alpha\beta'$, where α and β have rank $g-1$ where g is the dimension of the vector y_t. Then the matrix A has one unit root, and its column eigenvector is $[p' \; p'A_2']'$, where p is orthogonal to β, i.e., the expression $\beta'p$ vanishes. Suppose that all other eigenvalues of the matrix A are less than one in magnitude. Then in the representation (12) the data vector is expressible as $y_t = p\,\xi_t + [I \;\; 0]Q\,\eta_t + u_t$, where the matrix Q is chosen as the matrix of row eigenvectors of the matrix A as described above. Here the combination $\beta' y_t$ has no dynamic mode with unit root because p is orthogonal to β, i.e., the column space of β is the cointegration space in the sense of Granger [1983], as we describe the notion in Section 11.4.

Unit Roots

A unit root is also introduced if (first) differences of data are used to construct state space models, as mentioned, for example, in connection with the two alternative choices in time series modeling discussed by Nelson and Plosser [1982]. To illustrate this, consider

$$y_t = a + bt + u_t$$

where a and b are constant parameters and the exogenous disturbance, u_t, is modeled as

$$\phi(L)\, u_t = \theta(L)\, \varepsilon_t \tag{14}$$

where ε_t is the basic exogenous stationary serially-uncorrelated stochastic process.

An alternative model treats the difference of y_t as stationary

$$(1 - L)\, y_t = b + v_t \tag{15}$$

where v_t is modeled as

$$\mu(L)\, v_t = \nu(L)\, \varepsilon_t . \tag{16}$$

Extending slightly the definition given by Granger [1983], the relation (15) may be called co-integrated of order 1. To simplify exposition, assume that (14) and (16) define invertible stationary processes. The unit root is then found only in (15). Substition of (16) into (15) yields

$$\mu(L)\, (1-L)\, y_t = b\, \mu(1) + \nu(L)\, \varepsilon_t . \tag{17}$$

This representation, when converted to state space form, produces a p-dimensional state space model for $(1-L)y_t$, in (16) where p=deg $\mu(L)$, while the dimension of the state vector of a model for y_t will be $p+1$. The dimension is one less if the state space model is constructed for the first difference, and the model dynamic matrix does not contain any unit root, whereas that for the model of y_t clearly contains one unit root. Slightly generalizing Granger [1983] again, the expression $(1-L)y_t$ can be regarded as a special case of co-integration in which the two series, $\{y_t\}$ and $\{y_{t-1}\}$ are co-integrated.

The presence of unit roots in traditional ARMA type representations causes some difficulty, because the representations are neither invertible nor stationary. The method of dynamic aggregation of the preceding section effectively removes unit roots by deflating the dynamic matrix appropriately. See Householder [1964, p.168] on deflation. Aoki [1971] provides some link between deflation with aggregation of dynamic systems. Box and Tiao [1977] offers a related discussion for autoregressive processes.

If one does not wish to decompose time series into deterministic trends plus disturbances, but rather regards the time series as containing random growth components, one may wish to model the time series directly without preliminary processing such as differencing. Suppose that one considers a state space model of a vector-valued time series where its dynamic matrix has one or more unit roots. The earlier examples suggest that a suitable redefinition of state vectors could yield a lower-dimensional state space model. This is indeed true, and the process of deflating the dynamic matrix by removing unit eigenvalues is related to aggregation of the original state vector into a lower-dimensional one. The notion of co-integration introduced by Granger [1983] can also be regarded as a way of deflating the dynamic matrix.

Regime Shifts and Structural Changes

Time series may fail to be weakly stationary for a variety of reasons. Their statistical properties may suddenly or gradually be altered by external events, such as oil price rises, strikes, changes of governments, or modifications of operating procedures by central banks. Time series may have time-varying statistics, such as variances changing slowly with time. When parameter changes are discontinous, the time instants may be known or unknown. The magnitudes of changes are usually unknown. There may be no well-marked point of discontinuity when changes are gradual. Statistical, econometric and engineering literatures have many approaches to offer, such as the switching regression analysis of Goldfeld and Quandt [1973], the Markov models of Hamilton [1989], or the fault detection approaches in Basseville and Benviniste [1980] and more recently Basseville et al. [1987] and Moustakides and Benveniste [1986]. Turumi [1988] summarizes model stability works in econometrics. A small example is later discussed in Sections 11.9 and 12.3.

3.4 Minimum-Phase Transfer Function Representation

A linear state space model of time series $\{y_t\}$ is given by a pair of equations

$$\begin{cases} x_{t+1} = A_t x_t + B_t \xi_t \\ y_t = C_t x_t + D_t \eta_t \end{cases} , \tag{18}$$

where the first is called the state transition or dynamic equation and the second the data or observation equation. In these equations the vector x_t is the state vector at time t, and ξ_t and η_t are exogenous noise vectors. Time subscripts on the matrices A, B, C, and D indicate the possibility that they may be changing with time. This is one way to represent non-stationary time series. In most of the book, these matrices are taken to be constant, so that time series are modeled by constant coefficient dynamics, and the exogenous processes possibly have time-varying means or variances. The use of state space representation to model nonstationary time series has been proposed by Kitagawa [1981], for example.

This model is an example of the time domain description of time series. An alternative description is the transfer function representation used in discussing spectral density decomposition in Chapter 3. It is convenient to have a symbol for designating lags or leads in a dynamic model. In the time series literature the symbol B is usually used, $Bx_t = x_{t-1}$, for example. In the econometric literature the letter L is used for the same purpose. In the engineering literature the symbol for the lag is either z^{-1} or q^{-1}. In this book we use both L and q^{-1} for lag and q for lead, i.e., $qx_t = x_{t+1}$. Then the the state transition equation in (1) is rendered as $(qI - A)x_t = B\xi_t$. Assuming that ξ and η are the same, then the transfer function from the exogenous noise to the data is

$$y_t = G(q)x_t ,$$

with the transfer function given by

$$G(q) = I + C(qI - A)^{-1}B ,$$

assuming that the system matrices are constant. This transfer function can also be written as

$$G(q) = d(q)^{-1}N(q) ,$$

where the polynominal $d(q)$ is the characteristic polynomial of the matrix A. This is called the left matrix-fraction representation in the systems literature.

The zeros of the matrix $N(q)$ are called the zeros of the transfer function, and are important determinants of how well the simulations of the model perform, because they affect what is called, in the engineering literature, the signal-to-noise ratio. The zeros can be calculated as the eigenvalues of the matrix pencil

$$\lambda\begin{bmatrix} I & 0 \\ 0 & 0 \end{bmatrix} - \begin{bmatrix} A & -B \\ C & -I \end{bmatrix} .$$

Transfer functions with zeros inside the unit circle are called minimal phase. Models in this book are all estimated as minimal-phase stable models. Zeros on the unit circle represent a peculiar case of unit canonical correlations and are not likely to arise in time series models, see Hannan and Poskitt [1989]. However zeros close to the unit circle of the numerators and denominators of transfer functions present difficult modeling problems. We return to this topic later in the book.

Conversion of ARMA models into state space representation is extensively explained in Chapter 4. Because Markovian models use auxiliary variables, x_t , in relating observed data y_t and the exogenous disturbance process $\{n_t\}$, one may wonder about the wisdom of such an indirect way of describing time series. As is made clear in this book, however, this new representation has much to recommend it. The structure of information transmission is very simple in state space models, because only a single lag is involved in describing dynamics, which in turn makes calculations of impulse responses very straightforwardi--as we will see in Chapter 7. Likelihood functions are also easier to write out, and we are able to construct a model estimation algorithm which has nested estimates and guarantees asymptotic stability of estimated submodels as detailed in Chapter 9. A particular state space form called innovation representation is used throughout this book.

4 State Space and ARMA Models

This chapter establishes the theoretical equivalence of time series model descriptions in terms of well-known ARMA models and less familiar Markovian (state space) models, introduces the notion of minimal dimensional models and the associated minimal dimensional state vectors, and presents several methods for putting models into state space forms in general and into the observable or observability canonical form in particular. Although econometricians and statisticians are perhaps less accustomed to the state space representation of time series, this representation is quite useful in building models of times series for the purposes of either forecasting or analyzing dynamic interdependence of components of the series.

This chapter provides a short but self-contained background account and an introductory bridge to Chapter 5, and will be generally useful to the reader who is less familiar with state space representation of dynamic phenomena.

4.1 State Space Models

A state space model is specified by three matrices A, B, and C, called system matrices, plus the covariance matrix of the exogenous noise vectors. Within the framework of state space representation, there are several variants. One that we use most often takes the form

$$x_{t+1} = Ax_t + Be_t, \text{ and } \quad y_t = Cx_t + e_t \ , \tag{1}$$

where the noise vector e_t is mean zero with a constant covariance matrix $Ee_t e_s' = \Delta\delta_{t,s}$. This model is named as the innovation model because the vector e_t will later be shown to be the innovation component of the data vector y_t. The vector x is called the state vector and the subscript t refers to the time instant. The first equation specifies how this vector--which is a theoretical construct and, to the uninitiated, may appear to be only tenuously connected with the data $\{y_t\}$--evolves with time. Later we will be more definite about the relation. The matrix A is the transition or dynamic matrix, and its eigenvalues are important in determining the way the data behave. The second equation is the observation or data generation equation which specifies the relation between the data and the newly-introduced auxiliary vector x_t. See Chapters 6 and 7, where more will be said about innovation models. Another common state space representation takes the form

$$x_{t+1} = Ax_t + u_t, \text{ and } \quad y_t = Cx_t + v_t \ , \tag{2}$$

where the exogenous disturbances u_t and v_t appearing in the state transition and observation equations are possibly mutually correlated, although serially uncorrelated. An alternative to (2) is the model in which u_{t+1} rather than u_t appears in the dynamic equation and no noise is present in the observation equation. Akaike [1975] uses a model in this alternative form. Pavon [1980] examines both model representations and suggests that Akaike's representation leads to cumbersome operations in some applications. We prefer the form of (1) in which the observation equation exhibits explicitly the innovation because this form leads us directly to a simple model identification scheme discussed in Chapter 7.

The three system matrices determine the three important dynamic properties of state space models. First, stability, which is of foremost concern, depends only on the matrix A. The system (1) or (2) is asymptotically stable if all the eigenvalues of the matrix A lie strictly inside the unit disc in the complex plane. The algorithm for model building we develop in this book requires asymptotic stability. Stable innovation models generate weakly stationary stochastic sequences. Recall that weak stationarity is one of the key assumptions imposed on (transformed) data used to build models. Stability also appears as one of the technical conditions in solving certain matrix Riccati equations to determine innovation vector covariance matrices in Chapter 9. The only exception in this book is when the matrix A is allowed to have eigenvalues of unit magnitude for series containing random walk or seasonal components.

As far as the observed data are concerned, the state vector is an extraneous theoretical construct and is usually not observed directly. In Chapter 5 we consider the question of estimating it or the "observability" of the state vector in the precise technical sense of system theory, which depends on the matrices A and C jointly. The pair of matrices A and B determines another system property bearing on the question of singularity of the probability distribution induced by the exogenous stochastic processes. This is the controllability property discussed in Chapter 5.

The system matrices A, B, C and the covariance matrix of the exogenous noise are usually all constant. In addition, A is a stable matrix, so that the model (1) generates a weakly stationary time series. In some cases, some of these system matrices or the noise covariance matrix may vary with time; for example, in models of time series with heteroskedastic variances, the covariance matrix will changes with time. The state space representation can deal with such nonstationarity without much added complexity as will be seen in Chapters 6 and 7, for example. The third system property of controllability is taken up in Chapter 5.

4.2 Conversion to State Space Representation

Models in traditional AR, MA, or ARMA representation can be put into state space form in a simple and systematic way. Conversely, models in state space representation can be put into ARMA representation. Theoretically, then, models in these two alternative representations are equivalent. This section describes how models in one representation may be put into the other. This conversion employs changes of variables that may, at first sight, appear to be arbitrary, but are in fact quite natural once the procedures are understood.

Although quite simple, and capable of being executed in several ways, the procedures for converting traditional models into state space form described here are guaranteed to achieve "minimal dimensional" Markovian models. Other seemingly-simpler procedures may yield state space models with too many dimensions--i.e., redundant information or irrelevant information embedded in their state vectors. Generally speaking, non-minimal dimensional models are to be avoided because such representations may effectively prevent efficient optimization computations because of needlessly high dimensions. If information redundancy were the only problem associated with non-minimal dimensional models, then non-minimality might not be frowned upon too much. However, they usually suffer from other technical deficiencies that may not be obvious at first sight. For example, algorithms for filtering, estimation and control are usually stated for models in minimal-dimensional representation, and may break down, or require special handling if applied to non-minimal dimensional representations, because of certain matrices becoming singular or for other technical reasons. The model construction methods of this book all result in minimal dimensional representation.

Models with scalar endogenous and exogenous variables are discussed first. Because the issue of minimal representation is less likely to arise in such cases, we can better concentrate on methods for converting model representations into state space forms. Next, a special but useful class of vector-valued models is considered. The methods for scalar systems generalize naturally to this class of vector-valued models. Conversion of more general systems into state space representation may be found in Guidorzi [1975], for example. After models are thus converted, their observability and reachability should always be verified to ensure minimal representation, since these are necessary and sufficient for minimality.

Observability Canonical Form

Conversion into state space representation is the simplest and easiest to grasp for AR models, because the conversion is done by using the current data y_t and its past (or future) y's as components of a state vector. Here no misgivings can possibly arise about introducing extraneous variables with no intrinsic economic interpretation. In this section, the symbols y_t and u_t are used to denote scalar endogenous and exogenous variables in this section.

A special state space representation called observable or observability canonical form is sometimes useful in theoretical analysis. To illustrate, consider a scalar-valued ARMA(p, q) model, $\phi(L)y_t = \theta(L)u_t$, where the AR part is a p-th order polynomial in the lag operator L, and the MA part is a q-th order polynomial. Our notation is that $Ly_t = y_{t-1}$ and $L^{-1}y_t = y_{t+1}$. For the time being, assume that q is not greater than p. Let the polynomical in the AR part be given by $1 + \phi_1 L + \phi_2 L^2 + \cdots + \phi_p L^p$. The observability canonical form for a scalar time series is given by the state space model of the form (1), with vectors b and c instead of the corresponding matrices. The system matrix is put in a phase canonical form. There are four possible canonical forms, depending upon where the coefficients of the polynomial in the AR part of the model are displayed. The one which displays them as the last row vector is

$$A = \begin{bmatrix} 0 & 1 & \cdot & 0 \\ 0 & 0 & \cdot & \cdot \\ \cdot & \cdot & \cdot & \cdot \\ \cdot & \cdot & \cdot & \cdot \\ \cdot & \cdot & \cdot & 1 \\ \phi_p & \phi_{p-1} & \cdot & \phi_1 \end{bmatrix} .$$

The vector in the observation equation takes the form $c = [1\,0\,0\ \cdots\ 0]$. With these A and c, the observability matrix of this representation, to be introduced in Chapter 5, is simply the p-dimensional identity matrix. The elements of the vector b, $b_1, b_2, \ldots, b_p$ is chosen to satisfy the condition that the model's transfer function is the same as $\theta(L)/\phi(L)$. This is so because the state variables are mathematical constructs internal to the representation, which leaves the external description of the model such as the transfer function unchanged.

To determine the vector b, first note that the transfer function is given by

$$\theta(L)/\phi(L) = \theta_0 + Lh(L)$$

where

$$h(L) = \gamma(L)/\phi(L) = h_1 + h_2 L + h_3 L^2 + \cdots \tag{3}$$

where

$$\gamma(L) = \gamma_1 + \gamma_2 L + \cdots + \gamma_p L^{p-1},$$

and where

$$\gamma_i = \theta_i - \theta_0\,\phi_i\ ,\ i=1,\ldots,p\ .$$

Set θ_i to zero in the above for i greater than p. Here these h's in (3) are the impulse responses of the model. Set h_0 equal to θ_0. Then the data are expressed as a convolution of the impulse response sequence and the sequence of the exogenous noises

$$y_t = h_0 u_t + h_1 u_{t-1} + h_2 u_{t-2} + \cdots\ .$$

This expression is known as the Wold decomposition of the data y_t. The coefficients in (3) can be obtained by long division, or by solving an algebraic matrix equation

$$\begin{bmatrix} 1 & 0 & \cdot & 0 \\ \phi_1 & 1 & \cdot & 0 \\ \phi_2 & \phi_1 & \cdot & 0 \\ \cdot & \cdot & \cdot & \cdot \\ \cdot & \cdot & \cdot & \cdot \\ \cdot & \cdot & \cdot & \cdot \\ \phi_{p-1} & \phi_{p-2} & \cdot & 1 \end{bmatrix} \begin{bmatrix} h_1 \\ h_2 \\ \cdot \\ \cdot \\ \cdot \\ h_p \end{bmatrix} = \begin{bmatrix} \gamma_1 \\ \cdot \\ \cdot \\ \cdot \\ \gamma_p \end{bmatrix}. \tag{4}$$

Since $h_0 + Lc(I_p - AL)^{-1}b$ is the transfer function of the state space model, the impulse response h_i is given by $cA^{i-1}b$ for $i=1,2,...,$ which is simply b_i for i from 1 to p.

The first state vector component, $x_1(t)$ is equal to $y_t - u_t$ directly from the observation equation, $y_t = cx_t + u_t$. To see how the rest of the state vector components are related to data, advance the time index to $t+j$ in the observation equation, use the dynamic matrix to relate it to x_t

$$\begin{aligned} y_{t+j} &= cx_{t+j} + u_{t+j} \\ &= c(A^j x_t + bu_{t+j-1} + Abu_{t+j-2} + \cdots + A^{j-1}bu_t) + u_{t+j} \\ &= x_{j+1}(t) + u_{t+j} + h_1 u_{t+j-1} + h_2 u_{t+j-2} + \cdots + h_j u_t \,. \end{aligned}$$

Next we advance the time index to $t+j$ of the Wold decomposition, and use the notion of orthogonal projection, formally introduced in the next Chapter, to observe that the projected value of y_{t+j} based on the information available at time $t-1$ is given by

$$y_{t+j|t-1} = h_{j+1}u_{t-1} + h_{j+2}u_{t-2} + \cdots,$$

because the first $(j+1)$ terms in the Wold decomposition--namely $u_{t+j} + h_1 u_{t+j-1} + \cdots + h_j u_t$--are uncorrelated with the information set used in forming the forecast, i.e., the set containing $y_{t-1}, y_{t-2}, \cdots$. From these we conclude that

$$x_{j+1}(t) = y_{t+j|t-1}\,, \quad j = 0, 1, \ldots p-1\,.$$

Since the orthogonally-projected values are the best in the sense of least squares, the components of the state vector in this coordinate system are the best predicted values of p consecutive future values of the series at time $t-1$.

By using another phase canonical form of the matrix A, another observability canonical form is obtained. This alternative one is simpler to obtain than the one we just discussed, because the vector b is more directly determined, even though the first one has the advantage that the first p impulse responses are displayed as the elements of the vector b. To derive this second canonical form we first rewrite the ARMA model as a sequence of nested lag operators:

$$y_t - \theta_0 u_t = L[\phi_1 y_t + \theta_1 u_t + L[\phi_2 y_t + \theta_2 u_t + L[\cdots + \phi_p y_t + \theta_p u_t]\cdots] . \quad (5)$$

State variables are introduced as we undo or remove the lag operators. Define the first state variable by

$$x_1(t) = y_t - \theta_0 u_t .$$

Remove one lag operator and introduce the second state variable $x_2(t)$ to represent the remainder of the nested sequence of the lag operators

$$L^{-1}x_1(t) = \phi_1 y_t + \theta_1 u_t + x_2(t) .$$

Now repeat the process of removing one lag operator and introducing a new state variable

$$L^{-1}x_k(t) = \phi_k y_t + \theta_k u_t + x_{k+1}(t) , \; k=2,\cdots,p-1 .$$

Finally the last lag operator is removed to define its dynamics as

$$L^{-1}x_p(t) = \phi_p y_t + \theta_p u_t .$$

Collect the above after substituting y_t out by $x_1(t) + \theta_0 u_t$. We now have another observable phase canonical form in which the elements of the vector b are $b_i = \phi_i \theta_0 + \theta_i$, $i = 1,\cdots p$.

By specializing the above developments we can put *AR* or *MA* processes in state space representation. Note that the dynamic matrix of an *MA* process becomes all zeros except for ones along a superdiagonal in canonical forms.

Both to illustrate the procedure and to be more definite about the meanings of the state variables we give several examples next. We also introduce another canonical form called the controllable or controllability phase canonical form in these examples, although these forms are not used in our model algorithm.

Example 1. Given

$$y_t = \phi_1 y_{t-1} + \phi_2 y_{t-2} + u_t + \theta_1 u_{t-1} + \theta_2 u_{t-2} ,$$

write it as a nested sequence of lag operations

$$y_t - u_t = L[\phi_1 y_t + \theta_1 u_t + L(\phi_2 y_t + \theta_2 u_t)] .$$

Name the left-hand side above as x_{1t}. Then

$$L^{-1}x_{1t} = \phi_1 y_t + \theta_1 u_t + L(\phi_2 y_t + \theta_2 u_t)$$

Introduce x_{2t} as the last term in the right-hand side, and undo the L operator

$$L^{-1}x_{2t} = \phi_2 y_t + \theta_2 u_t \quad .$$

Substitute y_t out by $x_{1t} + u_t$. Then these relations define a dynamic model for x_{1t} and x_{2t}

$$x_{1t+1} = \phi_1 x_{1t} + \gamma_1 u_t + x_{1t}$$

and

$$x_{2t+1} = \phi_2 x_{1t} + \gamma_2 u_t$$

with

$$\gamma_i = \phi_i + \theta_i \quad , \quad i = 1, 2 \, .$$

This model is called an observable canonical form.

$$x_{t+1} = \begin{bmatrix} \phi_1 & 1 \\ \phi_2 & 0 \end{bmatrix} x_t + \begin{bmatrix} \gamma_1 \\ \gamma_2 \end{bmatrix} u_t \quad , \quad y_t = (1 \;\; 0) x_t + u_t \tag{6}$$

Here the state vector components are related to the series as

$$x_{1t} = y_t - u_t$$

and

$$x_{2t} = \phi_2 y_{t-1} + \theta_2 u_{t-1}$$

The first component is defined in the same way, but the second is not in this coordinate system.

The same model can be put into another canonical form called controllable canonical form.

Rewrite the example as

$$L^{-2}(y_t - u_t) = L^{-1}(\phi_1 y_t + \theta_1 u_t) + \phi_2 y_t + \theta_2 u_t \quad ,$$

or

$$(L^{-2} - \phi_1 L^{-1} - \phi_2)\,(y_t - u_t) = (\gamma_1 L^{-1} + \gamma_2) u_t \quad .$$

Define x_{1t} by

$$y_t - u_t = L\,(\gamma_1 L^{-1} + \gamma_2) x_{1t}$$

The dynamics are described by

$$[(L^{-1} - \phi_1 L - \phi_2 L) x_{1t} - u_t] = 0$$

and

$$y_t = (\gamma_1 + \gamma_2 L) x_{1t} + u_t$$

Now let $x_{2t} = Lx_{1t}$. Then

$$y_t = \gamma_1 x_{1t} + \gamma_2 x_{2t} + u_t ,$$

and

$$x_{1t+1} = \phi_1 x_{1t} + \phi_2 x_{2t} + u_t .$$

This produces another phase canonical form

$$\begin{bmatrix} x_1 \\ x_2 \end{bmatrix}_{t+1} = \begin{bmatrix} \phi_1 & \phi_2 \\ 1 & 0 \end{bmatrix} \begin{bmatrix} x_1 \\ x_2 \end{bmatrix}_{t-1} + \begin{bmatrix} 0 \\ 1 \end{bmatrix} u_t$$

$$y_t = (\gamma_1 \ \ \gamma_2) \begin{bmatrix} x_1 \\ x_2 \end{bmatrix}_t + u_t .$$

This model is dual of (6) in the sense that the dynamic matrix is the transpose and the vectors multiplying u_t and x_t are interchanged. Here x_{1t} is not directly tied to $y_t - u_t$ but through a dynamic relation

$$\gamma_1 x_{1t} + \gamma_2 x_{1t-1} = y_t - u_t .$$

Example 2. This example introduces state variables by undoing the nested sequence of lag operators so that the observability canonical representation results. In addition the polynomial for the *AR* part contains a factor with a unit root

$$(1 + a\ L + b\ L^2)\, y_t = \beta + \theta_0\, u_t + \theta_1\, u_{t-1} \tag{7}$$

where $1+a+b=0$, indicating the presence of the unit root in the AR part of the equation. Because of the presence of two lags in the AR part of this example, a two-dimensional state space model is constructed by defining the first component of the state vector as

$$z_{1,t} = y_t - \theta_0\, u_t , \tag{8}$$

and the second component by

$$z_{2,t-1} = z_{1,t} + a y_{t-1} - \theta_1 u_{t-1}.$$

This second equation is rewritten by substituting for y_t from (8)

$$z_{1,t} = z_{2,t-1} - a\,(z_{1,t-1} + \theta_0\, u_{t-1}) + \theta_1 u_{t-1} \tag{9}$$

$$= z_{2,t-1} - a\ z_{1,t-1} + (\theta_1 - a\ \theta_0)\, u_{t-1} .$$

Advance t by one in the second relation to derive

$$z_{2,t} = z_{1,t+1} + a\ y_t - \theta_1 u_t$$

$$= y_{t+1} + a\ y_t - \theta_0 u_{t+1} - \theta_1 u_t = -b\ y_{t-1} + \beta\,,$$

where (7) is used to obtain the last line. Eliminating y_{t-1} by (8), the relation becomes

$$z_{2,t} = -b\ (z_{1,t-1} + \theta_0 u_{t-1}) + \beta\,, \tag{10}$$

which, when combined with (9), gives its state space representation. The characteristic polynomial of the dynamic matrix has roots 1 and b.

When a new variable z_t is defined as $(1-L)\,y_t$ in (7), then the dynamic equation for it reduces to

$$z_t = bz_{t-1} + \beta + \theta_0 u_t + \theta_1\, u_{t-1}\,.$$

This fact can also be deduced by noting the factorization of the AR part as $(1-bL)(1-L)$. Note that the eigenvalue of these dynamics, b, is the non-unit root of the original dynamic system. This method of removing the unit root is a special case of deflation, which was discussed near the beginning of this chapter.

Example 3. Suppose that the model

$$y_t + 1.2y_{t-1} + 0.35y_{t-2} = u_t$$

is to be converted into controllability canonical state space representation. A two-dimensional state vector is defined by

$$x_t = \begin{bmatrix} y_{t-1} \\ y_t \end{bmatrix},$$

and is governed by the state transition equation

$$x_{t+1} = A\ x_t + \begin{bmatrix} 0 \\ 1 \end{bmatrix} u_{t+1}\,,$$

and the data equation

$$y_t = (0\ \ 1)\, x_t\,,$$

where the matrix A is given by

$$A = \begin{bmatrix} 0 & 1 \\ -0.35 & -1.2 \end{bmatrix}.$$

A different definition of the state vector components

$$x_{1t} = y_{t-1}\ , \text{ and } x_{2t} = y_t - u_t$$

produces a state space representation with the same A matrix, but the vector b is changed to $b = (1\ \ -5)'$, and u_t is added to the data equation.

Example 4. Next, we put an MA(q) model into the controllability canonical form

$$y_t = \theta(L)u_t\ , \tag{11}$$

where

$$\theta(L) = \theta_0 + \theta_1 L + \cdot + \cdot + \cdot + \theta_q L^q \quad \theta_q \neq 0\ ..$$

Introduce state vector components by

$$x_{k,t} = u_{t-k}\ ,\ k=1,...,q\ .$$

The state vector x_t with these q components evolves with time according to the state transition equation with the $q \times q$ matrix A and the q-dimensional vector b

$$A = \begin{bmatrix} 0 & 0 & 0 & \cdot & 0 & 0 \\ 1 & 0 & 0 & \cdot & 0 & 0 \\ 0 & 1 & \cdot & \cdot & 0 & 0 \\ \cdot & 0 & \cdot & \cdot & \cdot & \cdot \\ \cdot & \cdot & \cdot & \cdot & \cdot & \cdot \\ \cdot & \cdot & 0 & \cdot & \cdot & \cdot \\ 0 & 0 & 0 & \cdot & 1 & 0 \end{bmatrix}, \text{ and } b = \begin{bmatrix} 1 \\ 0 \\ \cdot \\ \cdot \\ \cdot \\ 0 \end{bmatrix}, \tag{12}$$

and the data equation is given by

$$y_t = (\theta_1, \theta_2, \ldots, \theta_q)x_t + \theta_0 u_t\ .$$

In this model, we know that auto-covariance matrices of lags equal or greater than q all vanish. We can verify this also by noting that the matrix A has only zero eigenvalues and A raised to the q-th power vanishes because of the relation (3) in Chapter 3.

Note that the same number of parameters appears in the state space representation as in the original MA(q) representation. This model is said to be in the controllable canonical form, since the controllability or reachability matrix $[\, b\ Ab\ A^2b\ \cdots\ A^{q-1}b\,]$ is the q-dimensional identity matrix. We return to this topic in Chapter 5.

The nested sequence representation we introduced in (5) can be used for (11) to put it into the observability canonical form. Then, the roles of the vectors b

and c are reversed, i.e., the vector b displays the parameter θs and components of the vector c are all zero except for the first element which is 1.

Direct Sum Representation. Suppose now that $q>p$ in $ARMA(p, q)$. The transfer function is the sum of a $(q-p)$-degree polynomial and a rational function in which the numerator degree is strictly less than p,

$$\theta(L)/\phi(L) = \alpha(L) + L\gamma(L)/\phi(L) . \tag{13}$$

The endogenous variable y_t is now expressed as a sum of two variables y_t^i, $i=1, 2$. The first one is modeled as a MA$(q-p)$,

$$y_t^1 = (\alpha_1, \ldots, \alpha_{q-p})x_t^1 + \alpha_0 n_t ,$$

where the $(q-p)$-dimensional vector x_t^1 is governed by the dynamics of the form (7), with the matrix A_1 and the vector b as defined in (12). The second dynamic variable comes from the second term in (13). When put into the observable canonical form, the coefficients of the polynomial $\phi(L)$ are displayed as the last row of the dynamic matrix A_2, and the vector multiplying n_{t-1} is composed of the first p components of the transfer function $\gamma(L)/\phi(L)$, denoted by h. When these two subsystems are combined, the state space representation becomes

$$\begin{bmatrix} x_t^1 \\ x_t^2 \end{bmatrix} = \begin{bmatrix} A_1 & 0 \\ 0 & A_2 \end{bmatrix} \begin{bmatrix} x_{t-1}^1 \\ x_{t-1}^2 \end{bmatrix} + \begin{bmatrix} 1 \\ 0 \\ \cdot \\ \cdot \\ \cdot \\ h \end{bmatrix} u_{t-1} ,$$

$$y_t = \alpha_0 u_t + (\alpha_1 \ \cdots \ \alpha_{q-p} \ 1 \, 0 \ \cdots \ 0) \begin{bmatrix} x_t^1 \\ x_t^2 \end{bmatrix} .$$

At most, a total of $p+q+1$ parameters completely specify this state space model. Note that the number of parameters is the same as in the original ARMA(p, q) model.

$$y_t = (1 \, 0 \cdot \cdots \cdot) \, x_t + \theta_0 \, n_t .$$

Vector Models

Suppose now that $\{y_t\}$ and $\{u_t\}$ are vector-valued time series related to each other by a transfer function matrix

$$y_t = H(L)\, u_t ,$$

where elements of the matrix $H(L)$ are rational functions in L--i.e., ratios of two polynomials in L. Such a transfer function matrix may arise from a vector-valued ARMA(p, q) process where the coefficients are matrices rather than scalars. For example, the model

$$y_t + A_1 y_{t-1} \cdots + A_p y_{t-p} = B_0 u_t + \cdots + B_q u_{t-q}$$

has the transfer function matrix

$$H(L) = (I + A_1 L + \cdots + A_p L^p)^{-1}(B_0 + B_1 L + \cdots + B_q L^q) .$$

This is an example of the description of dynamic relations called a (left) matrix fraction description--i.e., the transfer function matrix is expressed as $D(L)^{-1} N(L)$, where D and N are two (relatively prime) matrices whose elements are polynomials in L. Instead of pursuing the most general case, consider the transfer function matrix written as

$$H(L) = \frac{1}{d(L)} N(L) ,$$

where $d(L)$ is a scalar polynomial in L (the minimal polynomial of the matrix A in the corresponding state space representation). For cases in which the roots of this polynomial are distinct and known, a simple construction of a state space representation has been given by Gilbert [1963]. Guidorzi [1975] has developed an algorithm which does not require the information on the roots of the denominator polynomial. The next example, Example 5, points out that the state space conversion methods for scalar models of the previous section do not generalize straightforwardly to vector-valued processes. More examples are found in Chapter 9.

Minimal and Nonminimal Representations. Models we build should be as simple as is compatible with the dynamics contained in the data and consistent with data accuracy. Thus, the dimension of the state vector of a model should not be unnecessarily large. The minimality of the state vector in representation of the dynamics of the vector-valued time series depends on the two additional properties of the model, referred to as observability and reachability. Unlike stability, each of these two properties depends on a pair of matrices. The pair of matrices A and B jointly determine the reachability property; and the pair of A and C, the observability property. When the model possesses both of these properties, then the minimality of the state vector dimension is assured (see Kalman [1960], or Kailath [1980, p.363]).

To illustrate these ideas simply, suppose that the state vector is artificially inflated by appending some arbitrary vector z_t to x_t in the model (1) and rewrite it as

$$\begin{bmatrix} x_{t+1} \\ z_{t+1} \end{bmatrix} = \begin{bmatrix} A & 0 \\ 0 & I \end{bmatrix} \begin{bmatrix} x_t \\ z_t \end{bmatrix} + \begin{bmatrix} B \\ 0 \end{bmatrix} e_t \text{ , and } y_t = (C \ \ 0) \begin{bmatrix} x_t \\ z_t \end{bmatrix} + e_t \ .$$

This artificially enlarged state vector, of course, does nothing but increase the size of the model. In this present representation, one can readily see that the vector z_t is redundant. However, if some coordinate transformation is performed to hide this obvious redundancy of the vector z_t in this inflated model, it may not be readily detected. That the state vector is not of minimal dimension can be deduced, however, by appealing to the concepts of observability and reachability. This model is not observable, as will be seen in the next Chapter. (Common factors in the AR and MA parts of ARMA model representations imply that the corresponding state space models are uncontrollable. See Appendix 11.) Here are two less trivial examples.

Example 5. This bivariate example shows that the minimal state vector dimension need not be an integral multiple of the data vector dimension.

A two-dimensional data vector is generated by AR(2)

$$y_t = A_1 y_{t-1} + A_2 y_{t-2} + u_t$$

with

$$A_1 = \begin{bmatrix} \phi_1 & \phi_2 \\ \phi_4 & \phi_3 \end{bmatrix} \text{ and } A_2 = \begin{bmatrix} \phi_5 & 0 \\ \phi_6 & 0 \end{bmatrix}, \ \phi_5 \neq 0.$$

Note that A_1 is full rank, but rank A_2 is 1.

Applying the standard procedure to obtain observable forms

$$x_{1t} = y_t - u_t .$$

Then

$$x_{1t} = A_1 d_t + L A_2 d_t$$

$$= A_1 (x_{1t} + u_t) + x_{2t}$$

where

$$x_{2t} = LA_2 y_t$$

or

$$L^{-1} x_{2t} = A_2 y_t$$

$$= A_2(x_{1t} + u_t).$$

This state space representation is

$$\begin{bmatrix} x_1 \\ x_2 \end{bmatrix}_{t+1} = \begin{bmatrix} A_1 & I_2 \\ A_2 & 0 \end{bmatrix} \begin{bmatrix} x_1 \\ x_2 \end{bmatrix}_t + \begin{bmatrix} A_1 \\ A_2 \end{bmatrix} u_t,$$

$$y_t = [I_2 \;\; 0] x_t + u_t.$$

The rank of the observability matrix is four, but that of the controllability matrix is only three as a consequence of the deficient rank of the matrix A_2.

By writing out the AR(2) in components, it is clear that the dynamics are of the order three, because only $y_{1,t-2}$ is present and not $y_{2,t-2}$:

$$y_{1t} = \phi_1 y_{1t-1} + \phi_2 y_{2t-1} + \phi_5 y_{1t-2} + u_{1t},$$

and

$$y_{2t} = \phi_4 y_{1t} + \phi_3 y_{2t-1} + \phi_6 y_{1t-2} + u_{2t}.$$

Assume that $\phi_5 \neq 0$.

Introduce

$$x_{1t} = y_{1t} - u_{1t}.$$

Then from the first equation

$$L^{-1} x_{1t} = \phi_1 y_{1t} + \phi_2 y_{2t} + L \, \phi_5 y_{1t}.$$

Introduce $x_{2t} = L \, \phi_5 y_{1t}$ to keep the maximum lag to be one in the above equation,

$$x_{1t+1} = \phi_1(x_{1t} + u_{1t}) + \phi_2 y_{2t} + x_{2t}.$$

In the second equation, define

$$x_{3t} = y_{2t} - u_{2t}.$$

Then, unravelling the nested lag structure of the second equation

$$L^{-1} x_{3t} = \phi_4 y_{1t} + \phi_3 y_{2t} + L \, \phi_6 y_{1t}$$

$$= \phi_4(x_{1t} + u_{1t}) + \phi_3 (x_3 t + u_{2t}) + (\phi_6/\phi_5) x_{2t}.$$

Normally, the last term $L\phi_6 y_{1t}$ calls for a new state variable indexed as t. Here it is proportional to $L\phi y_{1t}$ which has already been defined as x_{2t}. This is why, in this example, the state vector dimension is three rather than four. The result, after substituting $y_{2t} = x_{3t} + u_2 t$, is

$$\begin{pmatrix} x_1 \\ x_2 \\ x_3 \end{pmatrix}_{t+1} = \begin{pmatrix} \phi_1 & 1 & \phi_2 \\ \phi_5 & 0 & 0 \\ \phi_4 & \phi_6 / \phi_5 & \phi_3 \end{pmatrix} \begin{pmatrix} x_1 \\ x_2 \\ x_3 \end{pmatrix}_t + \begin{pmatrix} \phi_1 & \phi_2 \\ \phi_2 & 0 \\ \phi_4 & \phi_3 \end{pmatrix} u_t$$

$$y_t = \begin{pmatrix} 1\,0\,0 \\ 0\,0\,1 \end{pmatrix} \begin{bmatrix} x_1 \\ x_2 \\ x_3 \end{bmatrix}_t + u_t$$

We can easily verify that the ranks of the observability and the controllability matrices are three, if the vectors $(\phi_2, \ \phi_3)$ and $(\phi_5 \ \phi_6)$ are not proportional to each other.

Example 6. Consider a bivariate AR(2) process given by

$$y_t = \phi_1 y_{t-1} + \phi_2 y_{t-1} + u_t$$

with $dim\, y_t = 2$ and

$$\phi_2 = \begin{pmatrix} 0 & 0 \\ \alpha & \beta \end{pmatrix} .$$

Assume that the rank of the matrix ϕ_1 is 2. The usual procedure will produce a 4-dimensional model,

$$x_t = \begin{pmatrix} \phi_1 & I_2 \\ \phi_2 & 0 \end{pmatrix} x_{t-1} + \begin{pmatrix} \phi_1 \\ \phi_2 \end{pmatrix} u_{t-1} ,$$

and

$$y_t = (I_2 \ \ 0)\, x_t + u_t .$$

Note, however, that the third component of x_t can be eliminated because it is zero and the resulting state transition equation

$$\begin{pmatrix} x_1 \\ x_2 \\ x_4 \end{pmatrix}_t = \begin{pmatrix} \phi_1 & I_2 \\ \alpha\beta & 0 \end{pmatrix} \begin{pmatrix} x_1 \\ x_2 \\ x_4 \end{pmatrix}_{t-1} + \begin{bmatrix} \phi_1 \\ \alpha\beta \end{bmatrix} u_{t-1}$$

has dimension 3 rather than 4. With this redefined state vector, the observation equation remains the same as above. The the state vector dimension can also be reduced in less obvious ways. Here is a simple example. (See Chapter 9 for a more involved example.)

Suppose now that

$$\phi_1 = diag\ (\lambda_1, \lambda_2) \quad and$$

$$\phi_2 = \begin{bmatrix} \alpha \\ \beta \end{bmatrix} (0\ \ 1) \ , \quad \beta \neq 0 \ ,$$

so that in components the model is given by

$$y_{1t} = \lambda_1 y_{1t-1} + \alpha y_{2t-2} + u_{1t}$$

and

$$y_{2t} = \lambda_2 y_{2t-1} + \beta y_{2t-2} + u_{2t} \ .$$

Treating y_{2t} first, let

$$x_{1t} = y_{2t} - u_{2t} \ ,$$

and define

$$x_{2t-1} = x_{1t} - \lambda_2 y_{2t-1}$$
$$= x_{1t} - \lambda_2 (x_{1t-1} + u_{2t-1}) \ ,$$

which becomes

$$x_{2t} = y_{2t+1} - u_{2t+1} - \lambda_2 y_{2t}$$
$$= \beta y_{2t-1} \ .$$

Introduce x_{3t} by $x_{3t} = y_{1t} - u_{1t}$. It evolves with time according to

$$x_{3t} = \lambda_1 y_{1t-1} + \alpha y_{2t-2}$$
$$= \lambda_1 (x_{3t-1} + u_{1t-1}) + (\alpha / \beta) x_{2t-1} \ .$$

Written jointly, the dynamics are three-dimensional,

$$\begin{bmatrix} x_1 \\ x_2 \\ x_3 \end{bmatrix}_t = \begin{bmatrix} \lambda_2 & 1 & 0 \\ \beta & 0 & 0 \\ 0 & \alpha / \beta & \lambda_1 \end{bmatrix} \begin{bmatrix} x_1 \\ x_2 \\ x_3 \end{bmatrix}_{t-1} + \begin{bmatrix} 0 & \lambda_2 \\ 0 & \beta \\ \lambda_1 & 0 \end{bmatrix} u_{t-1} \ ,$$

$$y_t = \begin{bmatrix} 0 & 0 & 1 \\ 0 & 1 & 0 \end{bmatrix} x_t + u_t \ .$$

A state vector model is called minimal when its state vector dimension can't be reduced further. How can we be sure that we have a minimal dimensional model? To answer this question, we need to introduce the notion of controllability and observability. Then a short answer is that the model is minimal if and only if it is controllable and observable.

Gilbert's Method Gilbert gave a general procedure for constructing state space models with minimal state vector dimension when the roots of the polynomial $d(L)$ are all distinct, Gilbert [1963]. First, factor the polynomial $d(L)$

$$d(L) = \prod_{1}^{r}(\lambda - \lambda_i)$$

where r is the degree of $d(L)$. Next, expand the transfer matrix into partial fractions

$$H(L) = \sum_{1}^{r} \frac{R_i}{\lambda - \lambda_i} ,$$

where the coefficient matrix has rank ρ_i. Third, factor the coefficient matrix into the product of two full rank matrices as

$$R_i = C_i\, B_i \text{ , where } rank\ B_i = rank\ C_i .$$

Finally define the system matrices by

$$A = diag\ [\lambda_i I_{\rho_i}] ,$$

$$B = [B_1' \cdots B_r']' \text{ , and } C = [C_1 \cdots C_r] .$$

The state space model thus has the dimension n, which is equal to $\sum_1^r \rho_i$. To verify that this n is the least dimension, note that

$$C\,A^k = [\lambda_i^k C_1 \cdots \lambda_r^k C_r\,] .$$

The observability matrix of this model is then expressible as the product of two full-rank matrices

$$\begin{bmatrix} I & I & \cdot & I \\ \lambda_1 I & \lambda_2 I & \cdot & \lambda_r I \\ \cdot & \cdot & \cdot & \cdot \\ \cdot & \cdot & \cdot & \cdot \\ \cdot & \cdot & \cdot & \cdot \\ \lambda_1^{n-1} I & \lambda_2^{n-1} I & \cdot & \lambda_r^{n-1} I \end{bmatrix} \begin{bmatrix} C_1 & 0 & 0 & 0 \\ 0 & C_2 & \cdot & \cdot \\ \cdot & \cdot & \cdot & \cdot \\ \cdot & \cdot & \cdot & \cdot \\ \cdot & \cdot & \cdot & \cdot \\ 0 & 0 & 0 & C_r \end{bmatrix} .$$

The first matrix is of full rank because λ's are all distinct by assumption, and the second is of full rank by construction.

4.3 Conversion of State Space Models into ARMA Models

The innovation models are such that the noise e_t in (1) is equal to $y_t - y_{t|t-1}$, where $y_{t|t-1}$ is the orthogonal projection of y_t on the subspace spanned by $y_{t-1}, y_{t-2}, \ldots, y_1$. A model of the form (1) is completely specified once its system matrices A, B, C and the covariance matrix of the innovation process, Δ, are determined.

The innovation model (1) of Chapter 4 implies a relation between the state vector, x_t, and stacked vectors y_t^+ and e_t^+ of the form

$$y_t^+ = \boldsymbol{O} x_t + \boldsymbol{G} e_t^+ , \tag{14}$$

where $\boldsymbol{O}$ is the observability matrix and $\boldsymbol{G}$ is the infinite matrix with the $Jp \times Jp$ upper left-hand section given by

$$\boldsymbol{G}_J = \begin{bmatrix} I & 0 & \cdot\ \cdot \\ H_1 & I & \cdot\ \cdot \\ H_2 & H_1 & \cdot\ \cdot \\ \cdot & \cdot & \cdot\ \cdot \\ \cdot & \cdot & \cdot\ \cdot \\ H_{J-1} & H_{J-1} & \cdot\ I \end{bmatrix} .$$

Suppose the minimal polynomial of the transition matrix A is $A^n - a_1 A^{n-1} - \cdots - a_n I$. Then $y_{t+n}, y_{t+n-1}, \ldots, y_t$ are related by

$$y_{t+n} - a_1 y_{t+n-1} - \cdots - a_n y_t = Q_0 e_{t+n} + \cdots + Q_n e_t ,$$

where

$$[\, Q_n \ Q_{n-1}, \ldots, Q_0] = [-a_n I \ -a_{n-1} I \ \ -a_1 I \ I\,] \boldsymbol{G}_n .$$

In the above equation $Q_0 = I$. These relations show that the innovation model in the time domain representation has the transfer function matrix given by

$$(I - a_1 L - \cdots - a_n L^n)^{-1} (I + Q_1 L + \cdots + Q_n L^n) .$$

These matrices (Q's) minimize the expression $E|y_t - \hat{y}_t|^2$, where $\hat{y}_t = a_1 y_{t-1} + \cdots + a_n y_{t-n} + e_t + Q_1 e_{t-1} + \cdots + Q_n e_{t-n}$, if $e_t, \ldots, e_{t-n}$ are available.

A more explicit relation exists between the transfer matrix of the innovation model for a time series and the covariance matrix generating function--i.e., the spectral density matrix of the series. This topic was discussed in Section 3.2.

5 Properties of State Space Models

This chapter briefly discusses three properties of dynamic systems in state space form that are important in our model-building procedure. They are stability, observability, and reachability (controllability). The notion of stability is well known. A dynamic system is asymptotically stable if the effects of initial conditions vanish asymptotically over time. This property holds if and only if all the eigenvalues of the transition matrix A of a state space model are less than one in magnitude. The other two properties are less familiar to statisticians and econometricians. They may be motivated by requiring that state space model representations be parsimonious--i.e., by the "minimality" of state vector dimension. We briefly mentioned observability in connection with the observability canonical form in Chapter 4. The pair of matrices A and C in a state space model determines this property. More on these properties, as well as economic examples of state space models, can be found in Aoki [1976]. The pair of matrices A and B jointly determines the reachability property. When the model possesses both of these properties, then the minimality of the state vector dimension is assured (see Kalman [1960], or Lindquist and Picci [1979]).

Since orthogonal projection is our main tool in estimating models, it is described briefly in this chapter also.

5.1 Observability

We now consider a state space model for the time series in innovation representation (1) of Chapter 4. For it, define the observability matrix by

$$\boldsymbol{O} = [C' \ A'C' \ A'^2C' \ \cdots \ A'^{n-1}C' \ \ldots]' , \tag{1}$$

and the reachability matrix by

$$\boldsymbol{C} = [B \ AB \ A^2B \ \cdots \ A^{n-1}B \ , \ldots] . \tag{2}$$

(This matrix is sometimes called the controllability matrix in the systems literature. When the matrix A is nonsingular, the distinction between controllability and reachability disappears.)

A Markov model is observable, or a pair of matrices A and C is an observable pair, if the rank of $\boldsymbol{O}$ is n, where n is the dimension of the state vector of the model. Similarly, the matrix pair (A, B) is called a reachable pair if the rank of the reachability matrix is equal to the dimension of the state vector.

These two system properties guarantee that the models behave normally--i.e., they exhibit no pathological properties. For example, if the pair (A, B) is reachable, then the probability distribution of the state vector induced by the innovation vectors has a nonsingular second moment matrix. Appendix 7 demonstrates the last assertion for a scalar model. The next section establishes a relation between the consistency of least squares estimates and observability.

Observability and Consistency of Least Squares Estimates

One can introduce the notion of observability by posing a question: When can the state vector x_t be reconstructed exactly from future observations if no noise is present? Or, more generally, what ensures that some estimators of x_t (ordinary least squares, conditional mean, etc.), constructed from $y_t, y_{t+1}, \ldots$, are unbiased or efficient if the data are corrupted by noise? By reversing the flow of data, if x_t can be exactly reconstructed from past data $y_t, y_{t-1}, \ldots$, then the system is said to be reconstructible. With nonsingular A matrices, this distinction also disappears.

Using the innovation model and the observability matrix equation, we can relate x_0 to $y_0, y_1, \ldots, y_T$ by

$$Y_T = O_T x_0 + \eta_T \tag{3}$$

where Y_T is a stacked vector and O_T is a finite section of $\boldsymbol{O}$ consisting of T submatrices, i.e., the stacked matrix

$$Y_T = \begin{bmatrix} y_0 \\ y_1 \\ \cdot \\ \cdot \\ \cdot \\ y_T \end{bmatrix}, \quad O_T = \begin{bmatrix} C \\ CA \\ \cdot \\ \cdot \\ \cdot \\ CA^{T-1} \end{bmatrix}$$

and η_T is given by a linear combination of e's:

$$\eta_T = \begin{bmatrix} D \; O \; \ldots\ldots\ldots\ldots \; O \\ H_1 \, D \; O \ldots\ldots\ldots \; O \\ H_2 \, H_1 \, D \; O \ldots. \; O \\ \cdot \\ \cdot \\ \cdot \\ H_T \; H_{T-1} \ldots. \; H_1 D \end{bmatrix} \begin{bmatrix} e_0 \\ e_1 \\ \cdot \\ \cdot \\ \cdot \\ e_T \end{bmatrix},$$

where H_l stands for $CA^{l-1}B$, $l = 1, 2, \ldots$ The vector x_0 can be solved from (3) if rank O_T is n:

$$x_0 = (O_T{}'O_T)^{-1} \; O_T{}'(Y_T - \eta_T)\,. \tag{4}$$

In the absence of noise, x_0 is exactly reconstructed as a linear combination of $y_0, \ldots, y_T$. Here T can be set to n, because a larger T does not increase the rank of O_T, as guaranteed by the Cayley-Hamilton theorem. If es are zero mean i.i.d.

Gaussian noise, then the conditional mean $E(x_0 | Y_T)$ equals $(O_T'O_T)^{-1}$ times $O_T'Y_T$.

Alternatively, consider

$$O_T'Y_T = O_T'O_T x_0 + O_T'\eta_T .$$

When the matrix

$$\Sigma = O_T'E(\eta_T \eta_T')\, O_T$$

is invertible, then the minimization of

$$| O_T'Y_T - O_T'O_T x_0 |_{\Sigma^{-1}}^2$$

yields the same estimate as when $O_T'O_T$ is invertible, i.e., as when the model is observable.

Lyapunov Equations

The rank condition for the observability matrix can be stated equivalently as the positive definiteness condition for the product of the transpose of the obervability matrix with itself, $\boldsymbol{O'O}$. This product is important in the model construction method discussed later in the book, and is called the observability Gramian, because all the elements of the matrix are inner products of suitable vectors. (The name "Gramian" usually refers to the determinant of such a matrix, Gantmacher [1959, p.247]. In the systems literature, the matrix itself is called Grammian.) By carrying out the multiplication and summation operations, the observability Gramian is equal to to

$$G_0 = \sum_{k=0}^{\infty} (A')^k C'CA^k \quad . \tag{5}$$

By the Calyley-Hamilton theorem,

$$\text{rank}\, \boldsymbol{O} = n \text{ if and only if } \boldsymbol{O}'\boldsymbol{O} = G_o > 0 . \tag{6}$$

From this infinite summation expression, it is easy to see that the Gramian satisfies a matrix algebraic equation called the Lyapunov equation

$$A'G_oA - G_o = -C'C . \tag{7}$$

A useful alternative characterization of observability is that the pair (A, C) is observable if and only if for any eigenvector x of the matrix A the condition that $Cx = 0$ implies that $x = 0$, Hautus [1969] or Aoki [1989, Sec.1.4]. Suppose u is an eigenvector of A with the eigenvalue λ. From (7), $(1 - \lambda^2)u'G_o u = u'C'Cu$, which is positive by the Hautus condition. Since the matrix A is asymptotically stable, it follows that the observability Gramian is positive definite.

The reachability Gramian, denoted by G_c, is analogously defined by the infinite sum $\sum_{k=0}^{\infty} A^k BB' (A')^k$. As with the observability matrix, note that

$$\text{rank } \mathbf{C} = n \text{ if and only if } \mathbf{CC}' = G_c > 0 .$$

This Gramian solves another Layapunov equation,

$$AG_c A' - G_c = -BB'. \tag{8}$$

Analogous to the test for observability, the pair (A , B) is a controllable or reachable pair if and only if for any eigenvector u' of the matrix A such that $u'A = \lambda u'$, the condition $u'B = 0$ implies that u is zero. Applying this Hautus condition to (8), we conclude that the controllability Gramian is positive definite if and only if the pair (A , B) is controllable. By rewriting the Lyapunov functions using the Kronecker product, their solutions are unique if and only the product of two eigenvalues of the transition matrix A is never equal to one. The condition that the matrix A is asymptotically stable is sufficient.

We show later that there exists a coordinate system in which these two matrix equations can be simultaneously satisfied by the same diagonal matrix. This happens when a state space model is expressed in a particular coordinate system called the (internally) balanced representation. In other words, the reachability and observability Gramians of a balanced model are equal to each other and simultaneously satisfy (7) and (8). More on the balanced representation will be found later in Chapter 9. In Chapter 6 we encounter another Gramian and the Lyapunov equation it satisfies. It has the same form as (8) except for the fact that another matrix M appears instead of B

$$AG_\Omega A' - G_\Omega = -MM' \tag{9}$$

We call this Ω Gramian. It is important in our modeling algorithm as described in Chapters 6 and 9.

5.2 Orthogonal Projections

The notion of orthogonal projection has been implicitly employed when we mentioned the innovation vector in the state space model. We now make it explicit and use it to introduce a certain Hankel matrix which is the basis for our estimation algorithm for state space models. The Hankel matrix and its relation to the state space models and state vectors are developed fully in the next section. A new piece of data brings news, contains surprises, or is useful if it can't be predicted exactly from past data. We define the innovation component of data y_t

to be the difference $y_t - y_{t|t-1}$ where $y_{t|t-1}$ is the image of y_t orthogonally projected onto the manifold spanned by the past data y_{t-1}^-, which is the stacked vector of past observations $y_{t-1}, y_{t-2}, \ldots,$. We use the orthogonal projection because it coincides with the conditional expectation, when the data are jointly normally distributed. We use the notation $\hat{E}(x \mid y)$ to denote the image of x orthogonally projected onto the manifold spanned by y. By definition, the residual $x - \hat{E}(x \mid y)$ and y are uncorrelated. Assume that the means are zero. Noting that the orthogonally-projected image is proportional to y, $x - Sy$ is uncorrelated with y for some S. This shows that

$$S = \Sigma_{xy}^{-1} \Sigma_{yy},$$

where

$$\Sigma_{yy} = E(yy'), \text{ and } \Sigma_{xy} = E(xy'),$$

when the indicated inverse exists. When means are not zero, subtract them from x and y.

The orthogonal projection itself is given by

$$\hat{E}(x \mid y) = \Sigma_{xy}{}' \Sigma_{yy}^{-1} y .$$

When the indicated inverse does not exist, it is replaced with the pseudo-inverse. Denote either of these expressions by $\hat{E}(x \mid y)$ and call it the wide-sense conditional expectation or the best linear (least squares) prediction of x given the data vector y. When x and y are jointly normally distributed, then the wide-sense conditional expectation is exactly the conditional expectation. This is easily checked by calculating the exponents for $p(x, y) / p(y)$ where $p(\cdot)$ is the Guassian density expression.

When a new piece of data η is added to the existing data set y, the weights on the data and the least squares estimates all change. However, by extracting a component from the new piece of data that is uncorrelated with the existing set of data, one can compute the orthogonal projection $\hat{E}(x \mid y, \eta)$ very easily. The component of η that is uncorrelated with y is simply that part of η that is orthogonal to the subspace spanned by y, i.e., the innovation component of η relative to y

$$e = \eta - \hat{E}(\eta \mid y) = \eta - E(\eta y')E(yy')^{-1} y . \tag{10}$$

Plainly e and y are uncorrelated.

The least squares estimate of x, given y and η, is the same as the least squares estimate of x, given y and e. To see this, first calculate

$$\hat{E}(x \mid y, e) = E[x(y'e')] \; diag \; [E(yy')^{-1}, E(ee')^{-1}] \begin{pmatrix} y \\ e \end{pmatrix} \tag{11}$$

$$= \hat{E}(x \mid y) + \hat{E}(x \mid e).$$

Because of the uncorrelatedness of y and e, their covariance matrix becomes diagonal. To establish the equality of the conditional expectation of x given y and e and that given y and η, note that they are related by the transformation

$$\begin{bmatrix} y \\ \eta \end{bmatrix} = \begin{bmatrix} I & 0 \\ E(\eta y')E(yy')^{-1} & I \end{bmatrix} \begin{bmatrix} y \\ e \end{bmatrix}.$$

Using this transformation, the equality of these two conditional expectations is verified. The inverse is replaced with the pseudo-inverse if it does not exist.

Call e constructed by (10) the (wide-sense) innovation of the new data. Now introduce explicitly a time index and denote the data vectors as $y_1, y_2, \cdots$, and construct the innovation vectors exactly by the Cholesky decomposition

$$e_1 = y_1 \text{ , and } \quad e_i = y_i - \hat{E}(y_i \mid y_{i-1}, \ldots, y_1), \; i=2, 3, \cdots .$$

Then the conditional expectation is expressible as

$$\hat{E}(x \mid y_n, y_{n-1}, \ldots, y_1) = \sum_{i=1}^{n} \hat{E}(x \mid e_i) = \sum_{i=1}^{n} \Theta_i \Delta_i^{-1} e_i , \tag{12}$$

where

$$\Theta_i = E(xe_i'), \text{ and } \Delta_i = E(e_i e_i') .$$

Eq.(12) shows that the best estimate or prediction of x, given data, is expressible by a moving-average model of innovations. By construction, these innovations are uncorrelated. Let $E(e_i e_i') = \Delta_i$, i=1,...n, and define $\Delta = diag\ (\Delta_1, \Delta_2, \ldots, \Delta_n)$, and perform the Cholesky decomposition of Σ_{yy} as $W \Delta W'$, where W is a block lower triangular matrix. Then the vector of innovations is related to the data by

$$[e_1 \ \cdots \ e_n] = W^{-1} y.$$

The covariance matrix $E(ee') = \Delta$, where $e' = (e_1, \ldots, e_n)$. Because the matrix W is lower triangular, the operation can be reversed to express y_j in terms of $e_1, \ldots, e_j$ for every positive $j \leq n$. The two sequences $\{y_i\}$ and $\{e_i\}$ are therefore said to be causally equivalent or invertible, because they span the same subspace in the Hilbert space.

Later more general innovation models are discussed in conjuction with Kalman filters, and the notion of causal invertibility is amplified. However, as an important example of the formula (11), suppose we have a state space model of the form (2) of Chapter 4. Let z_t be the estimate of x_t, given the stacked data vector y_{t-1}^-,

$$z_t = \hat{E}(x_t \mid y_{t-1}^-) . \tag{13}$$

Then the estimate is propagated forward in time by

$$z_{t+1} = \hat{E}(x_{t+1} \mid y_t^-),$$

which by (11) is equal to $\hat{E}(x_{t+1} \mid y_{t-1}^-) + \hat{E}(x_{t+1} \mid e_t)$, where $e_t = y_t - \hat{E}(y_t \mid y_{t-1}^-)$. Because the disturbances are serially uncorrelated their orthogonal projection on the manifold spanned by y_{t-1}^- vanishes and we obtain

$$z_{t+1} = Az_t + Ke_t, \tag{14}$$

where the matrix K is given by

$$K = \Sigma_{xe}\Sigma_{ee}^{-1}.$$

We defer the discussion of how to actually calculate the matrix K until we discuss Kalman filters.

Example: Kalman Filters Almost all materials in this book address the question of estimating state space models given a set of data, without a priori knowledge of process dynamics and second-order information on the noise processes--i.e., covariance matrices of noise processes driving the dynamic equations, or in the measurement equations. Kalman filters are designed on the assumption that this covariance information, and the system matrices such as the transition matrix and the matrix C in the observation equation, are known. As an example of the use of the orthogonal projection, this section assumes the information to derive the filter for a model of time series given by

$$x_{t+1} = Ax_t + u_t, \text{ and } y_t = Cx_t + v_t,$$

where the noise processes are both mean zero, serially uncorrelated with the covariance matrix

$$cov\begin{bmatrix} u_t \\ v_t \end{bmatrix} = \begin{bmatrix} Q_t & N_t \\ N_t' & S_t \end{bmatrix},$$

and uncorrelated with the state vector at some initial time. Since the orthogonal projection does not require that the covariances are constant, we allow for time-varying covariance matrices in this example.

Suppose that data from time 1 up to the current time t are available, and denote this set by y_t^-. (Actually this is a finite segment of y_t^-, which we define later. The same symbol is used because no confusion will arise.) We use the notation

$$x_{t+1|t} = \hat{E}(x_{t+1} \mid y_t^-) = \hat{E}(x_{t+1} \mid y_{t-1}^-, e_t),$$

where $e_t = y_t - \hat{E}(y_t \mid y_{t-1})$. Because this residual of the orthogonal projection of

y_t on the subspace spanned by y_{t-1} is uncorrelated with y_{t-1}^- by construction, we can write the orthogonal projection relation as

$$\hat{E}(x_{t+1} | y_{t-1}^-, e_t) = x_{t+1|t-1} + \Theta_t \Delta_t^{-1} e_t \ , \tag{15}$$

where

$$\Theta_t = E(x_{t+1} e_t'), \text{ and } \Delta_t = E(e_t e_t') .$$

Recall that $\{e_t\}$ has been shown to be serially uncorrelated in the previous section. So far, the Markovian structure of the model has not been used. Now, use it to relate x_{t+1} to x_t. To lighten notation write z_t for $x_{t|t-1}$ and obtain the relation

$$z_{+1} = x_{t+1|t-1} = Az_t + u_{t|t-1} = Az_t \ ,$$

because $E(y_{t-1}^- u_t')$ is zero, and hence $u_{t|t-1}$ vanishes. The Kalman filter for the process is

$$z_{t+1} = Az_t + \Theta_t \Delta_t^{-1} e_t \ , \tag{16}$$

where

$$e_t = y_t - y_{t|t-1} = C(x_t - z_t) + v_t \ ,$$

because $\hat{E}(v_t | y_{t-1}^-)$ is also zero.

To relate the matrices Θ_t and Δ_t to the basic noise covariances and system matrices, first define the prediction error covariance matrix by

$$\begin{aligned} \Xi_t &= E(x_t - z_t)(x_t - z_t)' \\ &= \Sigma_t - \Pi_t \geq 0 \ , \end{aligned} \tag{17}$$

where Σ_t and Π_t are covariance matrices of x_t and z_t, respectively;

$$\Sigma_t = E(x_t x_t') \ ,$$

and

$$\Pi_t = E(z_t \ z_t') \ .$$

The recursion (16) shows that the dynamics for Π's are

$$\Pi_{t+1} = A \Pi_t A' + \Theta_t \Delta_t^{-1} \Theta_t' \ , \tag{18}$$

because $E(z_t \ e_t') = 0$. The model yields the recursion of Σ_t

$$\Sigma_{t+1} = A \ \Sigma_t A' + Q \ , \tag{19}$$

because u_t is uncorrelated with the initial condition state vector by assumption.

The matrix Θ_t in (15) becomes, after z_{t+1} is substituted out by the model dynamics and e_t by (16),

$$\begin{aligned}\Theta_t &= E(x_{t+1}e_t') \\ &= E[(Ax_t + u_t)\{C(x_t - z_t) + v_t\}]' \\ &= A\,\Xi_t C' + N\ ,\end{aligned} \tag{20}$$

where the fact that $x_t - z_t$ is uncorrelated with, i.e., orthogonal to, z_t is used. Similarly, (15) and (16) lead to an important relationship between the covariance matrices for the innovation vector and the prediction error vector

$$\begin{aligned}\Delta_t &= E(e_t e_t') \\ &= C\,\Xi_t C' + S\ .\end{aligned} \tag{21}$$

Advancing t by one in (17) and taking the difference of the two recursion relations (18) and (19), the recursion formula for Ξ_t is deduced to be

$$\begin{aligned}\Xi_{t+1} &= \Sigma_{t+1} - \Pi_{t+1} \\ &= A\,\Xi_t A' + Q - \Theta_t \Delta_t^{-1} \Theta_t'\ .\end{aligned} \tag{22}$$

This important recursion can also be alternatively obtained by direct calculations. Since this is a useful exercise, it is done next. From (13) and (16), the filter error vector is equal to

$$\begin{aligned}x_{t+1} - z_{t+1} &= A(x_t - z_t) + u_t - \Theta_t \Delta_t^{-1} e_t \\ &= (A - \Theta_t \Delta_t^{-1} C)(z_t - z_{t|t-1}) + u_t - \Theta_t \Delta_t^{-1} v_t\ .\end{aligned}$$

The covariance matrix of the filter error is thus expressible as

$$\begin{aligned}\Xi_{t+1} = {}& (A - \Theta_t \Delta_t^{-1} C)\,\Xi_t (A - \Theta_t \Delta_t^{-1} C)' + Q \\ & - N\Delta_t^{-1}\Theta_t' - \Theta_t \Delta_t^{-1} N' + \Theta_t \Delta_t^{-1} S \Delta_t^{-1} \Theta_t'\ .\end{aligned}$$

Use the identity (21) to collect the quadratic terms in Θ_t in the above relation, which reduce to $\Theta_t \Delta_t^{-1} \Theta_t'$. The terms linear in Θ_t are collected as

$$-\Theta_t \Delta_t^{-1}(A\,\Xi_t C' + N)' = -\Theta_t \Delta_t^{-1} \Theta_t'\ ,$$

where (20) is used. Finally, (22) follows because

$$\Xi_{t+1} = A\,\Xi_t A' + Q - 2\Theta_t \Delta_t^{-1} \Theta_t' + \Theta_t \Delta_t^{-1} \Theta_t'$$

$$= A\,\Xi_t A' + Q - \Theta_t \Delta_t^{-1} \Theta_t' .$$

The covariance matrices of $\{y_t\}$ are related to these matrices by

$$\Lambda_0 = E y_t y_t'$$

$$= C\Pi_t C' + S \tag{23}$$

and by

$$\Lambda_{t,s} = E(y_t y_s') = CA^{t-s-1} M_s ,$$

where

$$M_s = A\Pi_s C' + N , \qquad t > s . \tag{24}$$

The derivations so far show that Kalman filters can deal with non-stationary noise (and time-varying dynamics, although this fact has not been emphasized). In covariance stationary time series, the time subscripts t and s enter only in the combination of $t - s$.

Suppose now that the Kalman filter is activated at some point in the remote past. First, replace the subscript t by $t - t_0$ to indicate that the filter is switched on at time t_0. If the noise sequences are wide-sense stationary and A is asymptotically stable, then letting t_0 recede into the infinite past has the same effect on the recursions for $\{\Pi_t\}$ as letting t approach infinity, and similarly for $\{\Sigma_t\}$, and $\{\Xi_t\}$. Denoting the steady state (limiting) value of Π_t by Π_*, (19) shows that it satisfies an algebraic matrix equation

$$\Pi_* = A\Pi_* A' + Q ,$$

and from (23), the $\{y_t\}$ sequence becomes stationary with the covariance matrices

$$\Lambda_0 = C\Pi_* C' + S , \tag{23a}$$

while (24) shows that

$$E(y_t y_s') = \Lambda_{t-s} = CA^{t-s-1}(A\Pi_* C' + N) , \qquad t > s .$$

For the existence of the limit, see the Lyapunov Theorem in Appendix 4. Similarly, (22) becomes an algebraic equation for the limiting matrix of Ξ_t denoted by Ξ:

$$\Xi = A\Xi A' + Q - \Theta\Delta^{-1}\Theta' , \tag{25}$$

where (20) shows that

$$\Theta = A\Xi C' + N \tag{26}$$

and the expression for the limiting innovation covariance matrix, (21), becomes

$$\Delta = C\Xi C' + S \ . \tag{27}$$

Furthermore, $\Xi = \Pi - \Sigma$ also holds, and (18) shows that $\Sigma = A\Sigma A' + \Theta\Delta^{-1}\Theta'$.

From (23a) and (27), the matrix Δ is alternatively given by

$$\Delta = \Lambda_0 - C\Sigma C' .$$

Letting $t_0 \to -\infty$, (17) shows that $\Pi = \Sigma + \Xi \geq \Sigma$, hence $\Pi_* = \Sigma_*$, i.e., the covariance matrix associated with the Kalman filter (16) attains the minimum among all Π's, i.e., among the covariance matrix of the state vector z_t. From (18), (25) and (27), Π_* satisfies the algebraic Riccati equation

$$\Pi_* = A\Pi_* A' + (M - A\Pi_* C')(\Lambda_0 - C\Pi_* C')^{-1}(M - A\Pi_* C')' . \tag{28}$$

Recursive and non-recursive solution methods are discussed in Chapter 7.

Assuming that A, C, and M are known, the noise covariance matrices are related to the matrix Π_* by

$$S = \Lambda_0 - C\Pi_* C' ,$$

$$Q = \Pi_* - A\Pi_* A' ,$$

and

$$N = M - A\Pi_* C' .$$

The covariance matrix of the noise vector in (14) is positive definite if the model is controllable and observable, and if the regularity condition is met. Indeed it is easy to see that the covariance matrix (14) also has the factored expression (13), with

$$B = (M - A\,\Pi_*\,C')\,\Delta^{-1} , \quad \Delta = \Lambda_0 - C\,\Pi_*\,C' .$$

6. Hankel Matrix and Singular Value Decomposition

This chapter constructs a Hankel matrix from auto-covariance matrices of time series which are used in Chapter 9 to estimate system matrices in the state space models for the data-generating processes. This chapter describes several concepts and relations needed in the later chapters.

6.1 The Hankel Matrix

Suppose we forecast the future data stream by orthogonally projecting it onto the manifold spanned by the past data. Define first the stacked vector of future observations and stacked vector of past data vectors as follows:

$$y_t^+ = \begin{bmatrix} y_t \\ y_{t+1} \\ \cdot \\ \cdot \\ \cdot \end{bmatrix}, \text{ and } y_{t-1}^- = \begin{bmatrix} y_{t-1} \\ y_{t-2} \\ \cdot \\ \cdot \\ \cdot \end{bmatrix}.$$

By design, time runs in the opposite directions in these two stacked vectors. By relating the stacked vector y_t^+ to the state vector at time t of the state space model (2) of Chapter 4, we see that it is equal to

$$y_t^+ = \boldsymbol{O} x_t + G(\xi_t^+, \eta_t^+), \tag{1}$$

where the stacked vector ξ_t^+ and η_t^+ are defined analogously to the stacked future observation vector, and $\boldsymbol{O}$ is the observability matrix introduced in Chapter 4. The term G is of no particular concern here. It depends on the future noise realizations, which are uncorrelated with the stacked data vector, because these noises are serially uncorrelated by assumption. For the moment assume that the stacked vectors are infinite-dimensional. We discuss the effects of finite truncation later.

Project y_t^+ orthogonally onto the subspace spanned by y_t^-. It produces a vector in this data subspace

$$\hat{E}(y_t^+ | y_{t-1}^-) = \boldsymbol{H} R^{-1} y_{t-1}^- \tag{2}$$

where

$$\boldsymbol{H} = E(y_t^+ y_{t-1}^{-\prime}) \tag{3}$$

is called a Hankel matrix expressing covariance between the future and past stacked vectors. It is an infinite-dimensional matrix with the same submatrices arranged along counter-diagonal lines--i.e., its (i, j)-th submatrix is given by $\Lambda_{1+|i-j|}$. The covariance matrix of the stacked data vector is given by

$$R = E(y_{t-1}^- y_{t-1}^{-\prime}).$$

We next use (1) to derive the alternative expression for the orthogonal projection. Recall that the noises are serially uncorrelated by assumption. We obtain

$$\hat{E}(y_t^+ \mid y_{t-1}^-) = \boldsymbol{O}\hat{E}(x_t \mid y_{t-1}^-) \tag{4}$$

$$= \boldsymbol{O}\,\Omega R^{-1} y_{t-1}^- ,$$

where the cross covariance matrix between x_t and the stacked data vector is denoted by

$$\Omega = E(x_t y_{t-1}^{-\prime}) . \tag{5}$$

Now, introduce a new state vector z_t by

$$z_t = \hat{E}(x_t \mid y_{t-1}^-) . \tag{6}$$

Note that the vector z_t differs from x_t by some vector which is orthogonal to the stacked data vector. Therefore, the cross covariance matrix Ω given in (5) can be equivalently defined by $E(z_t y_{t-1}^{-\prime})$. Recall how the matrix M is defined as the cross covariance matrix between z_t and y_{t-1} by (4) in Chapter 3. In (14) of Chapter 5 we have seen that the vector z_t obeys the same transition equation as the vector x_t. Consequently, $E(z_t y_{t-2}')$ is the same as AM. (This point is further elaborated in Section 7.1.) Proceeding analogously we deduce that

$$\Omega = [\, M \;\; AM \;\; A^2M \;\; \cdots \,] . \tag{7}$$

Using (6), the two right-hand sides in (4) lead to $\boldsymbol{O}(z_t - \Omega R^{-1} y_{t-1}^-) = 0$, or to an important expression which shows that the state vector z_t is a summary statistic of the data vector y_{t-1}^-

$$z_t = \Omega R^{-1} y_{t-1}^- , \tag{8}$$

if the observability matrix is full rank. The state vector as statistic of the data vector is taken up again in Chapter 8 when we discuss how best to summarize past information for a given measure of optimality.

The matrix Ω defined in (21) is very similar to the controllability matrix introduced earlier. Only the matrix M replaces the matrix B. As in the case of the controllability Gramian we define the Ω Gramian, G_Ω by the product $\Omega\Omega'$. It satisfies a Lyapunov equation (8) when M is used instead of B,

$$AG_\Omega A' - G_\Omega = -MM' . \tag{9}$$

This was introduced in Chapter 5. We will call the pair (A, M) reachable or controllable if this Gramian is positive definite. The Hautus characterization says that this pair is controllable if and only if for any row eigenvector, u' of A, the condition that $u'M = 0$ implies that u is zero. We speak of M-controllability when we wish to distinguish the notion of controllability with the matrix M rather than with the matrix B. This characterization is due to Hautus, see Aoki [1989, p.15].

Directly from (3) and (4), and also from (3) of Chapter 3, where we deduced the structure of the covariance matrices for the mean zero weakly stationary process parametrized by an innovation state space model, we immediately see that the Hankel matrix in (3) can be written as the product of two matrices, the observability matrix, $\boldsymbol{O}$, and another matrix, Ω, given in (7),

$$H = \boldsymbol{O}\Omega . \tag{10}$$

In general, Hankel matrices do not admit such a neat factorization. Only Hankel matrices constructed from auto-covariance matrices of weakly stationary mean zero processes with rational spectral densities have this type of factorization. Put differently, all data-generating processes we consider are such that their auto-covariance matrices are expressible as (3) of Chapter 3, and hence this factorization.

The rank of this Hankel matrix is not greater than the smaller of the ranks of the observability Gramian and Ω Gramian. For a model which is observable and M-controllable, the ranks are both equal to the dimension of the state vector. We speak of an M-minimal model when the above condition holds. For M-minimal model, the rank of the Hankel matrix is equal to the dimension of the state vector.

From this factored expression, we immediately note that

$$C\Omega = [\Lambda_1 \ \Lambda_2 \ \cdots \ ] \tag{11}$$

where the right-hand side is the first submatrix row of the Hankel matrix. Similarly the product $\boldsymbol{O}M$ is equal to the first submatrix column of the same Hankel matrix. Further the covariance matrix between y_{t+1}^{+} and the data vector is

$$E(y_{t+1}^{+} y_{t-1}^{-}{}') = \boldsymbol{O}A\Omega \tag{12}$$

These relations are used to recover estimates of the matrices A, C, and M later, in Chapter 9.

Because of the parallelism between the covariance matrices and the impulse response matrices, we can construct another Hankel matrix by substituting the impulse response matrices for the covariance matrices. This Hankel matrix is the cross covariance matrix between the stacked future vector y_t^{+} and the stacked past innovation vector e_{t-1}^{-} or the stacked exogenous input (decision) vector, rather than the stacked data vector, using the innovation model representation of (1) of Chapter 4. It can be factored as the product of the observability matrix and the controllability matrix

$$C = [B \;\; AB \;\; A^2B \;\; \cdots \;] .$$

Given a finite number of impulse responses, a method called the Padé approximation attempts to match this finite number of impulse response matrices with those of the model. Impulse responses, however, are not directly available from the observed data for economic time series in general. To take this approach, the impulse response matrices must be estimated by relating them to the observed data or sample covariance matrices. In engineering and other areas where data can be generated in experiments, impulse responses may be the raw data from which models are to be built. An algorithm which parallels the one described in this book is possible, but is not explicitly developed, since our primary interest is in series in which such experiments are not possible.

The matrices M and B are related by (4) of Chapter 3. Using (5) of Chapter 3 to substitute out the matrix Π repeatedly, we see that the range space of the matrix Ω is contained in that of the controllability matrix C. Substituting out the matrix B by (4) in (5), we see that the range space of the controllability matrix is contained in that of the Ω, i.e., we establish

Fact The ranks of the controllability Gramian and Ω Gramian are the same.

Suppose now that the noise in the state-space representation is specified slightly differently, as in (2) of Chapter 4 with the noise covariance matrix denoted by Σ_{uu}, Σ_{uv}, and Σ_{vv}. Then the covariance matrices of the data are related to those of the noises and the system matrices by $\Lambda_0 = C\Pi C' + \Sigma_{vv}$, $\Lambda_k = CA^{k-1}M$, $k \geq 1, j$ where the matrix M is now redefined to be $A\Pi C' + \Sigma_{uv}$. On the assumed weak stationarity of the process $\{x_t\}$, the matrix Π now satisfies the relation $\Pi = A\Pi A' + \Sigma_{uu}$. Since the joint noise covariance matrix Σ is symmetric and at least positive semi-definite, the matrix Π must exist to satisfy these conditions. This condition is related to the concept of positive realness of transfer function (for further detail, see Anderson et al. [1974] and Appendix 10).

As will be established later, this model is related to the model (1), in which an innovation sequence appears instead, where $\{e_t\}$ is mean zero, weakly stationary with covariance matrix Δ, and satisfies the relation

$$\begin{bmatrix} \Sigma_{uu} & \Sigma_{uv} \\ \Sigma_{vu} & \Sigma_{vv} \end{bmatrix} = \begin{bmatrix} B \\ I \end{bmatrix} \Delta \, (B' \;\; I) \; .$$

If we truncate the future stack at $t+J$ for some positive integer J and truncate the data vector at some past $t-K+1$, then we obtain the J by K upper left-hand submatrix block of the Hankel matrix in the orthogonal projection, and the covariance matrix R will be a $K \times K$ block Toeplitz matrix,

$$H_{J,K} = E \begin{bmatrix} y_{t+1} \\ y_{t+2} \\ y_{t+3} \\ \cdot \\ \cdot \\ \cdot \\ y_{t+J} \end{bmatrix} (y_t' \, y_{t-1}' \cdots y_{t-K+1}') = \begin{bmatrix} \Lambda_1 & \Lambda_2 & & \Lambda_K \\ \Lambda_2 & \Lambda_3 & \cdot & \Lambda_{K+1} \\ \cdot & \cdot & \cdot & \cdot \\ \cdot & \cdot & \cdot & \cdot \\ \cdot & \cdot & \cdot & \cdot \\ \Lambda_J & \Lambda_{J+1} & \cdot & \Lambda_{J+K-1} \end{bmatrix} .$$

Given a finite number of covariance matrices a branch of system theory called realization theory examines the question of reproducing them or generating data sequences with the given covariance matrices. It also examines how to extend finite sequences so that the extendted sequences are still covariance matrices of some weakly stationary processes (see Kimura [1986] or Byrnes and Lindquist [1988]). With a finite number of impulse responses given, a method called the Padé approximation attempts to match a finite number of impulse response matrices with those of the model. Impulse responses, however, are not directly available from the observed data. To take this approach, the impulse response matrices must be estimated by relating them to the observed data or sample covariance matrices. For economic time series, at least, use of covariance matrices are advantageous.

The rank of the Hankel matrix H of (10) is not greater than the minimum of those of the two factors. For a minimal state space model these ranks are both equal to the dimension of the state vector. Later in this chapter, the rank of any matrix is shown to be equal to the number of its non-zero singular values. Numbers of non-zero singular values thus theoretically determine the dimension of state space models. We return to this important topic in Chapter 9.

6.2 Singular Value Decomposition

In Section 1 we argued that the rank of the Hankel matrix constructed there is equal to the dimension of the state space model for the time series. We now describe a numerical way of reliably determining the rank of a matrix.

Numerical analysts recommend determining ranks of matrices using the following

Theorem. An $m \times n$ real matrix A can be written as $A = U \Sigma V' = U_r \Sigma_r V'_r$, where $U'U = I_m$, $V'V = I_n$, and

$$\Sigma = \begin{bmatrix} \Sigma_r & 0 \\ 0 & 0 \end{bmatrix} ,$$

where $\Sigma_r = diag(\sigma_1, \sigma_2, \cdots \sigma_r)$, and r is the rank of the matrix A.

The σ's are called singular values. They are all positive and arranged in decreasing order. (The set of matrices with some repeated singular values is 'negligible' in a precise technical sense.) Since $AA'U_r = U_r \Sigma_r^2$, the matrix U_r is constructed by using orthonormal eigenvectors with positive eigenvalues of $A\ A'$ in decreasing order. The matrix V'_r is then calculated as $\Sigma_r^{-1}U'_rA$ (see Stewart [1973]). Without loss of generality, we partition the matrices U and V conformably with the matrix Σ and retain only U_r, V_r and Σ_r, and drop subscript r. The number of positive singular values equals the rank of the matrix. Golub and van Laon [1983] state that this method is numerically most reliable. Computer algorithms for the singular value decomposition can be found in Forsythe, Malcolm, and Moler [1977], Golub and van Loan [1983], or in IMSL.

The number of singular values which are "significantly" different from zero determines the numerical rank of A. Unfortunately, large singular values and small singular values do not always separate neatly, making numerical determination of rank somewhat arbitrary. Put differently, judging a singular value to be effectively zero is somewhat arbitrary, and must be done relative to the sizes of other singular values, as well as relative to the statistical accuracy of the elements making up the original Hankel matrix. Construction of a lower dimensional state space model is also based on the judgmental choice of the effective numerical rank--i.e., on the judgment of the separation of singular values into (1) larger ones to be retained in the approximate models and (2) smaller ones to be ignored. Various error norms may be used in reaching such judgments. Estimates of error magnitudes from a variety of sources and statistical accuracy of data need to be evaluated for tentative choices of the singular values to be retained later in Chapter 9 and 10.

Appendix 8 lists some useful properties of the singular value decomposition. One that is relevant to our application is that the best approximation for a matrix in the 2-norm (i.e., the matrix norm induced by the usual Euclidean norm of a vector)--subject to the constraint that approximating matrices have a given rank, i for example--is given by taking the first i largest singular values and corresponding submatrices of U and V. The error of the approximation is given by σ_{i+1} where

$$\text{Min}[\,|A - X|_2 : \text{rank}\ X = i\,] = \sigma_{i+1}\,,$$

where minimization is over all matrices of the same size as A and of rank i. One of the best X is given by $\sum_{j=1}^{i}\sigma_j u_j v'_j$, where u's and v's are the column vectors of U and V respectively. When A is a Hankel matrix, this X may not be a Hankel matrix. Certain other Xs give the same error norm. In building a lower-dimensional approximate model by approximating the Hankel matrix of the type (10), the approximating matrix must also be a Hankel matrix so that it can be realized by a lower-dimensional state space model. This is one of the topics discussed in Chapter 9. There are two related minimization problems. One minimizes the error norm, given the dimension of the approximate models. The other minimizes the dimension of the approximating model, given a upper bound on the error norm. Glover [1984] solved the former and also gave numerical algorithms for the latter. The latter is also discussed by Kung and Lin [1981]. The approximations are further discussed in Chapter 10.

Sensitivity of Singular Values

Singular values of submatrices obtained by column partitioning are interlaced. Let $A = [\, a_1, a_2, \ldots, a_n \,]$ be a column partitioning of an $m \times n$ matrix A, $m \geq n$. Let A_r be the submatrix of A containing the first r columns, and let $\sigma_k(A_r)$ denote the k-th singular value of A_r. Then they are related by the sequence of inequalities

$$\sigma_1(A_{r+1}) \geq \sigma_2(A_r) \geq \sigma_2(A_{r+1}) \geq \cdots \geq \sigma_r(A_r) \geq \sigma_{r+1}(A_{r+1}).$$

See Golub and van Loan [1983, Cor.8.3-3].

Another useful relation states that the difference of the singular value of $A + E$ and A is bounded by the largest singular value of E, where E is regarded as perturbation. See Golub and van Loan [1983, Cor.8.3-2]. For example, if a computed A differs from its theoretical expression by some matrix E embodying various sources of numerical errors of computation, then $\sigma_1(E)$ is a measure by which "significance" of $\sigma_1(A+E)$ can be judged. More generally, a singular value of the sum of two matrices is bounded by

$$\sigma_{\mu+\nu+1}(X+Y) \leq \sigma_{\mu+1}(X) + \sigma_{\nu+1}(Y)$$

for any two matrices X and Y and any two non-negative integers μ and ν, as shown in K. Fan [1951].

Singular values change when units of measuring data series are changed. They are discussed in the next section.

Rank and Singular Values

Consider a partition of the matrix U and V as $(U_1\ U_2)$ and $(V_1\ V_2)$ respectively. Partitioning the matrix Σ conformably, the Hankel matrix can then be expressed as the sum of two Hankel matrices $H = H_1 + H_2$, where $H_i = U_i \Sigma_i V_i'$, i=1,2, where these Hankel matrices correspond to the two subsystems of the original system obtained by the partition, if $A = diag\ (A_1, A_2)$, $B' = (B_1', B_2)$,and $C = (C_1, C_2)$.

Clearly, the approximation of the Hankel matrix H with H_1 implies that

$$|H - H_1| = |H_2|$$

in any matrix norm. If the 2-norm is used, then the right hand side equals the largest sigular value of H_2 or σ_{n^*+1} where n* is the dimension of the submatrix Σ_1. The relative error is given by $|H - H_1| / |H_1| = \sigma_{n^*+1}/\sigma_1$.

In theory, the rank of the Hankel matrix is equal to the number of its positive singular values. In practice, things are not so clear-cut. Typically, a number of singular values will have similar small positive numbers. Clearly some criteria or rules of thumb must be established for determining whether these small numbers are significantly different from zero. The following error bounds on the singular values serve as one of the guides.

Let H be a theoretical Hankel matrix with its theoretically exact singular value decomposition, $U \Sigma V'$. Suppose that a numerically constructed approximation to H is available as $\bar{H} = H + \Delta H$, where it is known that $|H - \bar{H}| \leq a|\bar{H}|$. Here the constant a represents a measure of data accuracy. A program to calculate singular values of $\bar{H}$ will compute singular values not of $\bar{H}$ but of $\hat{H}$, because of computer round-off errors--i.e., $|\bar{H} - \hat{H}| \leq b|\bar{H}|$, where the constant b measures relative precision of the computer. From these two inequalities we derive

$$|H - \hat{H}| \leq (a + b)|\bar{H}|.$$

From the second inequality it follows that $|\bar{H}| \leq (1 - b)^{-1} |\hat{H}|$. Hence the error between the matrix H and $\hat{H}$ can be bounded by $(a + b)(1 - b)^{-1}|\hat{H}|$. In terms of the singular values of H and $\hat{H}$, this inequality can be stated as

$$|\sigma_i - \hat{\sigma}_i| \leq c\,\hat{\sigma}_1,$$

where c is the constant (a + b)/(1 - b). In general, b is much smaller than a, hence c is nearly equal to a. This inequality is suggested by Paige [1981]. If $\hat{\sigma}_r$ is greater than $c\sigma_1$, but $\hat{\sigma}_{r+1}$ is less than this number, then clearly σ_r is positive. Hence the rank of the matrix is at least r. The next singular value, σ_{r+1}, possibly can be zero. Such an r may be tentatively chosen as the numerical rank of the true but unknown Hankel matrix $\boldsymbol{H}$.

Spurious disturbances affecting computations of sample covariances tend to increase the numerical rank of the Hankel matrix constructed from them. This fact is illustrated by using the data in Tiao and Box [1981]. They simulated a two-dimensional *MA* (1) model with 250 data points to calculate sample covariance matrices Λ_k for k =1,...,12. They use the model given by

$$y_t = n_t - \Theta n_{t-1},$$

where Θ is a 2×2 matrix. The theoretical rank of the Hankel matrix is two, because this model can equivalently be put as a two-dimensional state space model given by

$$x_{t+1} = n_t, \text{ and } \quad y_t = -\Theta x_t + n_t.$$

As a rough measure of signal-to-noise ratio, they use $2/\sqrt{N}$.
The singular values of the Hankel matrix $\boldsymbol{H}_{J,K}$ are examined for several combinations of J and K. Using the value of the largest singular value times the signal-to-noise ratio as a guide to the numerical rank determination, the numerical rank chosen by this cut-off measure monotonically increases as J or K increases.

To put this more dramatically, replace by zero all elements in the sample covariance matrices with entries less than this signal-to-noise ratio. Each sample covariance matrix of orders 2 through 6 becomes all zero and that of order 7 has a single non-zero entry as

$$\Lambda_7 \approx \begin{bmatrix} 0 & 0 \\ -.17 & 0 \end{bmatrix} .$$

Theoretically, Λ_7 is zero, of course. The only non-zero covariance matrix is Λ_1 given as the product of minus Θ and the noise covariance matrix which is

$$\Lambda_1 = -\begin{bmatrix} -1.1 & -.5 \\ 1.3 & -.5 \end{bmatrix} .$$

With these noisy sample covariance matrices, the Hankel matrix will have singular values contributed by $\Lambda_1 \Lambda_1'$ plus the repeated singular value .17 which comes from $\Lambda_7 \Lambda_7'$ if J or K is sufficiently large. This is easily seen, for example, by constructing the Hankel matrix $\boldsymbol{H}_{4,6}$ in which Λ_1 and Λ_7 are the only non-zero sample covariance matrices, and they contribute non-overlapping singular values.

Approximate Regression Analysis

The relative magnitudes of singular values of a Hankel matrix convey important information about the dimension of approximate models to explain the data structure, whether models are explicitly dynamic such as innovation models for time series, or are such that time does not play an explicit or essential role, as in regression models. The nature of the approximation by dynamic models is further discussed in Chapter 10.

As an example of the use of the singular value decomposition in static models, forecast errors in linear regression models are analyzed in this subsection, following Mandel [1982] in broad outline. Regression models explain N scalar variables, $y_1, y_2, ..., y_N$, ignoring possible intertemporal relationships among them, by a common parameter vector β (regression coefficients) as weighted sums of exogenous variables $x_{i,j}$ as

$$y_i = \sum_{j=1}^{q} x_{i,j} \beta_j + e_i \ , \ i=1, \cdots, N \ ,$$

where the vector e has mean zero and covariance $\sigma^2 I$. For definiteness take N to be greater than q. Denote the singular value decomposition of the matrix X made up of elements $x_{i,j}$ by $U \Sigma V'$, and assume that there is a wide gap between the r-th singular value, σ_r, and the (r+1)-th singular value, σ_{r+1}, i.e., σ_r / σ_{r+1} is very small. When the matrix is full rank, there are q positive singular values. Primary interest in this example lies in situations where, although the rank is full, the smallest q-r singular values are nearly zero.

Accordingly, assume that the rank of the matrix X is q, but that $\sigma_r \gg \sigma_{r+1} >, \ldots, > \sigma_q > 0$, where σ_i is the i-th largest singular value of X. Partition conformably the matrices U, Σ, and V as $(U_1\, U_2)$, $(\Sigma_1\, \Sigma_2)$, and $(V_1\, V_2)$, where U_1 is $N \times r$, and so on. Thus

$$XV = U\Sigma\,, \text{ where } \quad XV_2 = U_2\Sigma_2\ .$$

This equation shows that if rank X is r which is less than q, then $XV_2=0$ because Σ_2 is zero. This equation constrains X. The matrix X is said to be collinear when its rank is less than q. Given a new set of $x_{i,j}$'s, the regression model can be used to forecast y only when these new xs satisfy this constraint equation. With the elements of Σ_2 much smaller than those in Σ_1, the matrix X is approximately constrained by a linear relation, $XV_2 \approx 0$.

Partitioning X as $(X_1\, X_2)$, and V_2 as $(V_{21}'\, V_{22}')'$, and noting that V_{22} is nonsingular, the constraint equation can be solved for X_2 when the matrix X is collinear

$$X_2 = -X_1 V_{21} V_{22}^{-1}$$

where V_{22} is $(q-r)\times(q-r)$. This equation explicitly expresses the collinearity in X.

By defining a new parameter vector by $\alpha = \Sigma V'\beta$, the regression model is reformulated as $y = U\alpha + e$. The least squares estimate of α is unbiased because

$$\hat{\alpha} = (U'U)^{-1}U'y = \alpha + U'e\ ,$$

and the covariance of $\hat{\alpha}$ is the same as that of e. The estimate of the original parameter β is recovered from that for α as $\hat{\beta} = V\Sigma^{-1}\hat{\alpha}$. The mean and covariance of the estimate is given by $E(\hat{\beta}) = V\Sigma^{-1}\alpha$, and cov $\hat{\beta} = \sigma^2 V\Sigma^{-2}V'$. Because the submatrix Σ_2 is much smaller than Σ_1, the inverse of Σ_2 in Σ^{-1} will be large. Such a large inverse tends to magnify the error in α, however.

6.3 Balanced Realization of State Space Model

A change of coordinate system causes the observability matrix, the matrix Ω, and the controllability matrix to change. Suppose that the model in some coordinate system is

$$x_{t+1} = Ax_t + B\xi_t\ , \text{ and, } \ y_t = Cx_t + \eta_t\ ,$$

Any nonsingular matrix W can be used to define a new state vector s_t from the old one x_t by $s_t = Wx_t$. In this new coordinate system, the model is

$$s_{t+1} = WAW^{-1}s_t + WB\xi_t\ , \text{ and, } \ y_t = CW^{-1}s_t + \eta_t\ .$$

The covariance matrix of the state vector is transformed by

$$cov(s_t) = W cov(x_t) W' .$$

The observability matrices in the old and new coordinate systems are related by

$$\boldsymbol{O}_{new} = \boldsymbol{O}_{old} W^{-1} . \tag{13}$$

The old and new observability Gramians are related by

$$W' G_{o,new} W = G_{o,old} .$$

Recall the definition of the matrix M as the cross covariance matrix between the state vector and the data vector with a single period lag given in (4) of Chapter 3. Therefore, this matrix is transformed into

$$M_{new} = E(s_t y_{t-1}') \tag{14}$$

$$= WE(x_t y_{t-1}') = WM_{old} .$$

Therefore the matrix Ω_{new} is equal to $W\Omega_{old}$. A probably easier way to see this is to use the decomposition of the Hankel matrix as noted in (10). Alternatively, we recognize that auto-covariance matrices are invariant with respect to any change of coordinate in which state vector is expressed, because this is purely internal to the model and has nothing to do with the data series and hence the Hankel matrix remains the same. Then the above follows immediately from (10). Arguing similarly, we conclude that the controllability matrix is transformed into $\boldsymbol{C}_{new} = W\boldsymbol{C}_{old}$, because the impulse response matrices given in (11) of Chapter 3 are also invariant with respect to coordinate changes. The controllability Gramian thus changes from G_c into WG_cW', and the product of $\Omega\Omega'$, i.e., Ω Gramian turns into $W\Omega\Omega'W'$ in the new coordinate system.

A particular coordinate system called "balanced" was introduced by Moore [1978]. He was concerned with choosing units to express input and output variables in control systems in order to avoid inappropriate choices for state variables. His proposal is to choose these units so that the controllability Gramian and the observability Gramian are equal to a same diagonal matrix. This in effect properly "balances" the scaling of input and output variables. In time series there is only one type of variable--i.e., the data vectors--unless one has, in addition, exogenous input or decision variables. Nevertheless, the question of proper choices of units to express data and state vectors is important.

Example. Here is an example of a poor choice of scaling in an extreme form. Suppose a two-dimensional model for a univariate series has the parameter

$$A = \begin{bmatrix} \alpha & 0 \\ 0 & \beta \end{bmatrix}, \quad C = (10^6 \ \ 10^{-6}), \text{ and } B' = (10^{-6}, \ 10^6)', \text{ with } \mathrm{E}e_t^2 = \sigma^2.$$

Choose α and β to be small in magnitude so that α^2, $\alpha\beta$, and β^2 are not close to 1. Since the matrices A and C are given in this example, we can construct the observability matrix and the Gramian directly. The observability matrix is equal to

$$O = \bar{O}D$$

with

$$\bar{O}' = \begin{bmatrix} 1 & \alpha & \alpha^2 & \cdot & \cdot \\ 1 & \beta & \beta^2 & \cdot & \cdot \end{bmatrix}, \quad \text{and,} \quad D = \mathrm{diag}(10^6 \ \ 10^{-6}) \ .$$

The observability Gramian, $G_0 = D\bar{O}'\bar{O}D$, is extremely ill-conditioned. The contour of $x'G_0x = const.$ is an extremely thin ellipse. Solving the Riccati equation, (3) of Chapter 3, the elements of the state vector covariance matrix are easily seen to be given by

$$\Pi_{11} = 10^{-12}\sigma^2 / (1-\alpha^2),$$

$$\Pi_{12} = \sigma^2 / (1-\alpha\beta),$$

and

$$\Pi_{22} = 10^{12}\sigma^2 / (1-\beta^2).$$

Note that this covariance matrix is also extremely ill-conditioned. The variance of the first state variable is practically zero, but that of the second state variable is huge, and these two are correlated. The variances of the data series can be calculated using (6) of Chapter 3,

$$\Lambda_0 = \sigma^2 + C\Pi C'$$

$$= \sigma^2 + \frac{\sigma^2}{1-\alpha^2} + \frac{2\sigma^2}{1-\alpha\beta} + \frac{\sigma^2}{1-\beta^2},$$

and from (4) of Chapter 3 we see that the elements of the matrix M are quite unbalanced as well,

$$M = \begin{bmatrix} m_1 10^{-6} \\ m_2 10^6 \end{bmatrix},$$

with

$$m_1 = \sigma^2\alpha\left[\frac{1}{1-\alpha^2} + \frac{1}{1-\alpha\beta}\right],$$

and

$$m_2 = \sigma^2\beta\left[\frac{1}{1-\alpha\beta} + \frac{1}{1-\beta^2}\right].$$

Then (3) of Chapter 3 yields

$$\Lambda_k = CA^{k-1} M = \alpha^{k-1} m_1 + \beta^{k-1} m_2 \quad , \quad k \geq 1.$$

This sequence of variances show no anomaly, because the extreme values of the elements of B and C cancel out. The matrix Ω is equal to $\Omega = D^{-1}\overline{\Omega}$, with

$$\overline{\Omega} = \begin{bmatrix} m_1 & m_1\alpha & m_1\alpha^2 \ldots \\ m_2 & m_2\beta & m_2\beta^2 \ldots \end{bmatrix} .$$

Here is an example of state space coordinates in which the Gramians are both ill-conditioned, even though the auto-covariance matrices show no abnormality.

These anomalies are of course due to the artificially introduced improper scalings of the state variables, and disappear once we change the coordinate to correct the unbalanced units with which they are expressed,

$$s_t = Dz_t \ .$$

Then

$$DAD^{-1} = A \ ,$$

$$DB = \begin{pmatrix} 1 \\ 1 \end{pmatrix} ,$$

and

$$CD^{-1} = (1 \ \ 1) \ .$$

The covariance matrix of this new state vector is given by

$$cov \ s_t = D \ \Pi \ D = \sigma^2 \begin{bmatrix} 1/(1-\alpha^2) & 1/(1-\alpha\rho) \\ 1/(1-\alpha\rho) & 1/(1-\beta^2) \end{bmatrix}$$

and it is well-conditioned, and the contour $x'D\Pi Dx$ is not extremely elongated as before, indicating that the variances of the two state variables are not extremely unbalanced, as before.

Based upon this example, we call a model balanced, if the observability Gramian and Ω Gramians are the same diagonal matrix (equal to the diagonal singular matrix) in the singular value decomposition of the Hankel matrix. This notion is distinct from that proposed by Moore, in which the observability Gramian and the Gramian G_c, i.e., the controllability Gramian, are equal to the singular matrix. From the Lyapunov equations (7), (8), and (9) in Chapter 5, three Gramians can't be simultaneously equal to the same diagonal matrix. We use the word "balanced" in the former sense.

Any state space model can be put in balanced form by equating the factorization of the Hankel matrix (10) with its singular value decomposition by setting

$$O = U\Sigma^{1/2}\ , \text{ and } \Omega = \Sigma^{1/2}V'\ ,$$

in the algorithm of Chapter 9.

Effects of Scaling

Some analysts prefer to use correlation matrices rather than covariance matrices to avoid the fact that singular values of the Hankel matrices are scale-dependent, i.e., are not invariant to changes in measurement units. Denote the covariance matrix of the stacked data vector by R_-. Scale the data vector as $R_-^{1/2} y_{t-1}^-$. Then this vector has the unit matrix as its covariance matrix. Scaling the stack of future observation by the square root of its covariance matrix, the Hankel matrix becomes $R_+^{1/2} H R_-^{1/2\,\prime}$ rather than H, where R_+ is the covariance matrix of y_t^+. We discuss this scaling, which leads to the introduction of canonical variates, in Chapter 8.

Although singular values of Hankel matrices are not invariant to units of measurement of data series, model sizes and state vectors in some coordinate systems are not overly sensitive to unit changes. In some coordinate systems, state vectors remain the same. To see this, let D be a p-dimensional diagonal matrix and consider models for the data series Dy_t and how they are related to the innovation model for the orignal data series, y_t, which is given as

$$z_{t+1}=Az_t+Be_t\ ,\ y_t=Cz_t+e_t\ .$$

Multiply the observation equation from the left by D and rewrite the model as

$$Dy_t=DCz_t+De_t\ ,\ z_{t+1}=Az_t+BD^{-1}De_t\ . \tag{*}$$

This equation indicates that there is a coordinate system (which is not balanced unless the matrix D is a scalar multiple of the identity matrix) in which the state vector of the original model can serve as that of this rescaled data series. We can see this by noting that the covariance matrices of the rescaled series are the original covariance matrices pre- and post-multiplied by D. Let $D \otimes I$ be denoted by $\mathcal{D}$. The new Hankel matrix is given by $\mathcal{D} H \mathcal{D}$. One factorization of this new Hankel matrix is $\mathcal{D}$ O-by- $\Omega\mathcal{D}$. The state space model with respect to this coordinate system has z_t as the state vector because of (6) and

$$\Omega\mathcal{D}(\mathcal{D}R_-\mathcal{D})^{-1}\mathcal{D}y_{t-1}^- = \Omega R_-^{-1} y_{t-1}^-\ .$$

We later see that the estimator for the dynamic matrix A is invariant to this type of scale changes.

The determinant of the new singular values is $|D|^{2K}$ times the determinant of the original singular values when the Hankel matrix is $Kp \times Kp$. When D is d times the identity matrix, all the singular values are multiplied by d^2. By putting (9) into a balanced form we can obtain bounds on the new singular values in terms of the old. Follow the coordinate changes described in Aoki (1987 p.75), applied to the observability gramian and the other gramian $\Omega \mathcal{D}^2 \Omega'$. The required non-singular transformation to put (9) into the balanced form is $\Delta^{1/2} \Gamma_o' W \Lambda^{-1/2}$, where the singular value decomposition of $\Theta' \mathcal{D}^2 \Theta$ is $\Gamma_o \Delta_o \Gamma_o'$, and that of $\Delta_o^{1/2} \Gamma'_o \Omega \mathcal{D}^2 \Omega' \Gamma_o \Delta_o^{1/2}$ is $W \Lambda^2 W'$. Denoting the new singular values by $\sigma_i(D)$ we have

$$\sigma_1(D) \leq \kappa \sigma_1, \text{ and, } \quad \sigma_N(D) \geq \kappa_2 \sigma_N,$$

where $\kappa_1 = d^2_{\max}$, and $\kappa_2 = d^2_{\min}$. From these we know that $\sigma_N(D)/\sigma_1(D) \geq (\kappa_1/\kappa_2)(\sigma_N/\sigma_1)$. Here we use the inequality which states that the largest eigenvalue of any matrix of the form $K'NK$ where N is symmetric positive definite and $K'K$ is positive definite is not greater than the product of the maximum eigenvalue of N and that of $K'K$.

Any state space model can be put into the balanced form in two steps. When the Hankel matrix composed of covariance or impulse response matrices is available, we can decompose it by the singular value decomposition. This method is described fully in Chapter 9. Here we give an alternative method without recourse to the Hankel matrix. First, express the observability Gramian in a factored form, $G_o = P'P$, where $P = \Delta_o^{1/2} \Gamma_o'$, in terms of the orthonormal eigenvector matrix Γ_o of the observability Gramian and its associated eigenvalue matrix Δ_o. Then, change the state vector from x_t to $\hat{x}_t = Px_t$, with the corresponding changes in the system matrices, $\hat{A} = PAP^{-1}$, $\hat{B} = PB$, $\hat{C} = CP^{-1}$. The new observability Gramian becomes an identity matrix, I, and the new reachability Gramian is $\hat{G}_c = PG_cP'$. Next, decompose $\hat{G}_c$ as $W\Lambda^2 W'$, and define S_1 to be $\Lambda^{-1/2}W'$ in order to change the state vector once more from $\hat{x}_t$ to $\bar{x}_t = S_1 \hat{x}_t$. The new system matrices are $\bar{A} = S_1 \hat{A} S_1^{-1}$, $\bar{B} = S_1 \hat{B}$, and $\bar{C} = \hat{C} S_1^{-1}$. In this new representation, the reachability Gramian is $\bar{G}_c = S_1 \hat{G}_c S_1^{-1} = \Lambda$, and the observability Gramian becomes $\bar{G}_o = S_1^{-T} \hat{G}_o S_1^{-1} = \Lambda$. The combined transformation from x_t to $\bar{x}_t$ puts the original model in the balanced form. It is given, therefore, by $\bar{x}_t = Sx_t$ where $S = S_1 P = \Lambda^{-1/2} W'P$.

This conversion method works provided that the matrix S is invertible, i.e., if the system is observable, which ensures the existence of the inverse of the matrix P, and if the reachability Gramian is positive definite to ensure the existence of the inverse of Λ. The set of observable and reachable asymptotically stable dynamic systems with distinct singular values can be mapped continously into the balanced forms by the above procedure. The parametrization of balanced models is treated more fully in the following subsection.

Parametrization

This subsection discusses how to parametrize the class of all asymptotically stable, reachable, and observable discrete-time models. (This section may be skipped without loss of continuity.) A balanced model has a diagonal Gramian X that satisfies the two matrix Lyapunov equations

$$A'XA-X=-C'C \text{ , and } AXA'-X=-MM' \text{ .} \tag{15}$$

Recall that if all the eigenvalues of A have modulus less than one, i.e., if A is asymptotically stable, and if the ranks of the observability matrix O and the matrix Ω, which has the same structure as the reachability matrix, are equal to the dimension of the state vector, then X is symmetric and positive definite. (Actually any two of these three properties imply the third.)

Consider the class of all triples (A, C, M) satisfying these two matrix equations, where X is positive definite and A is asymptotically stable. This class is the same as the class of all asymptotically stable, observable and reachable systems. Since A is asymptotically stable, its singular value decomposition is given by $U\Lambda V'$, where Λ is diagonal with entries all less than one in magnitude. These three matrices uniquely determine the matrix A. The two equations in (15) are then uniquely solved to determine C, M, and a positive definite X. To see this, multiply the first equation of (15) from the left by V' and by V from the right. Do likewise with its second equation from the left by U' and by U from the right. These operations produce

$$\Lambda T\Lambda-S=-(CV)'CV \text{ , and } \Lambda S\Lambda-T=-(M'U)M'U \text{ ,} \tag{16}$$

where $T=U'XU$ and $S=V'XV$. To simplify the notation let Y stand for CV, and Z for $M'U$. Multiply the two equations, given in (16), by Λ from the left and right, and use (15) to convert them to

$$\Lambda^2S\Lambda^2-S=-Y'Y-\Lambda Z'Z\Lambda \text{ , and } \Lambda^2T\Lambda^2-T=-\Lambda Y'Y\Lambda-Z'Z \text{ .}$$

They are uniquely solved for S and T, because Λ's elements are all less than one in magnitude. As a result of the choice of both Y and Z to be of full rank, the right-hand sides of the above two equations are both negative definite. Therefore, the matrices S and T are both positive definite. (See Lyapunov equations in Appendix 4.) The solution of the first equation is expressed explicitly as

$$vecS=(I-\Lambda^2\times\Lambda^2)^{-1}vec\,(Y'Y+\Lambda Z'Z\Lambda) \text{ ,}$$

with a similar expression for $vecT$.

Let the singular value decomposition of the symmetric matrix T be $W\Gamma W'$. Then from the uniqueness of representation up to sign changes of the elements, the matrix X is the same as Γ, and U is equal to W'. Similarly, from the singular value decomposition of S, the matrix V can be obtained. Then, the matrix C is recovered as $C=YV'$, M as UZ', and A as $U\Lambda V'$. This solution procedure establishes that a balanced model can be completely parametrized by specifying a p *by* n matrix Y of rank p, a r *by* n matrix Z of rank r, a set of n numbers with magnitude less than one to specify Λ. In any innovation model introduced in this book, the dimension of the innovation vector is that of the data vector, i.e., $p = r$. In balanced innovation models then, there are $n(2p + 1)$ free parameters. After these free parameters are specified, the matrices A, C, and M are specified as shown above. Then the balanced innovation model can be constructed by solving the Riccati equation as shown in Chapter 7. We must specify a symmetric matrix Λ_0, i.e., equivalently a matrix Δ which requires $p(p + 1)/2$ parameters, because it determines the matrix B. In total, then, we need $n(2p +1) +p(p + 1)/2$ parameters for a complete specification of the data-generating processes.

6.4 Examples with Exact Covariance Matrices

This section presents several analytical examples to illustrate the preceding developments. In the first example, the dynamic system is of the second-order, and y and e are both scalar time series. Let the model be given by

$$x_{t+1} = Ax_t + be_t, \quad y_t = cx_t + e_t.$$

Here we use the lower-case letters b and c to denote two-dimensional vectors. The two Lyapunov equations associated with this model, (7) and (13), are

$$A\Sigma A' - \Sigma = -mm', \text{ and } A'\Sigma A - \Sigma = -c'c.$$

These equations are not independent, because both are of rank one. They differ only in the signs of the components. Let D be a diagonal matrix with entries 1 or -1. Then from the second equation we obtain $DA'\Sigma AD - D\Sigma D = -Dc'cD$. Then note that $D\Sigma D = \Sigma$ to see that $DA'D = A$, and $m = Dc'$. In our two-dimensional system, these relations imply that the matrix A is either

$$A = \begin{bmatrix} \alpha & \beta \\ \beta & \delta \end{bmatrix}$$

or the (2, 1) element of A is replaced by -β. In other words, the matrix A must either be symmetric or anti-symmetric, and the components of m are those of c or their negatives. In our applications, the parameters of the matrix A are constrained to render it asymptotically stable. Other examples of balanced models are found in Chapters 9 and 10.

Next, a special case of this balanced model is used to obtain a closed-form solution for the associated Riccati equation (9). The matrix A is taken to be anti-symmetric and the vector $c = m' = (1 \ \ 0)$ (the second element necessarily being zero to satisfy the equation for the observability Gramian),

$$A = \begin{bmatrix} 0 & \gamma \\ -\gamma & 0 \end{bmatrix},$$

The elements of the matrix Σ are $(1-\gamma^4)^{-1}$, called σ_1, and σ_2 which is γ^2 times σ_1. From (9a), the matrix Δ is a scalar δ in this example, given by $\lambda_0 - \pi_1$, where π_1 is the (1, 1) element of the matrix Π. The parameter λ_0 is the variance of the data $\{y_t\}$. Naming the off-diagonal element π_2 and the (2, 2) element as π_3, the Riccati equation produces the next three equations

$$\pi_1 = \gamma^2\pi_3 + (1-\gamma\pi_2)^2 / \delta,$$

$$\pi_2 = -\gamma^2\pi_2 + \gamma\pi_1(1-\gamma\pi_2) / \delta,$$

and

$$\pi_3 = \gamma^2\pi_1 + (\gamma\pi_1)^2 / \delta.$$

The equation for π_1 becomes quadratic after π_3 is substituted out, since π_1 must be positive. When solved, π_1 is equal to $\lambda_0 - \gamma^2\sigma_1^{1/2}$, assuming that this expression is positive, i.e., that λ_0 is large enough. Then the remaining two equations can be solved separately for π_2 and π_3. Note that π_2 is proportional to δ. From these, the components of the vector b are found to be given by $(1-\gamma\pi_1)/\delta$ and $\gamma\pi_2/\delta$.

The odd- and even-lagged covariances take on different forms for this example system. From $cm = 1$, $cAm = 0$, and the fact that $A^{2k} = (-\gamma^2)^k I$, k=1, 2, ..., it follows that λ_{2k} is zero and $\lambda_{2k+1} = (-\gamma^2)^k$.

The singular value decomposition of this Hankel matrix is given by

$$\boldsymbol{H} = \sigma_1 u_1 v_1' + \sigma_2 u_2 v_2',$$

where the singular values are the same as the elements of Σ. (This general fact is proved in the next section.) The infinite-dimensional vectors are defined by

$$u_1 = v_1 = (1-\gamma^2)^{1/2}[1\ 0\ {-\gamma^2}\ 0\ \gamma^4\ 0\ \ldots]',$$

and

$$u_2 = -v_2 = (1-\gamma^2)[0\ 1\ 0\ {-\gamma^2}\ 0\ \gamma^4\ 0\ \ldots]'.$$

The transfer function $T(z)$ in the sum form of the spectral decompositon (8) of Chapter 8 can be directly shown to be equal to $\lambda_0/2$ plus $c(zI - A)^{-1}m$, since $\lambda_{2k+1}z^{-(2k+1)}$ is equal to z^{-1} times $cA^{2k}z^{-2k}$. The matrix W in the spectral decomposition (9) of that chapter can be similarly derived.

6.5 Hankel Norm of a Transfer Function

The Hankel singular value of a transfer function $G(z)$ is defined as

$$\sigma_i[G(z)] = \lambda_i^{1/2}(G_o G_\Omega), \ i=1, 2,...,$$

where G_o and G_Ω are the observability and Ω Gramians of the system with the transfer function $G(z)$. These singular values are coordinate-free, i.e., are invariants of the models, because the auto-covariance matrices are free of the coordinate system chosen to express the state vectors. They are also independent of constant matrices of the transfer functions, such as Λ_0 in $T(z)$ or the identity matrix in $W(z)$ of (8) of Chapter 3, since they do not enter into the Hankel matrices.

The i-th eigenvalue of the matrix $G_o G_\Omega$ is the same as that of the matrix $\boldsymbol{H}'\boldsymbol{H}$ and $\boldsymbol{H}\boldsymbol{H}'$, where $\boldsymbol{H} = \boldsymbol{O}\Omega$ is the Hankel matrix constructed from the auto-covariance matrices--i.e., $CA^{i-1}M$. Hence the name "Hankel singular value". To establish this assertion, suppose that u is an eigenvector of the matrix product $G_o G_\Omega$ with an eigenvlaue λ. Then, $\Omega' G_o G_\Omega u = \lambda \Omega' u$. Recalling that $\boldsymbol{H} = \boldsymbol{O}\Omega$, this equation can be rewritten as $\boldsymbol{H}'\boldsymbol{H}(\Omega' u) = \lambda(\Omega' u)$. Conversely, starting from $\boldsymbol{H}\boldsymbol{H}'v = \lambda v$ and multiplying this equation from the left by $\boldsymbol{O}'$, it can be rewritten as $G_o G_\Omega(\boldsymbol{O}'v) = \lambda(\boldsymbol{O}'v)$. This establishes the claim.

When the dynamic system matrix A is stable, one can define the Hankel norm of a transfer function as the largest Hankel singular value,

$$\|G(z)\|_H = \sigma_1[G(z)] = \sqrt{\lambda_{\max}(G_o G_c)} = \sigma_1(\boldsymbol{H}),$$

where these equalities hold by definition.

In the balanced representation of the innovation model, the Gramians G_0 and G_Ω are both equal to a diagonal matrix Σ, which appears as a part of the singular value decomposition of the Hankel matrix. Therefore the Hankel norm of $T(z)$ in the sum decomposition of the spectral density is equal to σ_1, which is the largest singular value of Σ.

6.6 Singular Value Decomposition in the z-Domain

One way to express the singular value decomposition relation for a Hankel matrix, $H = U\Sigma V'$, is to write it as

$$H v^i = \sigma_i u^i , \tag{17}$$

where u^i and v^i are the i-th column vectors of the matrix U and V respectively. The pair of vectors u^i and v^i satisfying (17) is called the Schmidt pair. When the elements of the Hankel matrix are scalars--e.g., when the sequences of covariances or impulse responses that form the elements of the Hankel matrices are scalar-valued--(17) can be easily restated in the z-transform form. (See Adamjan et al. [1968, 1971] for a more complete analysis.) This alternative expression has practical convenience when dealing with infinite-dimensional vectors, and yields insight into the nature of lower-dimensional modeling of a dynamic relation.

Aside from a constant term, the transfer function of the innovation model is

$$G(z) = c(zI - A)^{-1}m = n(z) / d(z) = \sum_{i=1}^{\infty} \lambda_i z^{-i} \tag{18}$$

where $d(z)$ is a polynomial in z of degree n which is also the dimension of the state space model, and $n(z)$ is also a polynomial in z of degree less than n. Assume that the matrix A is asymptotically stable. The roots of $d(z)$ all lie strictly inside the unit disc in the complex plane.

Define two functions of z by

$$\mu_i(z) = \sum_{j=1}^{\infty} u_j^i z^{-j}, \tag{19}$$

and

$$\nu_i(z) = \sum_{j=1}^{\infty} v_j^i z^{j-1}. \tag{20}$$

Here, the subscript refers to the component of the vector indicated by superscript. Note that these vectors have finite 2-norm, e.g., $\sum_j (u_j^i)^2$ is finite, when the dynamic system that underlies the Hankel matrix is l_2-stable.

The different ways of constructing functions of z in (18) and (19) come from the fact that the Hankel matrix transforms a sequence extending to $-\infty$ into another sequence extending to $+\infty$. An explicit expression is found in Section 8.3. Eq.(19) yields a function which is analytic outside the unit disc, while (20) produces a function of z which is analytic inside the unit disc.

The relation (17) is then expressible as

$$[\, G(z)\nu_i(z)]_- = \sigma_i \mu_i(z),$$

where $[\cdot]_-$ denotes the operation of retaining only that part of the infinite series with negative powers in z, i.e., the causal part.

We next relate μ_i and ν_i explicitly to the system matrix and vectors following Silverman and Bettayeb [1980], by first putting the model in the balanced representation. In the balanced form, the matrix U is given by $\boldsymbol{O}\Sigma^{1/2}$. Therefore the vector u^i is given by $\boldsymbol{O}\Sigma^{-1/2} e_i$, where e_i is the (infinite) column vector with its only non-zero entry of 1 at the i-th position. Recalling our definition of the observability matrix , we obtain

$$\mu_i(z) = \sigma_i^{-1/2} c(zI - A)^{-1} e_i \, . \tag{21}$$

Denote this as $m_i(z)/d(z)$, where the degree of the numerator is less than n.

Similarly, the vector v^i is $\Omega'\Sigma^{-1/2} e_i$. Hence

$$\nu_i(z) = \sigma_i^{-1/2} e_i' (I - zA)^{-1} m. \tag{22}$$

To relate (22) explicitly to a rational function, use the facts that $DAD = A'$ and $m = Dc'$ in a balanced form, where D is a diagonal matrix with entries ± 1. Using these we deduce that $\nu_i(z)$ can be written as

$$\sigma_i^{-1/2} e_i'(I - zA)^{-1} b = \sigma_i^{-1/2} z^{-1} c (z^{-1} I - A)^{-1} De_i = \pm z^{-1} \mu_i(1/z). \tag{23}$$

Define $m^*(z) = z^{n-1} m(1/z)$ and $d^*(z) = z^n d(1/z)$. Then, $\nu_i(z)$ is equal to $\pm m^*(z)/d^*(z)$.

Now define an approximation to G(z) by

$$G_a(z) = G(z) - \phi_i(z),$$

where $\phi_i(z)$ is defined to be $\sigma_{i+1}\mu_{i+1}(z)/\nu_{i+1}(z)$. We note first from (23) that the function ϕ_i takes on the magnitude 1 on the unit circle. In other words, the error of this approximation has magnitude one for $z = e^{j\omega}$. Second, this function $G_a(z)$ is expressible as the ratio of the polynomial $p(z)$ over $d(z)$, where

$$p(z) = [n(z)m^*(z) - \sigma d^*(z)m(z)]/d(z)$$

in which σ is either of $\pm\sigma_{i+1}$. To see that $p(z)$ is indeed a polynomial, suppose the opposite. Then because the zeros of the polynomial $d(z)$ all lie inside the unit disc, therefore $[p(z)]_-$ is not zero. Because $d^*(z)$ has all its zeros outside the unit disc, $[p(z)/d^*(z)]_-$ is not zero. But,

$$[n(z)m^*(z) - \sigma d^*(z)m(z)]/d(z)d^*(z) = G(z)\,\nu_i(z) - \sigma\mu_i(z).$$

Taking the part of negative powers, we note it is not zero. However this contradicts (17). Thus, p(z) is, as asserted, a polynomial.

As an example, consider the one in Section 6. The function $\mu_1(z) = \kappa z/(z^2 + \gamma^2)$, where $\kappa = (1 - \gamma^4)^{1/2}$. The function $\nu_1(z) = \kappa/(1 + \gamma^2 z^2)$, $\mu_2(z) = \kappa/(z^2 + \gamma^2)$, and $\nu_2(z) = -\kappa z/(1 + \gamma^2 z^2)$. The approximation to $G(z) = z/(z^2 + \gamma^2)$ is then given by $G_a(z) = 1/(1 - \gamma^4)z$. The rank-one best approximation to this rank-two Hankel matrix is given by the matrix, with its only non-zero entry at (1, 1) given by $(1-\gamma^4)^{-1}$.

7 Innovation Models, Riccati Equations, and Multiplier Analysis

7.1 Innovation Models

Forward Innovation Models

Suppose that a representation in the form of (2) of Chapter 4 is adopted for time series models. The orthogonally-projected image of the state vector, given in (8) of Chapter 6, showes that the filtered estimate of the state vector evolves dynamically with the same dynamic matrix A as the original model. This vector is a vector of statistics which summarizes the information carried by the stacked data vector. Now, we show by a different reasoning that this vector is updated by exactly the same dynamics as the original model. To be explicit about the timing, and to show that the derivation is valid also for time-varying processes, we add time subscripts to the matrices Ω and R. Advance t by one unit in (8) of Chapter 6:

$$z_{t+1} = \Omega_{t+1} R_t^{-1} y_t^- . \tag{1}$$

Because the stacked data vector is partitioned conformably as

$$y_t^- = \begin{bmatrix} y_t \\ y_{t-1}^- \end{bmatrix} ,$$

the matrices Ω_{t+1} and R_t are related to those at the previous time by

$$\Omega_{t+1} = [M \quad A\,\Omega_t] , \tag{2}$$

and

$$R_t = \begin{bmatrix} \Lambda_0 & C\,\Omega_t \\ \Omega_t' C' & R_{t-1} \end{bmatrix} ,$$

where $E(y_t y_{t-1}^{-\prime})$ is the matrix with row submatrices Λ_1, Λ_2, ..., i.e., $C\,\Omega_t$. Express R_t^{-1} in the partitioned form as

$$R_t^{-1} = \begin{bmatrix} R^{11} & -R^{12} \\ -R^{21} & R^{22} \end{bmatrix} , \tag{3}$$

where

$$R^{11} = (\Lambda_0 - C\,\Pi_t C')^{-1},\ R^{12} = R^{11} C\,\Omega_t R_{t-1}^{-1},\ R^{21} = (R^{12})' ,$$

and

$$R^{22} = R_{t-1}^{-1} + R_{t-1}^{-1} \Omega'_t C' R^{11} C\,\Omega_t R_{t-1}^{-1}.$$

From (8) of Chapter 6, the covariance matrix of z_t is equal to

$$\Pi_t = \Omega_t R_{t-1}^{-1} \Omega_t', \tag{4}$$

which is used in the above to simplify the algebraic expression. From the observation equation of the state space model we also note that the matrix R^{11} is equal to the innovation noise covariance matrix Δ_t. Time subscript is added to indicate the possibly time-varying nature of the innovation noise covariance matrix. Using (2) and (3), (1) is rewritten as

$$z_{t+1} = M(R^{11}y_t - R^{12}y_{t-1}^-) + A\,\Omega_t(-R^{21}y_t + R^{22}y_{t-1}^-),$$

where (4) and the definition of z_t are used to note that $R^{12}y_{t-1}^-$ is the same as $R^{11}Cz_t$, that $\Omega_t R^{21} y_t$ is equal to $\Sigma_t C' R^{11} y_t$, and that $\Omega_t R^{22} y_{t-1}^-$ is the same as $z_t + \Sigma_t C' R^{11} C z_t$. Substituting these relations into the above, (1) becomes

$$z_{t+1} = Az_t + B_t(y_t - Cz_t), \tag{5}$$

where

$$B_t = (M - A\,\Pi_t C')R^{11},$$

which is the same as (4) of Chapter 3 except for the presence of time subscripts we added in this section. Recalling that z_t is the same as $x_{t|t-1}$, the expression Cz_t is equal to $y_{t|t-1}$, i.e., the difference $y_t - Cz_t$ is the innovation in the latest data vector y_t, i.e., that part of y_t which is orthogonal to the subspace spanned by y_{t-1}^-. Denote this difference by e_t, and call it the innovation vector. Since the covariance of this vector is $\Lambda_0 - C\,\Pi_t C'$, it is at least positive semi-definite. We require that it be positive definite. This condition is called the regularity condition by Faurre et al.[1979]. Eq. (5) is the Kalman filter for the model given in (2) of Chapter 4 if the system matrices and the noise covariance matrix are known.

In the steady state, Π_t can be identified, under some technical conditions, with Π, and B_t becomes the matrix B.

To summarize, suppose that the Hankel matrix constructed from the covariance matrices of the data vectors is decomposed into two full rank factors as in (10) of Chapter 5, and one of the factors, Ω, changes with time according to (2) for some matrices Λ_0, A, M, and C. Then, the vector defined by the right-hand side of (8) of Chapter 5 evolves over time according to (5), where the matrix B and the vector e_t are as defined above.

For an alternative way to see that the vector defined by (8) of Chapter 5 indeed evolves over time according to (5) (given the dynamic and the observation equations of (2) of Chapter 4, in which the disturbance vectors are serially uncorrelated by assumption), note first that z_t is uncorrelated with $y_t - y_{t|t-1}$ where the second term refers to the orthogonal projection of the first term onto the subspace spanned by y_{t-1}^-. Then the orthogonal projection of x_{t+1} in the left-hand side of (2) of Chapter 4 onto the subspace spanned by y_t^- is simply z_{t+1} by definition. On the right-hand side, using a property of the orthogonal projection given in (11) of Chapter 5, note that

$$\hat{E}(x_{t+1} \mid y_t^-) = \hat{E}(x_{t+1} \mid y_{t-1}^-) + \hat{E}(x_{t+1} \mid e_t) ,$$

where the substitution for x_{t+1} from the dynamic equation in (2) of Chapter 4 yields the expression that the first term is equal to Az_t since ξ_t is uncorrelated with y_{t-1}^-. The second term is equal to Be_t, where B is given by $E(x_{t+1}e_t')\Delta^{-1}$. Note that $E(x_{t+1}e_t')$ is equal to $M - A\Pi C'$, as in (4) of Chapter 3.

From (4), (2), and (3), the matrix Π_t evolves with time according to

$$\begin{aligned} \Pi_{t+1} &= cov\ z_{t+1} \\ &= [M \quad A\Omega_t]R_t^{-1}[M \quad A\Omega_t]' \\ &= A\Pi_t A' + (M - A\Pi_t C')(\Lambda_0 - C\Pi_t C')^{-1}(M - A\Pi_t C')' \ . \end{aligned} \tag{6}$$

The covariance matrix of the innovation, Δ , is equal to $\Lambda_0 - C\Pi_t C'$. This expression also gives the covariance of the one-step-ahead prediction error, $y_t - y_{t|t-1}$. When the inverse does not exist, then the pseudo-inverse is used instead. Pseudo-inverses are discussed in Aoki [1967]. See also Appendix 8 on the singular value decomposition. From the interpretation of the innovation covariance matrix, positive semi-definiteness rather than strict positive definiteness implies that some linear combination of components of y_t can be predicted exactly, i.e., y_t is not purely random. By assumption, this case does not arise in time series treated in this book. Eq. (6) is the Riccati equation for the covariance matrices of the state vector. The steady-state covariance matrix satisfies the algebraic Riccati equation in which Σ_t and Σ_{t+1} are replaced by Σ_*. The matrix $M - A\Pi_t C'$ becomes B. By choice, Π_* is the minimal element of the set of symmetric solutions of the Riccati equation

$$X = AXA' + (M - AXC')\Delta^{-1}(M - AXC')'$$

in the sense that any positive definite solution satisfies the inequality $X \geq \Sigma_*$. See, for example, Faurre et al. [1979, chapter 4] on the class of all possible solutions of Riccati equations.

The innovation model (5) is special because the random disturbance term on the state vector dynamics, Be_t, and that on the observation vector y_t are such that their joint covariance matrix has the factored form

$$cov\begin{bmatrix} Be_t \\ e_t \end{bmatrix} = \begin{bmatrix} B \\ I \end{bmatrix} \Delta (B' \ I\ ) . \tag{7}$$

Such a factored expression for the random disturbances of a general state space models is not initially obvious.

Backward Innovation Models

We employ the procedure just described for the forward innovation models to derive a new representation, called backward innovation model, for the state space model given in (1) of Chapter 4, i.e., $z_{t+1} = Az_t + Be_t$, and $y_t = Cz_t + e_t$, by running time backward, i.e., the data set at time t is y_t^+ which becomes y_{t-1}^+ when a "new" data vector y_{t-1} is added. In a particular coordinate system we describe below, the backward innovation model is of the form

$$\zeta_t = A'\zeta_{t+1} + Nf_t \text{ ,and, } y_t = M'\zeta_{t+1} + f_t \ ,$$

where the backward innovation vector is defined by

$$f_t = y_t - \hat{E}(y_t \mid y_{t+1}^+) \ .$$

This is the one-step ahead forecast error when time runs backward. Note that we now project y_t^- onto the manifold spanned by y_{t+1}^+, while in the forward case, we project y_{t+1}^+ onto the manifold spanned by y_t^-. It is important to observe that the matrix M which does not appear explicitly in the forward innovation expression now appears in the backward one as the matrix in the observation equation, i.e., the one relating y_t to the backward state vector.

To derive this set of equations, we calculate the orthogonal projection of any state vector x_t onto y_t^+

$$\begin{aligned} s_t &= \hat{E}(x_t \mid y_t^+) \\ &= \Theta_t R_+(t)^{-1} y_t^+ \end{aligned}$$

with

$$\Theta_t' = E(y_t^+ x_t').$$

From the relation analogous to that in (1) of Chapter 6, note that Θ_t is equal to $\Pi_t O'$, with the covariance matrix of the original state vector at time t denoted by Π_t. Since time runs backward, y_t^+ becomes y_{t-1}^+ next period, where y_{t-1}^+ is the stacked vector composed of y_{t-1} above y_t^+. Hence

$$R_+(t-1) = \begin{bmatrix} \Lambda_0 & M'O' \\ OM & R_t(t) \end{bmatrix} \tag{8}$$

because the cross covariance matrix between y_{t-1} and y_t^+ is $M'O'$, and

$$\Theta_{t-1} = \Pi_{t-1}[\, C' \quad A'O' \,] \ . \tag{9}$$

Note that

$$R_+(t-1)^{-1} = \begin{bmatrix} R^{11} & -R^{12} \\ -R^{21} & R^{22} \end{bmatrix}$$

with

$$R^{11} = (\Lambda_0 - M'ZM)^{-1} = \Delta_f^{-1} ,$$

$$R^{12} = R^{11} M'O'R_+(t)^{-1} = (R^{21})' ,$$

$$R^{22} = R_+(t)^{-1} + R_+(t)^{-1} O\, M R^{11} M'O'R_+(t)^{-1}$$

with $Z_t = O'R_+^{-1}(t)O$, and Δ_f^{-1} is defined by the above equality. Define a new state vector, ζ_t by $\Pi_t^{-1} s_t$, i.e., by

$$\zeta_t = O'R_+^{-1} y_t^+ , \tag{10}$$

Collecting these terms exactly as in the case of the forward model derivation, we obtain

$$\zeta_{t-1} = A'\zeta_t + (C' - A'ZM)R^{11}(y_{t-1} - M'\zeta_t).$$

Now let

$$f_t = y_t - M'\,\zeta_{t+1}$$

and

$$N = (C' - A'ZM)R^{11}.$$

Then the backward innovation model is given by

$$\zeta_{t-1} = A'\zeta_t + N f_{t-1} \text{ ,and, } y_t = M'\zeta_{t+1} + f_t \ . \tag{11}$$

From the definitional relation, Z_t is the covariance matrix of this vector ζ_t. The matrix Z_{t-1} is then equal to

$$Z_{t-1} = [C' \ A'O'\,]R_+(t-1)[C' \ A'O']' .$$

This relation yields

$$Z_{t-1} = A'Z_t A + N_t \Delta_f N_t' ,$$

where we introduce a new matrix

$$N_t = (C' - A'Z_t M)R^{11}.$$

When the matrix A is asymptotically stable, this recursion equation for Z_t becomes an algebraic Riccati equation for Z,

$$Z = A'ZA + N\Delta_f N' . \tag{12}$$

In the backward model, given in (11), the matrix C plays the role of the matrix M in the forward model,

$$\begin{aligned} C' &= \mathrm{E}\zeta_t y_t' \\ &= A'SM + N\Delta_f . \end{aligned}$$

The expression for the covariance matrices are the same,

$$\begin{aligned} \Lambda_1 &= y_t y_{t-1}' \\ &= \mathrm{E}(M'\zeta_{t+1} + f_t)(\zeta_t'M + f_{t-1}) \\ &= \mathrm{E}(M'\zeta_{t+1} + f_t)\,\zeta_t'M \\ &= \mathrm{E}(M'\zeta_{t+1} + f_t)(\zeta_{t+1}'A + f_t'N')M \\ &= M'ZAM + \Delta_f N'M \\ &= CM , \end{aligned}$$

and similarly, $\Lambda = CA^{k-1}M$.

The observability of this model is the same as the M-controllability of the original forward innovation model. In this sense, the backward representation is "dual" to the forward one. In the forward model the matrix M does not appear explicitly, even though this matrix is an important stepping stone in deriving the matrix B. In the backward model, the matrix C does not appear explicitly in the representation, even though it is needed in calculating the matrix N, which is the "filter gain" in the backward model. In Chapter 9 we use the backward innovation model to propose an instrumental variables estimator for the matrix M.

7.2 Solving Riccati Equations

We first present a closed-form expression for the state vector covariance matrix for a class of VAR processes which satisfy certain technical conditions. For this class we can therefore bypass the solution of the Riccati equation for the state vector covariance matrices. We then present recursive and non-recursive solution methods of the Riccati equation for the state vector covariance matrix for a general class of data generating processes.

Closed Form Solutions for VAR Models

Following the general method for introducing state vectors, as given in Chapter 4, we note that a p-dimensional VAR(d) process given by

$$y_t = \phi_1 y_{t-1} + \phi_2 y_{t-2} + \cdots + \phi_d y_{t-d} + u_t \ ,$$

has a state vector

$$x_t = \Phi y_{t-1,d}^- \ , \tag{13}$$

where $y_{t-1,d}^-$ is a truncation of the stacked data vector y_{t-1}^- at y_{t-d}. The matrix Φ displays the coefficient matrices in the original VAR(d) model along the counter-diagonal line

$$\Phi = \begin{bmatrix} \phi_1 & \phi_2 & . & \phi_d \\ \phi_2 & \phi_3 & . & 0 \\ \cdot & \cdot & \cdot & \cdot \\ \cdot & \phi_d & 0 & . \\ \phi_d & 0 & 0 & 0 \end{bmatrix} .$$

Assume that this $dp \times dp$ matrix is non-singular, so that the vector x_t introduced above is minimal. This vector is, by construction, uncorrelated with the exogenous noise. Therefore the state vector x_t is the same as its orthogonal projection onto the manifold spanned by y_{t-1}^-. Consequently, this matrix Φ is the same as the expression in the orthogonal projection ΩR_-^{-1} given in (8) of Chapter 6, where R_- is the covariance matrix of y_{t-1}^-. We thus can state

Proposition When the matrix Φ above is nonsingular, then the VAR process has a minimal dimensional state space model with a dimension which is an integral multiple of the dimension of the data vector, and the Riccati equation for the state vector covariance matrix, given in (5) of Chapter 3, has a closed-form solution $\Omega R_-^{-1} \Omega'$ where Ω , which is equal to ΦR_-, is the second factor in the Hankel matrix decomposition, as given in (10) of Chapter 6. When the matrix Φ is singular, there exists a matrix $\hat{\Phi}$ of rank n, which is less than dp, such that ΩR_- is equal to $\hat{\Phi}$, i.e., $\hat{\Phi} R_- \hat{\Phi}'$ solves the Riccati equation.

Example. The following example illustrates the second half of this Proposition. Take d to be 2 and

$$\phi_1 = \begin{bmatrix} .2 & .1 \\ .3 & .5 \end{bmatrix} \ , \ \phi_2 = \begin{bmatrix} -.2 & -.2 \\ .1 & .1 \end{bmatrix}$$

and

$$cov \ u_t = \begin{bmatrix} 1 & .2 \\ .2 & 1 \end{bmatrix} .$$

Then the matrix Φ has rank 3, not 4. A three-dimensional state vector is defined by the method of Chapter 4, i.e., $x_{1t} = y_{1t} - u_{1t}$, $x_{2t} = y_{2t} - u_{2t}$, and $x_{3t} = y_{1t-1} + y_{2t-1}$, for example. Then matrix $\hat{\Phi}$ is defined as the 3×4 matrix with ϕ_1 and ϕ_2 as the top row submatrices, and the third row is a vector with elements 1,1, and 0. Some further examples are presented in Section 9.3.

Iterative Solution Algorithm

Solutions of a Riccati equation are non-unique. When ordered by using the partial ordering of symmetric positive semi-definite matrices, there is the minimal and maximal element in the solution set. We need the minimal solution because the state vector of the innovation model has the smallest covariance matrices of any state vector, since it is the orthogonally-projected image of the latter. In this subsection only, denote this covariance matrix by Π_*. This matrix Π_* may be iteratively calculated as follows:

$$\begin{cases} \Pi_0 = 0, \\ \Pi_{j+1} = A\Pi_j A' + (M - A\Pi_j C')(\Lambda_0 - C\Pi_j C')^{-1}(M - A\Pi_j C')' , \\ \Pi_* = \lim_{j\to\infty} \Pi_j , \end{cases} \quad (14)$$

or solved directly as discussed in the next subsection.

As is well-known concerining any Newton-Raphson type iterative scheme, there is always the danger that the iterates generated by the recursion will overshoot the true solution if the starting value is too 'far' from the true solution. To avoid this, it may be desirable to use a small scale factor to shorten each adjustment step. Let the right hand side of (14) be denoted by $f(PI_j)$. Then, instead of (14) use

$$\Pi_{j+1} = \Pi_j + \varepsilon[f(\Pi_j) - \Pi_j] ,$$

for some small positive value of ε.

The speed of convergence is governed by the eigenvalues of $F = A - BC$ rather than those of A itself in the neighborhood of Π_*. A straightforward demonstration of this point may be had by taking the variational expression of the Riccati equation and noting that deviations from the solution are governed by

$$Z_{j+1} = AZ_jA' - AZ_j(BC)' - BCZ_jA' + BCZ_j(BC)' ,$$

where $Z_j = \Pi_j - \Pi_*$, and we use a relation that $\delta\Delta^{-1}$ is equal to $-\Delta^{-1}\delta\Delta\Delta^{-1}$. Vectorizing the above we see that

$$vecZ_{j+1} = (F \otimes F)vecZ_j .$$

This iterative scheme fails to converge if the largest eigenvalue of F is not less than one in magnitude, and converges slowly if the largest eigenvalue is close to one. If the matrix A has an eigenvalue of unit magnitude and if it is not a controllable pair of (A, M) or equivalently such an eigenvalue is not an observable one of the pair (A', M') in the sense that there is an eigenvector q such that $A'q = \lambda q$ and $M'q = 0$, then this eigenvalue is inherited by F. To see this substitute the matrix B by its definitional relation and rewrite F as $A' - C'\Delta^{-1}(M' - C\Pi_* A') = (I + C'\Delta^{-1}C\Pi_*)A' - C'\Delta^{-1}M'$. Now multiply it from the right by q and note that $\Pi_* q$ is zero, because Π_* is equal to $\Omega R^{-1}\Omega'$, and that Ωq is zero, because $M'q = 0$, $M'A'q = 0$, and so on, leaving the relation $F'q = \lambda q$.

A Non-Iterative Solution Algorithm

A non-iterative algorithm for solving an algebraic Riccati equation is available by exploiting properties of symplectic matrices. The method described here adapts the techniques reported in Vaughan [1970] and Laub [1983]. (Although quite analogous to the symplectic matrices that appear in the Hamiltonian formulations of optimal control or usual Kalman filter gain calculations, there are some subtle differences in the Riccati equation that arise in the construction of the innovation models.)

Define three auxiliary $n \times n$ matrices by

$$\psi = A' - C'\Lambda_0^{-1}M', \quad \hat{Q} = C'\Lambda_0^{-1}C, \quad \text{and} \quad D = M\Lambda_0^{-1}M'.$$

Assume that the matrix ψ is nonsingular. This assumption is implied by the assumption that (A, C) is an observable pair from the Hautus' characterization of the observability. Construct a $2n \times 2n$ matrix Φ, where n is the dimension of the matrix A

$$\Phi = \begin{bmatrix} \psi - \hat{Q}\psi^{-T}D & \hat{Q}\psi^{-T} \\ -\psi^{-T}D & \psi^{-T} \end{bmatrix}, \tag{15}$$

where the symbol $-T$ stands for the inverse of transpose. This matrix is a product of three matrices

$$\begin{bmatrix} I & \hat{Q} \\ 0 & I \end{bmatrix}, \quad \begin{bmatrix} I & 0 \\ 0 & \psi^{-T} \end{bmatrix}, \quad \text{and} \quad \begin{bmatrix} \psi & 0 \\ -D & I \end{bmatrix}.$$

Alternatively, it is the product of the inverse of the product of the first two matrices and the third. Call the inverse of the product of the first two matrices Φ_1 and the third matrix Φ_2, i.e., $\Phi = \Phi_1^{-1}\Phi_2$.

The matrix Φ we constructed above is symplectic, i.e., satisfies the definitional relation $\Phi' J = J\ \Phi^{-1}$, where

$$J = \begin{bmatrix} 0 & I \\ -I & 0 \end{bmatrix} .$$

This can be verified directly using

$$\Phi^{-1} = \begin{bmatrix} \psi^{-1} & -\psi^{-1}\hat{Q} \\ D\psi^{-1} & \psi' - D\psi^{-1}\hat{Q} \end{bmatrix} .$$

The relation of the symplectic matrix, $\Phi' = J\ \Phi^{-1} J'$, implies that if λ is an eigenvalue of a symplectic matrix, then so is $1/\lambda$. A proof of this fact and other properties of symplectic matrices are collected in Appendix 6 on symplectic matrices. That the eigenvalues of a symplectic matrix come in pairs of λ and λ^{-1} is easily seen because the determinant of Φ is 1, as can be verified by its three-product form noted above.

Example. Let n be one. Then the three matrices are scalars, and we denote them by r, q and d in this example. The characteristic equation is the quadratic polynomial $x^2 - \alpha x + 1$, with $\alpha = \psi + (1-qd)/\psi$. In the univariate series the scalar r is the ratio of Λ_1/Λ_0. The product qd is equal to r^2. Then the characteristic polynomial has two equal roots, namely 1 and -1, because the product must be one if the dynamic parameter a is equal to one.

Next, transform the matrix Φ into the real Schur decomposition form (see Golub and van Loan [1983, p.219] for example)

$$W'\Phi W = \begin{bmatrix} \phi_{11} & \phi_{12} \\ 0 & \phi_{22} \end{bmatrix} ,$$

where the submatrix ϕ_{ii} is a block upper triangular $n \times n$ matrix,

$$\phi_{ii} = \begin{bmatrix} R_{11} & R_{12} & \cdot & R_{1m} \\ 0 & R_{22} & \cdot & \cdot \\ 0 & 0 & \cdot & \cdot \\ \cdot & \cdot & \cdot & \cdot \\ \cdot & \cdot & \cdot & \cdot \\ \cdot & \cdot & \cdot & \cdot \\ 0 & 0 & 0 & R_{mm} \end{bmatrix}$$

where R_{jj} is either a real number or a 2×2 matrix, $j=1, \ldots, m$ for some m, and $i=1, 2$.

Collect the eigenvalues with magnitudes less than one into the submatrix ϕ_{11}. (Since the solution is invariant with respect to any nonsingular transformation, the ordering of eigenvalues is immaterial so long as they all have magnitude less than one.) If the matrix pair (A, M) is reachable and the pair (A, C) is observable, then no eigenvalue of Φ will lie on the unit circle. This fact is established in Appendix 6. To see this heuristically, consider a special case in which the matrix $\hat{Q}$ is sufficiently "smaller" than other matrices. Then, the matrix Φ is nearly block triangular and its eigenvalues are those of the diagonal submatrices, ψ^{-T} and $\psi - \hat{Q}\psi^{-T}D$. The latter is approximately ψ when $\hat{Q}$ is sufficiently small. Next, note that the eigenvalues of ψ' are those of the Kalman filter predictor, $A - BC$, because $(M - A\Pi C')\Delta^{-1}C$ is nearly equal to $M\Lambda_0^{-1}C$, where the Woodbury identity, Householder [1964, p.124] is used to see that $C\Delta^{-1}C'$ is nearly zero under the current assumption. Therefore, Φ has eigenvalues of magnitude one if and only if $(A - BC)$ has eigenvalues of unit magnitude. This possibility is ruled out if the model is minimal dimensional, so that (A, B), or equivalently (A, M), is a reachable pair and (A, C) is an observable pair. Theoretically, these conditions may be satisfied, and yet numerically $\hat{Q}$ could be buried under the statistical noise, especially if Λ_0 is relatively larger than the entries in the other system matrices. Then eigenvalues of unit magnitudes could arise even if the model is theoretically minimal dimensional.

Partition the orthogonal matrix W conformably into $n \times n$ submatrices W_{ij}, and note from the orthogonality of the matrix W that $W_{21} W_{11}^{-1} = -W_{22}^{-T} W_{12}'$. Denote this matrix by X. The main result of this section is stated as

Proposition. The matrix $X = W_{21} W_{11}^{-1}$ of (30) solves the Riccati equation

$$X = AXA' + f(X),$$

where

$$f(X) = (M - AXC')(\Lambda_0 - CXC')^{-1}(M - AXC')'.$$

This proposition can be verified by direct substitution. First, the fact that the (2, 1) submatrix of $W'\Phi W$ is zero can be stated as

$$-X\Phi_{11} - X\Phi_{12}X + \Phi_{22}X + \Phi_{21} = 0,$$

which becomes upon substitution of the definitional expression

$$0 = X\psi + (-X\hat{Q} + I)\psi^{-T}(D - X).$$

This equation is the same as the Schur decomposition form written for the factors of the matrix Φ

$$\Phi_2\begin{pmatrix} I \\ \Pi \end{pmatrix} = \Phi_1\begin{pmatrix} I \\ \Pi \end{pmatrix}\Gamma,$$

for some matrix Γ.

Move the second term to the left and multiply from the left by $\psi'(-X\hat{Q}+I)^{-1}$ to obtain

$$X = \psi'(-\hat{Q}+X^{-1})^{-1}\psi + D.$$

Next, use the Woodbury identity

$$(X^{-1}-\hat{Q})^{-1} = (X^{-1}-C'\Lambda_0^{-1}C)^{-1} = X + XC'(\Lambda_0 - CXC')^{-1}CX,$$

and substitute ψ out by its definitional relation. Then the equation for X becomes

$$X = AXA' + (M-AXC')(\Lambda_0-CXC')^{-1}(M-AXC')'.$$

This last expression is obtained by grouping the terms in

$$X=\psi'X\psi+D+\psi'XC'\Delta^{-1}CX\psi$$

after moving the term AXA' to the left as

$$X-AXA' = -AXC'\Delta^{-1}[\Delta+CXC']\Lambda_0^{-1}M' - (\text{its transpose})$$

$$+M\Delta^{-1}M'+AXC'\Delta^{-1}CXA',$$

where $\Lambda_0 - CXC'$ is denoted by Δ. The terms on the right-hand side can be grouped as shown. Since X is the covariance matrix of state vector of the innovation model, the solution of the Riccati equation which is positive semi-definite must be used. Additionally, the solution must satisfy the inequality

$$\Lambda_0 > CXC'.$$

This is a condition to ensure that the innovation error covariance matrix is full rank, i.e., innovations are not confined to some subspace in the p-dimensional Euclidean space.

This procedure breaks down when the matrix A has one or more eigenvalues of unit magnitude. If the matrix Φ has an eigenvalue of magnitude one, then Appendix 6 establishes that the matrix A of the innovation model that underlies the matrix Φ is not asymptotically stable, that the pair (A, M) is not controllable or reachable, and that the pair (A, C) is not observable. For these reasons, the matrix Φ with an eigenvalue of unit magnitude is not useful in deriving asymptotically stable minimal dimensional innovation models.

Scalar Riccati equations are quadratic, and hence have two solutions, of which only one may satisfy the above inequality. If both do, then take the smaller one. More generally, this inequality must be imposed as a regularity condition on the solutions of Riccati equations.

Appendix 6 deduces some properties of symplectic matrices. One property important in the determination of the model dimension is that the matrix Φ has eigenvalues of unit magnitude if and only if they are either the uncontrollable or unobservable eigenvalues of the pair (A, M) or of (A, C). If the system matrix A of the innovation model has eigenvalues of unit magnitude, then the model is not asymptotically stable and the state vector will not be weakly stationary. We cannot construct innovation models in such cases. This situation arises if eigenvalues of unit magnitude are either uncontrollable or unobservable, i.e., when the model is not of minimal dimension.

7.3 Likelihood Functions

Pre-sample period disturbances and endogenous variables which serve as initial conditions are usually assumed to be constant, e.g., zero, or back-forecasted. More properly, they should be recognized as random vectors. As summarized in Ansley [1979], several derivations of the exact likelihood of ARMA processes are available in the literature. In state space representations the likelihood function conditional on the initial measurement is easily written down by inspection. One can also give the recursion for the expression with exponents of the likelihood function.

Recursive calculation of the exponential expression is stated as the next

Fact

$$y_T^{-\prime} R_T^{-1} y_T^- = \Sigma_1^T e_t{}' \Delta_t^{-1} e_t + y_0{}' R_0^{-1} y_0 \ . \tag{16}$$

To verify this relation, start with the last data period T. The data vector is y_T^- with the covariance matrix

$$R_T := \begin{bmatrix} \Lambda_0 & C\,\Omega_{T-1} \\ (C\,\Omega_{T-1})' & R_{T-1} \end{bmatrix}$$

with $R_{T-1} = cov\, y_{T-1}^-$ and $\Omega_{T-1} = E\,(z_T y_{T-1}^-)$. Dropping the subscript T, denote its inverse by

$$R_T^{-1} = \begin{bmatrix} R^{11} & -R^{12} \\ -R^{21} & R^{22} \end{bmatrix} \ .$$

Then calculate

$$y_T{}' R_T^{-1} y_T^- = (y_T{}' \ \ y_{T-1}^-{}') R_T^{-1} \begin{bmatrix} y_T \\ y_{T-1}^- \end{bmatrix} \ .$$

As in Section 1, we note that

$$R^{11} = \Delta$$

with

$$\Delta = \Lambda_0 - C\,\Omega_{T-1} R_{T-1}^{-1}\,\Omega_{T-1}{}'C' ,$$

$$= \Lambda_0 - C\,\Sigma_T C'$$

$$\Sigma_T = \Omega_{T-1} R_{T-1}^{-1}\,\Omega_{T-1}{}' ,$$

$$R^{12} y_{T-1}^{-} = \Delta_T^{-1} C\,\Omega_{T-1} R_{T-1}^{-1} y_{T-1}^{-} = \Delta_T^{-1} y_{T|T-1} ,$$

since

$$y_{T|T-1} = CE\,(z_T y_{T-1}{}') R_{T-1}^{-1} y_{T-1}^{-}$$

$$= C\,\Omega_{T-1} R_{T-1}^{-1} y_{T-1}^{-} \quad ,$$

and that

$$R^{22} = R_{T-1}^{-1} + R_{T-1}^{-1}\,(C\,\Omega_{T-1})'\Delta^{-1}(C\,\Omega_{T-1}) R_{T-1}^{-} \quad .$$

Therefore, we simplify the preceding expression as

$$y_T{}' R_T^{-1} y_T^{-} = e_T{}'\Delta^{-1} e_T + y_{T-1}^{-1} R_{T-1}^{-1} y_{T-1}^{-} \quad .$$

Iterating this reduction step, we obtain the Fact.

If $y_1, \cdots, y_T$ are generated by an AR(p) process, then the first p data points serve as the initial conditions for the remainder. Noting that

$$e_t = y_t - \Sigma_1^p \phi_j y_{t-j} \quad , \quad t \geq p+1$$

is independent of $y_1, \cdots, y_p$, the joint density of $e_{t+1}, \cdots, e_T$ has the expression

$$constant.\ |\Delta|^{-(T-p-1)/2}\ exp\ -\tfrac{1}{2}\Sigma_{p+1}^{T} e_t{}'\Delta^{-1} e_t.$$

The joint density of the first p data points is

$$const\ |R_p|^{-\frac{1}{2}}\ exp\ -\tfrac{1}{2}(y_p^{-})' R_p^{-1} y_p^{-} \quad .$$

Thus the joint density of y_T^{-} is the product of these two. The exponent is $-\frac{1}{2}$ times

$$(y_p^{-})' R_p^{-} y_p^{-} + \Sigma_{p+1}^{T} e_t{}'\Delta^{-1} e_t = tr\, R_p^{-} y_p^{-} y_p^{-\prime} + \Sigma\, tr\, \Delta^{-1} e_t e_t{}' .$$

Then, the logarithm of the likelihood function is

$$ln\, p(y_T^-) = -\frac{T}{2}\, ln\, 2\pi - \tfrac{1}{2}\, ln\, |R_N| - \tfrac{1}{2} y_T^{-\prime} R_N^{-1} y_T^-$$

$$= -\frac{T}{2}\, ln\, 2\pi - \tfrac{1}{2}\, ln\, |R_N| - \tfrac{1}{2}\, tr\, \Delta^{-1} S_T - \tfrac{1}{2} y_0' R_0^{-1} y_0$$

where

$$S_T = \sum_1^T e_t e_t' \quad .$$

The expression $|R_N|$ depends only on the data, and the last term $\Delta^{-1} S_T$ depends on both the model parameters and data.

When the model is AR(p), then

$$e_t = y_t - \Sigma_{j=1}^{p} \phi_j y_{t-j} \quad , \quad t \geq p+1$$

and

$$S_T = \Sigma y_t y_t' - \Sigma_j \phi_j y_{t-j} y_t' - \Sigma_i y_t y_{t-i}' \phi_i' + \Sigma_{i,j} \phi_i y_{t-i} y_{t-j}' \phi_j'$$

i.e., Λ_j, $j \leq p$ and the terms involving the initial and the last p data points are the sufficient statistics. This is the result obtained by Arato [1961] and Anderson [1963]. Dickinson [1983] provides a more refined analysis on the sufficient statistics. The main point of this type of argument is the proof that only AR processes have sufficient statistics, the number of which is independent of the data point, and potentially much smaller than the total number of data points.

The likelihood function is usually derived more directly. We follow Dent [1977] to illustrate. Ansley [1979] contains a short summary of the contributions by Newbold [1974] in addition to those by Dent. Given an ARMA (p, q) model for a mean zero y_t,

$$y_t = \Sigma_{i=1}^{p} \phi_i y_{t-1} + u_t - \Sigma_1 \theta_j u_{t-j} \ ,$$

let A_1 and A_2 be ($T \times T$) upper triangular matrices containing, respectively, only ϕs and θs, such that

$$A_1 y_T^- = A_2 u_T^- + A_b f \ , \tag{17}$$

where f is a vector containing only initial conditions

$$f' = (y_0, y_{-1}, \cdots, y_{1-p}, u_0, u_{-1}, \cdots, u_{1-p}) \quad .$$

The matrix A_1 is a Toeplitz matrix with $1, -\phi_1, \cdots, -\phi_p$ along the main and super diagonal lines and similarly for A_2 with $1, -\theta_1, \cdots, -\theta_p$. The matrix A_3 is such that

$$A_3 = \begin{bmatrix} O \\ I_r \end{bmatrix} \quad : T \times r^*$$

with $r^* = max\,(p, q)$.
Rewrite (17) as

$$y_T^- = A_1^{-1}(A_2, A_3)\begin{bmatrix} u_T^- \\ f \end{bmatrix} \quad .$$

Thus

$$R_T = A_1^{-1}(A_2 \;\; A_3)\begin{bmatrix} I_T \otimes \Delta & 0 \\ 0 & \Sigma_f \end{bmatrix}\begin{bmatrix} A_2' \\ A_3' \end{bmatrix}(A_1')^{-1}$$

$$= A_1^{-1}[A_2(I_T \otimes \Delta)A_2' + A_3\Sigma_f A_3'](A_1')^{-1} ,$$

where

$$cov\; u_T^- = I_T \otimes \Delta \quad ,$$

and

$$R_T^{-1} = A_1' G^{-1} A_1 ,$$

where

$$G = A_2(I_T \otimes \Delta)A_2' + A_3\Sigma_f A_3'$$

$$= A_2[I_T \otimes \Delta + A_2^{-1}A_3\Sigma_f A_3'(A_2')^{-1}]A_2' .$$

Next, rewrite G as

$$G = A_4[I_T + A_5A_5']A_4' ,$$

$$A_4 = A_2(I_T \otimes \Delta^{1/2}) ,$$

$$A_5 = (I_T \otimes \Delta^{-1/2})A_2^{-1}A_3\Sigma_t^{1/2}$$

$$= USV' ,$$

and USV' is the singular value decomposition of A_5. Since the rank of A_5 is r^*, we can take U to be $T \times r^*$, and S to be $r^* \times r^*$.
Then

$$G = A_4[I_T + US^2U']A_4$$

and

$$G^{-1} = A_4^{-1}{}'[I_T - US(I_r' + S^2)^{-1}SU']A_4^{-1} .$$

We arrive at the desired expression

$$y_T^{-\prime}R_T^{-1}y_T^{-} = \sum_1^T \eta_j^2 - \sum_1^r \delta_j \zeta_j^2$$

with

$$\eta = (I \otimes \Delta^{-1/2})A_1 A_2^{-1} y_T^{-}$$

and

$$\zeta = U'\eta \ , \quad \delta_i = s_i^2/(1+s_i^2) \ , \quad i = 1 \cdots r^* \ .$$

Identification

Outputs from Kalman filters can be used to construct the likelihood function of the observed data sequences, Schweppe [1965]. The time series need not be stationary. Applications to non-stationary cross-sectional data have been reported in Goodrich and Caines [1979], for example. Gardner et al. [1980] describe a computer algorithm for exact maximum likelihood estimation of ARMA models by Kalman filters. Denote the stacked vector consisting of $y_t, y_{t-1}, \cdots$ by y^t. The joint probability density function can be written as the product of successive conditional probability density functions

$$p(y^T \mid x_0) = \prod_0^{T-1} p(y_{t+1} \mid y^t, x_0)\, p(y_0 \mid x_0) \,, \tag{18}$$

where $p(y_{t+1} \mid y^t, x_0)$ is assumed to be approximately normal with mean $y_{t+1|t}$ and the covariance matrix Δ. For a reasonably large T, the initial condition vector x_0 may be put to zero for a causally invertible system. Then, the joint probability density function is approximately given by

$$(2\pi)^{-(T+1)p/2} det(\Delta)^{-(T+1)/2} \exp\frac{-1}{2}[\sum_0^T (y_t - y_{t|t-1})'\Delta^{-1}(y_t - y_{t|t-1})] \,.$$

This expression for the joint probability density function makes clear that two sets of model parameters (θ_1, Δ_1), and (θ_2, Δ_2), where θ stands for the collection of the matrices A, B, and C, are indistinguishable if and only if the two covariance matrices are equal, $\Delta_1 = \Delta_2$, and if the Markov parameters of the inverse system, H_k, are the same for each of the two parameters θ_1 and θ_2 for all $k \geq 1$. Since the latter condition on the Markov parameters of the inverse systems holds if and only if the Markov parameters of the original systems are identical, $H_k(\theta_1) = H_k(\theta_2)$, $k \geq 1$, two minimal innovation representations are indistinguishable if and only if the covariance matrices and the set of Markov parameters are the same.

Two notions of identifiability exist in the literature: one based on consistency, and the other on uniqueness. In the latter sense, different parameter values must produce different probability distributions of data for the model to be called identifiable (Solo [1986]). Suppose that for a suitable parametrization of models, the parameter vector θ uniquely specifies a model within a class. The parameter vector then has a one-to-one relation with each model in the class. Depending on the class of models under consideration, the "true" model, represented by the parameter θ_0, may or may not be in the class. When it is, the identifiability notion based on consistency is relevant, because the convergence of the estimated parameter to the true one becomes an important issue, Wertz [1981]. Even when the "true" model is not in the class of candidate models, it is useful to require that each model in the class generates a distinct output sequence so that only one model or its parametric representation of the equivalence class of models corresponds to a given joint probability density function. Two models are observationally equivalent if the probability distributions of data are the same. If two observationally equivalent models are indeed the same, then the model is said to be identifiable. A function $g(\theta)$ is identifiable if the equality of two probability distributions of data vectors, $p(y\,|\,\theta_1) = p(y\,|\,\theta_2)$, implies that $g(\theta_1) = g(\theta_2)$, i.e., observationally equivalent models assign the same value to $g(\theta)$.

7.4 Dynamic Multiplier Analysis and Structural Model Identification

This section discusses one of the important uses of the estimated model--i.e., impulse response or dynamic multiplier analysis by the estimated model--as a way of performing policy effectiveness studies.

Confidence Interval of Impulse Response Analysis

Perhaps, the most striking advantage of state space models is the ease of conducting impulse response, i.e., dynamic multiplier analysis. They are much easier to calculate when the model is in state space form than in ARMA form. The impulse response matrices of a state space model are simply given by CA^iB for $i = 0, 1, , \cdots$. Compare this simplicity with, for example, Brissimis [1976], who carries out cumbersome calculations to derive multipliers in ARMA models. The impulse response of the data vector to past innovations is exactly the Wold decomposition we mentioned in Chapter 3.

Calculation of impulse responses is one of the important products of modeling time series. Since the system parameters are only estimates, Runkel [1987] has stressed the importance of reporting confidence intervals for impulse responses. He calculates confidence intervals of impulse responses of a univariate model as follows. For the process modeled by $y_t = ay_{t-1} + e_t$, the impulse responses are a^j, $j = 0, 1, 2, \ldots$. Suppose that $T^{1/2}(\hat{a} - a)$ is normally distributed with mean zero and variance w. Then with any smooth function g, $T^{1/2}(g(\hat{a}) - g(a))$ is also normally distributed with mean zero and variance

wGG', where G is the derivative of g evaluated at $\hat{a}$. One standard deviation confidence interval of the k-th term of the impulse response sequence is $2k\hat{a}^{k-1}\sqrt{w}$. Since the impulse response itself is going to zero, the relative size of the confidence interval grows with k.

Models for economic time series usually do not have explicit exogenous variables and have only the innovations of the data as inputs in calculating impulse responses. (In Chapter 9 we discuss how to incorporate exogenous variables as inputs when they are available.) What we really want in macroeconomic time series modeling and other decision analysis modeling exercises is to estimate impulse responses of policy variables as a way of evaluating effects of actual or proposed policy changes. Some shocks in economic structural models, such as monetary shocks, may be interpreted as policy actions. In interpreting models in such cases, we need impulse responses with respect to structural shocks, or exogenous shocks in structural models, and not with respect to innovations. Clearly, we need a relation which connects innovations in data to structural shocks. Once this is done, the dynamic multiplier (i.e., impulse responses) may help weed out some structural models as not representative of reality if their impulse responses are contrary to economists' intuitions or to observed facts. This section illustrates a procedure by which estimated time series models may help identify structural models.

Variance Decomposition

A vector-valued autoregressive (VAR) model for a mean-zero weakly-stationary process $\{y_t\}$ is specified by

$$A(L)\,y_t = u_t\ ,$$

where $A(L)$ is a matrix polynomial in the lag operator L, $A(L) = I - A_1 L - \cdots - A_p L^p$, and $\{u_t\}$ is a vector-valued mean zero exogenous noise sequence with a constant covariance matrix

$$E(u_t\, u_s') = Q\ \delta_{t,s}\ .$$

The matrix Q need not be diagonal because of possible contemporaneous correlations among the components of the vector u_t. In this model u_t is the innovation vector e_t. If the constant term in $A(L)$ is A_0 rather than the identity matrix, then $A_0 e_t$ is equal to u_t.

When the matrix $A(L)$ is invertible, the model may be put into the $MA(\infty)$ form

$$y_t = A(L)^{-1} u_t\ .$$

A common practice in VAR literature is to transform u_t into white noise v_t with a diagonal covariance matrix Σ

$$u_t = W\ v_t\ ,$$

where W is a lower triangular matrix with the unit diagonal elements such that $W \Sigma W' = Q$, and then to rewrite the model as

$$y_t = A(L)^{-1} W v_t = G(L) v_t ,$$

where the transfer function matrix between the white noise sequence v_t and y_t is denoted by $G(L)$ as defined above. The identity relation $A(L) G(L) = W$ can be used to calculate the coefficient matrices of $G(L) = G_0 + G_1 L + \cdots$.

The same procedure can be applied to ARMA models. As an example consider

$$(I - A_1 L - A_2 L^2) y_t = u_t = (B_0 + B_1 L) v_t$$

where v_t is an innovation with mean zero and covariance Λ which is diagonal, then

$$G_0 = B_0,$$

$$G_1 = B_1 + A_1 G_0 ,$$

and

$$G_2 = A_1 G_1 + A_2 G_0 ,$$

and so on.

Given a VAR model driven by innovation disturbances, the predicted value of y_{t+k} for any positive integer k, given $y_t, y_{t-1}, \ldots$, which is denoted by $y_{t+k|t}$, can be calculated by

$$y_{t+k|t} = [\, G(L)/L^k]_+ \, v_t ,$$

where the symbol $[\ ]_+$ denotes that part of the expression with non-negative power in L. Now, the prediction error is expressible as

$$y_{t+k} - y_{t+k|t} = G_0 v_{t+k} + G_1 v_{t+k-1} + \cdots + G_{k-1} v_{t+1} .$$

The error covariance matrix is given as a sum $\sum_{i=0}^{k-1} G_i \Sigma G_i'$. This expression, often referred to as the variance decomposition, expresses the prediction error as a sum of components contributed by the elements of v_t. For example, with k=1, $G_0 \Sigma G_0'$ is equal to $\sum_{i=1}^{r} \sigma_i g_i g_i'$, where g_i is the i-th column vector of G_0 and σ_i is the i-th diagonal element of Σ.

The variance decomposition and the sequence of the impulse response matrices may be used to infer dynamic properties of the model and how innovations originating in a certain sector or variable affect the rest of the model or the time series components if clear interpretation of the components of the innovation is available. Often, however, no direct correspondence capable of economic or physical interpretation exists between y and v, because v is obtained by transforming u, which may have clearer identification with economic or physical disturbances. For example, u may be innovations of the money stock or the industrial production index. The variances of v, which are linear combinations of variances of these innovations, may not be associated with innovations of any clearly identifiable economic variable. We must assume a particular structural model to provide such an interpretation, as we next illustrate.

Identification Exercises

Economics structural models are usually stated as an ARMA model

$$A(L)y_t = B(L)u_t \ , \tag{19}$$

where

$$A(L) = A_0 + A_1 L + \cdots$$

$$B(L) = B_0 + B_1 L + \cdots$$

and u_t is a mean-zero, serially uncorrelated "structural" noise vector with finite covariance matrix. Then $\hat{\mathrm{E}}(u_t \mid y_{t-1}^{-}) = 0$, and we obtain from (19)

$$A_0 y_{t|t-1} + A_1 y_{t-1} + \cdots = B_1 \ u_{t-1} + \cdots \tag{20}$$

Subtracting (20) from (19) and denoting the innovation $y_t - y_{t|t-1}$ by e_t,

$$A_0 e_t = B_0 \, u_t \ .$$

More generally, the 'surprise' content $u_t - u_{t|t-1}$ is related to the innovation vector. Assume that A_0 is nonsingular. Then the innovation e_t and the structural shock u_t is related by

$$e_t = Du_t \tag{21}$$

with

$$D = A_0^{-1} B_{0.}$$

If D is nonsingular, the expression

$$u_t = D^{-1} e_t \tag{22}$$

provides structural relations of contemporaneous variables in the structural model and the innovations.

Example. As an example consider a 'toy' structural model specified by

Goods Demand Equation $Y_t = \gamma M_t - \alpha P_t + u_{1t}$,

Money Demand Equation $M_t - P_t = \beta Y_t + u_{2t}$,

Money Supply Equation $M_t = u_{3t}$,

where $y_t = (Y_t, P_t, M_t)'$ is observed. Here, Eq. (22) takes the form

$$\begin{bmatrix} u_{1t} \\ -u_{2t} \\ u_{3t} \end{bmatrix} = \begin{bmatrix} 1 & \alpha & -\gamma \\ \beta & 1 & -1 \\ 0 & 0 & 1 \end{bmatrix} \begin{bmatrix} e_y \\ e_p \\ e_m \end{bmatrix}_t \tag{23}$$

where the diagonal elements are normalized to be one without loss of generality. The fact that M_t in the money supply relation is independent of the contemporaneous Y_t and P_t may reflect the fact that the central bankers do not have contemporaneous information concerning these variables, an assumption of information patterns discussed, for example, by Sims [1986]. From (23), the covariance matrix of u_t is related to Δ by

$$cov\ \ u = Q = D^{-1}\,\Delta\, D^{-T} \ . \tag{24}$$

Sometimes its inverse relation is more convenient.

There are three unknown parameters in D^{-1}, as we specified the structural model. If we assume that the covariance matrix of u_t is diagonal with only three unknown variances, then the total number of unknowns is six and we have six relations from the six independent parameters in Δ. The unknowns can in principle be uniquely determined. This is the identification problem for this "toy" structural model.

Once we know D, then the impulse responses to structural shocks are calculated by

$$CA^kBD\,, \quad k = 0\ 1,\ \cdots$$

using the estimated system matrices for A, B, and C, in the innovation model for the data series $\{\ y_t\ \}$. Examples of the identification by imposing restrictions on the matrix D may be found in Bernanke [1986], Sims [1986], and Aoki [1989].

Common vs. Country (Site) Specific Shocks One way to achieve the just-identified condition is to decompose structural shock u into a shock common to all components of the series and mutually-uncorrelated component-specific shocks. Such a decomposition may be reasonable when the series is made up of the like components in different sites (countries), such as real GNP of several countries. The supposition is that each country is hit by shocks, some of which are world-wide, and others are specific to that country.

With this reasoning, decompose the innovation as

$$e = u_w \alpha + u_s ,$$

where u_w is a univariate random variable with unit variance, the vector α represents the weight assigned to the individual components, and the vector u_s is a vector of country specific shocks with $cov\ u_s = \Sigma$. These variances are related by $\Delta = \sigma_w^2 \alpha\alpha' + \Sigma$. There are $p(p+1)/2$ parameters in the innovation covariance matrix to determine the parameters in this decomposition. When $p = 3$ and Σ is diagonal, the just-identified condition is achieved. With $p = 4$, there are $(10 - 4)$ free parameters to be assigned to Σ.

Example: Real GNP of the USA, West Germany and Japan. Real GNP data from 1st quarter 1973 to third quarter 1989, a total of 66 data points, have been used. For reasons we discuss in chapter 11, each series is normalized by its initial data value so that each starts with 1. Because the series appears to be nearly integrated--i.e., nearly nonstationary--the two-step procedure described in Chapter 11 is used to estimate the innovation covariance matrix.

The set of algebraic equations is solved first for αs:

$$\alpha_1^2 = \Delta_{31}\Delta_{21} / \Delta_{32} , \quad \alpha_2^2 = \Delta_{32}\Delta_{21} / \Delta_{31} ,$$

and

$$\alpha_3^2 = \Delta_{31}\Delta_{32} / \Delta_{21} .$$

Then we solve for σs:

$$\sigma_1^2 = \Delta_{11} - \Delta_{31}\Delta_{21} / \Delta_{32} , \quad \sigma_2^2 = \Delta_{22} - \Delta_{32}\Delta_{21} / \Delta_{31} ,$$

and

$$\sigma_3^2 = \Delta_{33} - \Delta_{31}\Delta_{32} / \Delta_{21} .$$

The numerical values are

$$\alpha_{us} = .625 , \alpha_G = .682 , \alpha_J = .626 , \sigma_{us}^2 = .721 \times 10^{-4} , \sigma_{wg}^2 = .631 \times 10^{-4}$$

and

$$\sigma_J^2 = .718 \times 10^{-4} .$$

The common shock affects all three countries about equally. West Germany has the least country specific shock.

Example: A Structural Model. In the structural model approach, a matrix S, which is the inverse of the matrix D defined in (21), is hypothesized to relate structural shocks to the innovations by

$$Se_t = u_t \ ,$$

with $cov\ u =$ diagonal, where

$$e = \begin{bmatrix} e_{us} \\ e_G \\ e_J \end{bmatrix} \quad \text{and} \quad u = \begin{bmatrix} u_{us} \\ u_G \\ u_J \end{bmatrix} .$$

The diagonal elements of S are set to 1 without loss of generality. Suppose that it is given by

$$S = \begin{bmatrix} 1 & \alpha & \gamma \\ 0 & 1 & \beta \\ \gamma & 0 & 1 \end{bmatrix} .$$

This matrix S incorporates the assumption that contemporaneous interactions between the USA and West Germany are asymmetric, that interactions between Japan and the USA are symmetric, and that those between West Germany and Japan are negligible. Let $cov\ u = Q$. There are three parameters in S. Assume that the components of u are mutually uncorrelated. Then there are six parameters in total, and the model is just identified. Using the data for the period of 1973.1 to 1989.3, a total of 66 data points, the two-step procedure is used to estimate a three-dimensional model

$$y = \begin{bmatrix} .0485 & .250 & .978 \\ .1675 & .0976 & 1.793 \\ .3343 & .2054 & .795 \end{bmatrix} z_t + e_t ,$$

$$z_{t+1} = \begin{bmatrix} .971 & -.0026 & .006 \\ 0 & .887 & -.149 \\ 0 & -.004 & .832 \end{bmatrix} z_t + \begin{bmatrix} -3.063 & -3.738 & 5.352 \\ 2.843 & 2.006 & -1.235 \\ -0.027 & .499 & -.1807 \end{bmatrix} e_t \ ,$$

with

$$cov\ e_t = \begin{bmatrix} 1.789 & * & * \\ .557 & 1.842 & * \\ 1.285 & 1.560 & 2.964 \end{bmatrix} \times 10^{-3}.$$

It is convenient to use the relation in (24) in the inverse form (this step is not necessary if one uses software to handle symbol manipulations)

$$\Delta^{-1} = S'Q^{-1}S$$

to derive three equations

$$\begin{bmatrix} h_1 \\ h_2 \\ h_3 \end{bmatrix} = \begin{bmatrix} 1/\alpha & 0 & 0 \\ -\gamma/\beta & 1/\beta & 0 \\ -1/\alpha & 0 & 1/\gamma \end{bmatrix} \begin{bmatrix} \Delta^{12} \\ \Delta^{23} \\ \Delta^{13} \end{bmatrix}$$

where Δ^{ij} is the (i,j)-th component of Δ^{-1}, and

$$Q^{-1} = diag\,(h_1, h_2, h_3) \quad .$$

The remaining three equations provide the three nonlinear relations to determine the three parameters α, β, and γ. Use Δ^{11} and Δ^{22} to express α and β in terms of γ as

$$\alpha = (1-\gamma^2)\Delta^{12} / (\Delta^{11} - \gamma\Delta^{13})$$

and

$$\beta = (\Delta^{23} - \gamma\Delta^{12}) / (\Delta^{22} - \alpha\Delta^{12}) \quad .$$

Substitute these into Δ^{33}. We obtain a third order polynomial in γ. Of the three roots, only one is real, $\gamma = -.337$. The results are

$$\alpha = .124 \quad , \quad \sigma_1^2 = 1/h_1 = 1.296 \times 10^{-3}$$

$$\beta = -.542 \quad , \quad \sigma_2^2 = 1/h_2 = 1.022 \times 10^{-3}$$

and

$$\sigma_3^2 = 1/h_3 = 2.301 \times 10^{-3} \quad .$$

The sign of α seems wrong in this model. The inverse of the matrix S^{-1}(i.e., D) is equal to

$$D = \begin{bmatrix} 1.100 & -.136 & .297 \\ .200 & .975 & .594 \\ .371 & -.046 & 1.100 \end{bmatrix} \quad .$$

An alternative model with

$$S = \begin{bmatrix} 1 & 0 & \gamma \\ \alpha & 1 & 0 \\ \gamma & \beta & 1 \end{bmatrix}$$

produces the parameter values of $\alpha = -.096$, $\beta = -.745$, $\gamma = -.268$, $\sigma_1^2 = 1.313\times10^{-3}$, $\sigma_2^2 = 1.751\times10^{-3}$, and $\sigma_3^2 = 1.324\times10^{-3}$. The signs of the parameters are all reasonable. The inverse of this matrix is

$$D = \begin{bmatrix} 1.100 & .270 & .295 \\ .106 & 1.021 & .028 \\ .374 & .820 & 1.100 \end{bmatrix} \quad .$$

See Chapter 12 for the plots of impulse responses for this model. Another model with rather similar parameter values has the matrix S given by

$$S = \begin{pmatrix} 1 & \alpha & \gamma \\ \alpha & 1 & 0 \\ \gamma & \beta & 1 \end{pmatrix} .$$

Its parameter values are $\alpha = -.043$, $\beta = -.757$, $\gamma = -.267$, $\sigma_1^2 = 1.305 \times 10^{-3}$, $\sigma_2^2 = 1.797 \times 10^{-3}$, and $\sigma_3^2 = 1.324 \times 10^{-3}$.

The matrix

$$S = \begin{pmatrix} 1 & \gamma & \gamma \\ \alpha & 1 & 0 \\ 0 & \beta & 1 \end{pmatrix}$$

appears to produce too small effects than we would expect of exogenous shocks in the US on the Japanese real GNP.

When no structural form is posited, the matrix D is not available. Nevertheless, suppose we write

$$e_t = Du_t = \Sigma_{i=1}^{p} d_i u_{it} \text{ with } cov\ u_t = \Sigma ,$$

for some D, d_i being the i-th column vector of D. When u_{it}'s are uncorrelated

$$\Delta = D \Sigma D'$$

$$= \Sigma_i \sigma_i^2 d_i d_i' .$$

One way to select these d's is to use the eigenvectors of Δ. Its polar decomposition is

$$\Delta = \Sigma \mu_i v_i v_i' ,$$

where v_i is the i-th normalized column eigenvector $v_i' v_i = 1$ for the eigenvalue μ_i. Then $\mu_i = \sigma_i^2$, $d_i = v_i$, $i = 1 \cdots p$ achieves the just-identified condition. The multipliers are $CA^k BD$ with $D = [v_1 \cdots v_p]$.

This is an alternative to the VAR variance decomposition. Unlike the variance decomposition discussed above, this procedure is invariant to permutations of components of the y-vector.

We next allow for non-diagonal Σ. For easier exposition, take the case of $p = 2$ and drop the t subscript. Treat u_1 as a primary shock and write

$$e = d_1 u_1 + d_2 u_{2|1} + d_2(u_2 - u_{2|1})$$

$$= (d_1 + d_2 \sigma_{21} \sigma_{11}^{-1}) u_1 + d_2(u_2 - u_{2|1}) \, ,$$

where

$$u_{2|1} = \sigma_{21} \sigma_{11}^{-1} u_1 \quad .$$

Then because the last term is uncorrelated with the first, the first term is identified with $\mu_1 v_1$ and the second with $\mu_2 v_2$, where the eigenvalues of Δ is arranged in the non-decreasing order, $\mu_1 \geq \mu_2$. In the expression for the multipliers, Be is decomposed into $B\mu_1 v_1 + B\mu_2 v_2$. Then reversing the order u_2 is interpreted as primary.

When no such D is available, the multipliers must be solely based on the reponses to e. In some applications each component of the innovation vector may be treated as a primary source of disturbances, especially if the innovation covariance matrix is (nearly) diagonal. When no such D is available, the multipliers must be solely based on the responses to e. If Δ is diagonal, then the components of e are uncorrelated and $CA^k B_i$ (where B_i is the i-th column vector B) is the sequence of the impulse responses. When the matrix Δ is not diagonal, then Be may be decomposed into $B_i e_i + \hat{E}(Be - B_i e_i \mid e_i)$, i.e, each e_i may be regarded as primary cause for a shock represented by $B_i e_i$ and the induced or residual shocks, which is the second term above. By taking $i = 1, \cdots, p$ and examining these p impulse response vectors, one may gain some insight into the dynamic interactions. The procedure is equivalent to repeated applications of the Cholesky decomposition to data vectors after permuting the components. Such a procedure may be useful when each e_i is primarily associated with different site as in a multi-country study of interactions.

7.5 Out-of-Sample Forecasts

One of the ways to test a model constructed by the method of this book, given later in Chapter 9, is to use the model to produce out-of-sample forecasts. First, divide the total number of data points, N, into N_1 and N_2. The first N_1 points are used to construct the system matrices, A, B, and C. To be clear about the number of data points used in system matrix construction, we carry the number of data points as an argument of the matrices, as $A\,(50)$ for example.

These matrices are then sequentially updated as new data points from the remainder of the data set are presented to the model. Alternatively, the same set of estimates may be kept while additional data points are incorporated as shown below. Let F stand for the expression $(A - BC)$, and use H_1 for CB. Then, the first out-of-sample forecast is produced by

$$y_{N_1+1 \mid N_1} = CF\,(N_1) z_{N_1} + H_1(N_1) y_{N_1} \, .$$

Compare this forecast with the data y_{N_1+1}. Next, this data point is added to the existing pool of data points to update the system matrices to $A(N_1+1)$, and so on. In general, one-step-ahead forecasts are produced by

$$y_{N_1+k+1 \mid N_1+k} = CF(N_1+k) x_{N_1+k \mid N_1+k-1} + H_1(N_1+k) y_{N_1+k} .$$

When linear models are correctly specified and their system parameters estimated consistently and possibly asymptotically efficiently, there is no difference in performance between one k-step-ahead forecast and a sequence of k one-step-ahead forecasts, i.e., the model which minimizes a quadratic measure of one-step ahead forecast error also minimizes the similar measure for k-step forecast errors. In nonlinear models or possibly misspecified linear models, however, the use of different models (depending on forecast horizon) may be advantageous. This point has been illustrated by Findley [1985]. Here we give two alternative expressions for one-step- and k-step-ahead forecasts:

$$y_{t+k+1 \mid t} = CA^k (Fz_t + By_t), \quad k = 0, 1, \cdots ,$$

and

$$y_{t+k+1 \mid t} = CBy_{t+k \mid t} + CFBy_{t+k-1 \mid t} + \cdots + CF^k By_{t+1 \mid t} + CF^{k+1}(Fz_t + By_t) .$$

An alternative expression similar to the Kalman filter update equation is

$$y_{t+k \mid t+1} = y_{t+k \mid t} + CA^{k-1} B (y_{t+1} - y_{t+1 \mid t}) .$$

Simultaneous forecasts of y_t^+ is given by

$$\hat{E}(y_t^+ \mid y_{t-1}^-) = \boldsymbol{O}(Fz_t + By_t) .$$

These expressions require a knowledge of z_t. One may use the expression (8) in Chapter 6 for it, or some scheme of smoothing or back-forecasting may be used to estimate the initial condition vector z_0, which is in turn propagated forward to estimate z_t. The backward innovation model constructed in Section 1 may be used for this purpose.

8 STATE VECTORS AND OPTIMALITY MEASURES

The state vector introduced in Chapter 5 is by no means the only way of summarizing information in data sets. Two other choices, one used by Akaike and the other by Arun et al., are discussed in this chapter.

8.1 Canonical Variates

Akaike introduced the canonical variables as state vectors in his important pioneering works, Akaike [1974,1975,1976]. His state variables are therefore orthogonal to all previously introduced state variables. This choice of state variables, however, leads to non-nested estimates of the system matrices.

Denote the covariance matrices of the future and past stacked data vectors by R_+ and R_-, respectively. Then normalize the data vectors by the square-roots of their respective covariance matrices

$$d_t^+ = R_+^{-1/2} y_t^+ \text{ ,and } \quad d_{t-1}^- = R_-^{-1/2} y_{t-1}^- \text{ .}$$

The covariance matrix of these two normalized data vectors is a scaled version of the original Hankel matrix we used in all previous analysis:

$$\hat{E}(d_t^+ \mid d_{t-1}^-) = R_+^{-1/2} \boldsymbol{H} R_-^{-1/2} = P \Gamma Z' \text{ .} \tag{1}$$

In the above, the singular value decomposition of this scaled Hankel matrix is denoted by $P \Gamma Z'$. The matrix Γ is a diagonal matrix with diagonal elements γ_i arranged in decreasing order of magnitude, $i = 1, 2, \cdots$. They are the canonical correlation coefficients between the future and past stacked data vectors. Now the canonical variates are defined by rotating the scaled data vectors

$$u_t^f = P' d_t^+ \text{ , and } \quad u_{t-1}^p = Z' d_{t-1}^- \text{ .}$$

Here we use superscript f and p to refer to the canonical variates of the future and the past, because the dimension of these vectors is not the number of stacked vectors any more , but is equal to the rank of the scaled Hankel matrix which is the same as that of the Γ matrix, i.e., number of non-zero the canonical correlation coefficients retained in it. The scaled Hankel matrix can be factored just as the Hankel matrix can be factored into the observability and Ω-controllability matrix. The state vector in this framework can be chosen to be

$$x_t = \Gamma^{1/2} u_{t-1}^p \text{ ,} \tag{2}$$

so that its covariance matrix is Γ, which is diagonal--unlike our state vector covariance matrix Π.

Then, the developments in Section 7.1 can be repeated to derive the dynamics governing the time evolution of this state vector. See also Desai and Pal [1982], where a similar development is carried out for univariate data series.

Mutual Information

Gel'fand and Yaglom [1959] have shown that the mutual information between the canonical vectors u_t^+ and u_{t-1}^- is given by

$$I(u_t^+, u_{t-1}^-) = -\ln det(I - \Gamma^2)$$
$$= -\sum_{i=1}^{N} ln(1-\gamma_i^2) .$$

Based on this relation, Desai and Pal [1983] suggested the ratio

$$\sum_{i=1}^{n_1} ln(1-\gamma_i^2) / \sum_{i=1}^{N} ln(1-\gamma_i^2)$$

as a measure for selecting $n_1 < n$ as the dimension of an approximate state-space model. Having chosen n_1, rewrite the right-hand side of (1) as

$$[P_1, P_2] \begin{bmatrix} \Gamma_1 & 0 \\ 0 & \Gamma_2 \end{bmatrix} \begin{bmatrix} Z_1' \\ Z_2' \end{bmatrix} ,$$

and define the state vector of dimension n_1 by replacing Γ and Z by Γ_1 and Z_1 respectively, in their defining relation (2) --i.e.,

$$x_t = \Gamma_1 Z_1' L^{-1} y_t^- \ : \ n_1 \times 1 \ .$$

This state vector has mean 0 and covariance $\Gamma_1^{\ 2}$.

8.2 Prediction Error

A third way of summarizing the past data is used by Arun et al. [1983]. This approach chooses the subspace for which the prediction error covariance is minimized, i.e., the matrix S in the definition of the state vector $x_t = Sy_{t-1}^-$ is chosen by minimizing

$$\Xi = var\ [y_t^+ - \hat{E}(y_t^+ \,|\, x_t)] \ .$$

Because this matrix Ξ is equal to $R_+ - \boldsymbol{H} S'(SR_- S')^{-1} S \boldsymbol{H}'$, the optimal matrix is

$$S' = R_-^{-1} \boldsymbol{H}' W \ , \tag{3}$$

where W is the matrix made up of the eigenvectors of $\boldsymbol{H} R_-^{-1} \boldsymbol{H}'$, i.e.,

$$\boldsymbol{H} R_-^{-1} \boldsymbol{H}' Z = Z \Lambda \ ,$$

where the matrix $\Lambda = diag\,(\lambda_1, \lambda_2, \ldots)$ collects the eigenvalues of the matrix pencil

$\lambda R_- - \boldsymbol{H}'\boldsymbol{H}$. See Gantmacher [1959] or Rao [1964, 1979]. From (3), if the matrix V is chosen to be $(n_1 \times Np)$, then the variance Ξ is minimized by

$$S' = R_-^{-1} H' W_1 ,$$

where W_1 is the $Np \times n$ matrix made up of the eigenvectors corresponding to n largest eigenvalues of $\boldsymbol{H} R_-^{-1} \boldsymbol{H}'$. See Aoki and Havenner [1989] for further details and some numerical examples.

8.3 Singular Values and Canonical Correlation Coefficients

The construction of the state space model in this book chooses the state dimension based on the Hankel norm of the transfer function discussed in Section 5.7. Error analysis of this method of building models is conducted in Chapters 9 and 10.

We next make explicit the relation between the canonical correlation coefficients and the singular values by choosing a coordinate representation. It is convenient to slightly alter the notation for the innovation vector by assuming that innovation vector e_t has the covariance matrix I_p, $cov\ e_t = I_p$ rather than Δ. This change is reflected in the space presentation

$$z_{t+1} = Az_t + Be_t ,$$

and

$$y_t = Cz_t + De_t .$$

In other words, DD' equals Δ in the notation, and the matrix B here stands for $B\Delta^{1/2}$. We assume that the dimension of z_t is known to be n in this section, which is assumed to be the dimension of the minimal realization. That is, the reachability and observability gramians have rank n, and the matrix A is asymptotically stable.

As before, we stack the future and past innovation vectors of the data vector and denote them by e_t^+ and e_{t-1}^-. In particular, we focus on

$$e_o^+ = \begin{bmatrix} e_o \\ e_1 \\ \vdots \end{bmatrix} \quad and \quad e_{-1}^- = \begin{bmatrix} e_{-1} \\ e_{-2} \\ \vdots \end{bmatrix} .$$

The linear subspace $\overline{\text{span}}\, e_o^+$ is denoted by $\mathcal{E}^+$ and the linear subspace $\overline{\text{span}}\, e_{-1}^-$ by $\mathcal{E}^-$. Then the Hankel matrix H is a map from $\mathcal{E}^+$ to $\mathcal{E}^-$. For example, e_o is mapped into $e_{0|-1} = \hat{E}(e_o \mid e_{-1}^-)$. We write this using the operator notation

$$\mathcal{H}e_o = \Lambda_1 e_{-1} + \Lambda_2 e_{-2} + \cdots ,$$

where $\mathcal{H}$ is the projection operator with

$$\Lambda_{k+1} = CA^k B \quad , \; k = 1, 2, \cdots .$$

More generally,

$$\mathcal{H}e_t = \Sigma_{k=0}^{\infty} CA^{k+t} Be_{-1-k} \; , \; t > 0 . \tag{4}$$

It turns out to be useful to consider the dual map from $\mathcal{E}^-$ to $\mathcal{E}^+$. For example, it maps e_{-1} onto e_o^+. The projected image $e_{-1|0} = \hat{E}(e_{-1} | e_o^+)$ then is equal to $\Sigma_0^{\infty} \Lambda_{k+1}' e_k$, because $cov\; e_k = I$. Thus, this map for e_{-1} is given by

$$\mathcal{H}' e_{-1} = \Sigma_{k=0}^{\infty} B'(A')^k C' e_k \quad ,$$

and that for e_{-s} by

$$\mathcal{H}' e_{-s} = \Sigma_{k=0}^{\infty} B'(A')^{k+s-1} C' e_k \quad , \; s > 0.$$

From the model representation, the Wold-decomposition of the state vector z_0 is given by

$$z_0 = Be_{-1} + ABe_{-2} + \cdots$$

and

$$\mathcal{H}' z_0 = B\,(\Sigma_{k=0}^{\infty} B'(A')^k C' e_k) + AB\,(\Sigma_{k=0}^{\infty} B'(A')^{k+1} C' e_k) + \cdots \tag{5}$$

$$= G_c \Sigma_{k=0}^{\infty} (A')^k C' e_k \; ,$$

where G_c is the controllability Gramian. Eq. (5) shows that the backward dynamic model

$$s_t = A' s_{t+1} + C' e_t \quad , \; y_t = B' s_{t+1} + Re_t$$

naturally arises. This model may be thought of as a dynamic inverse regression curve, $y_{t-1}^- = \hat{E}(y_{t-1}^- | y_t^+) + v^+$, where v^+ is the residual vector. Just as the forward regression curve can be written as $y_t^+ = \boldsymbol{O}\, z_t +$ residuals, by defining the state vector z_t appropriately, i.e., by (20) of Chapter 5, the inverse regression model can be written as $y_{t-1}^- = \Omega' s_t + v^-$, where the state vector is $s_t = \boldsymbol{O}' R_+^{-1} y_t^+$. This has been shown in Section 1 of Chapter 7. Note that

$$s_0 = C' e_0 + A' C' e_1 + \cdots , \tag{6}$$

so that (4) is now

$$\mathcal{H}' z_0 = G_c s_0 \quad , \tag{7}$$

where $\mathcal{H}'$ is the adjoint operator of $\mathcal{H}$. The operator $\mathcal{H}$ is the Hilbert-Schmidt operator, and (6) is one of the pair of the Hilbert-Schmidt pair because (6) yields, in view of (4) and (6),

$$\mathcal{H}s_0 = G_0 \Sigma_{k=0}^{\infty} A^k Be_{-1-k} \tag{8}$$

$$= G_o z_0 ,$$

where G_o is the observability Gramian. Now put the model in balanced form. The relations of (7) and (8) show that the components of $\Sigma^{-1/2}s_0$ and $\Sigma^{-1/2}z_0$ are the Hilbert Schmidt pairs

$$\mathcal{H}(\sigma_i^{-1/2}s_0^i) = \sigma_i(\sigma_i^{-1/2}z_0^i)$$

and

$$\mathcal{H}'(\sigma_i^{-1/2}z_0^i) = \sigma_i(\sigma_i^{-1/2}s_0^i) \quad , \quad i = 1, \cdots, n,$$

with the normalization conditions

$$E(\sigma_i^{-1/2}s_0^i)^2 = 1 ,$$

and

$$E(\sigma_i^{-1/2}z_0^i)^2 = 1 \quad .$$

From (7) and (8),

$$\mathcal{H}\mathcal{H}'z_0 = \mathcal{H}G_c s_0 = G_c \mathcal{H}s_0 = G_c G_o z_0 = \Sigma^2 z_0 .$$

Note that

$$E[(\sigma_i^{-1/2}s_0^i)(\sigma_i^{-1/2}z_0^i)] = \sigma_i^{-1}E(s_o^i z_{-1}^i) = \sigma_i$$

because

$$< s_0^i, z_0^i > = < \mathcal{H}s_0^i, z_0^i > + < (I - \mathcal{H})s_0^i, z_0^i >$$

$$= \sigma_i < z_0^i, z_0^i >$$

$$= \sigma_i^2 ,$$

since $(I - \mathcal{H})s_0^i$ belongs to $\mathcal{E}^+$ and z_0^i to $\mathcal{E}^-$. Therefore in the balanced representation, the canonical correlation coefficients are the singular values and the canonical variates are $\Sigma^{-1/2}z_0$ and $\Sigma^{-1/2}s_0$ respectively. This relation is established in a more general context (i.e., not necessarily minimal relization) by Michaletzky (1989) using a different method.

We can generalize the above by using the notion of the maximal correlation coefficients, Renyi (1959). Suppose we postulate

$$y_t^+ = f(y_{t-1}^-, n_t)$$

for some unknown f and disturbance term n_t. In the state vector modeling approach, the state vector serves as "sufficient" summary of the data vector and the above is approximated by

$$y_t^+ \approx f(z_t, n_t),$$

which is the same as Li's approach* when we use only linear aggregation $z_t = S y_{t-1}^+$, where S has n rows, Li (1988). (In Li some scalar-valued function $\phi(y_t^+)$ is used as the left-hand side.) The maximal correlation coefficient is

$$R_b^2 = \max corr(h(y_t^+), b' y_{t-1}^-)^2$$

with a suitable normalization. The maximization is with respect to all Borel measureable functions with finite variances, and is achieved under some technical conditions (the joint probability measure of y_t^+ and y_{t-1}^- is absolutely continuous with respect to the product measure of the two component vectors) by the inverse regression, $h(y_t^+) = E(y_{t-1}^- \mid y_t^+)$ which is equal to $H' R_+^{-1} y_t^+$ in our linear model. The best linear aggregation is achieved by $\beta_i' y_t^+$, where β's are the eigenvectors of $H' R_+^{-1} H \beta_i = \lambda_i R_- \beta_i$ and the eigenvalues are the zeros of $|\lambda R_- - H' R_+^{-1} H|$, i.e., the canonical correlation coefficients.

9 Estimation of System Matrices

This chapter is central to this book. It describes two types of estimators for system matrices of state space models from given sets of time series data, and examines statistical properties of the proposed estimators. The estimators of the first type are based on the stochastic realization theory, and are suggested by the systems literature on the deterministic realization theory. The estimators of the second type are instrumental variable estimators often found in the econometric literature. We describe an important relation between these two estimators. The first type of estimators is computationally faster than the second, but the second is asymptotically more efficient than the first. There is an important class of VAR data-generating processes for which the two types of estimators are equally efficient asymptotically.

The dimensions of the models are determined by the numerical ranks of the Hankel matrices constructed from the sample covariance matrices of data sets. In this way the dimensions of the models are data-determined. Among theoretically-equivalent representations of models of given dimensions, the one called balanced is chosen for its various advantages. System matrices are estimated in a nested form in the sense described later in this section, and all estimated dynamic matrices of balanced lower-dimensional approximations to the original stable one remain stable. A test for model selection is also discussed.

First, the procedure for constructing a (hypothetical) full-dimensional model is described. Second, methods for constructing approximate models are considered, all based on relative sizes of singular values of the Hankel matrix. The chapter contains several small theoretical examples to illustrate some aspects of the estimators as well as results of some Monte Carlo simulation experiments. More extensive examples are found in Chapter 11. This chapter is based in part on Havenner and Aoki [1988], and Aoki [1989].

9.1 Two Classes of Estimators of System Matrices

As shown in (10) of Chapter 6, the Hankel matrix built up from the sample covariance matrices of the data set can be factored as the produce of the observability matrix $\boldsymbol{O}$ and another matrix Ω which has the same structure as the reachability matrix. This factored expression suggests a way of computing the system matrices A and C in the innovation representation (1) of Chapter 3. The matrix B and the noise covariance matrix are then computed after solving the matrix Riccati equation by one of the methods described in Section 7.2.

The basic steps in constructing innovation models are collected here for easy reference. Consider building an innovation model

$$z_{t+1} = Az_t + Be_t, \text{ and } \qquad y_t = Cz_t + e_t,$$

where e_t is a mean-zero, serially-uncorrelated covariance-stationary process with

$cov\ \ e_t = \Delta$. The covariances matrices of $\{y_t\}$ are related to the system matrices and the noise covariance matrix by the relations

$$\Lambda_k = E(y_{t+k}y_t{}') = CA^{k-1}M,\ k \geq 1, \text{ and } \Lambda_0 = C\Pi C' + \Delta, \tag{1}$$

where the matrix Π is the covariance matrix of the state vector,

$$\Pi = E(z_t z_t{}'),$$

and the matrix M denotes the cross-covariance matrix between the vectors z_{t+1} and y_t, and is related to the other system matrices by

$$M = A\ \Pi C' + B\ \Delta.$$

In this chapter, the matrix A is assumed to be stable, so that $\{z_t\}$ is covariance stationary, i.e., the matrix Π is constant. From weak stationarity, Π satisfies the matrix algebraic equation, called the Riccati equation

$$\Pi = A\ \Pi A' + B\ \Delta B'.$$

Both matrices B and Δ depend on Π, because

$$\Delta = \Lambda_0 - C\ \Pi C', \text{ and } B = (M - A\ \Pi C')\Delta^{-1}.$$

The matrix Δ must be positive definite for its inverse to exist. This condition is called the full-rank or regularity condition. We require that none of the components of the data vector $\{y_t\}$ be collinear at any time, so that the matrix Λ_0 is nonsingular.

The theoretical Hankel matrix is constructed from the covariance matrices for the data vector process given in (1). The sample Hankel matrix is similarly made using sample covariance matrices, which are denoted by a hat over the theoretical ones. The northwest section of size J sub-matrix rows by K sub-matrix columns of the Hankel matrix is

$$H_{J,K} = \begin{bmatrix} \Lambda_1 & \Lambda_2 & \cdot & \Lambda_K \\ \Lambda_2 & \Lambda_3 & \vdots & \Lambda_{K+1} \\ \cdot & \cdot & \vdots & \cdot \\ \cdot & \cdot & \vdots & \cdot \\ \cdot & \cdot & \vdots & \cdot \\ \Lambda_J & \Lambda_{J+1} & \vdots & \Lambda_{J+K-1} \end{bmatrix}.$$

As shown earlier in Sec.6.1, this matrix is the product of the observability matrix

$$O_J = \begin{bmatrix} C \\ CA \\ \cdot \\ \cdot \\ \cdot \\ CA^{J-1} \end{bmatrix}$$

with another matrix called Ω_K

$$\Omega_K = [M \ AM \ A^2 M \ \cdots A^{K-1} M],$$

because the covariance matrices of the weakly stationary mean zero data generating stochastic processes with rational spectral densities have the special structure shown in (1).

The sample Hankel matrix can actually be computed without calculating sample covariance matrices explicitly. Numerical accuracy of the singular value decomposition improves if the observed data are arranged in the way shown next, and the singular value decomposition is directly applied to it, thus skipping the intermediate calculations of sample covariance matrices. For this purpose arrange the data vector as follows:

$$Y_- = \begin{bmatrix} y_1 & y_2 & y_3 & \cdot & y_{N-1} \\ 0 & y_1 & y_2 & \cdot & y_{N-2} \\ 0 & 0 & y_1 & \cdot & \cdot \\ \cdot & \cdot & 0 & \cdot & \cdot \\ \cdot & \cdot & \cdot & \cdot & \cdot \\ \cdot & \cdot & \cdot & \cdot & y_{N-K+1} \\ 0 & 0 & 0 & \cdot & y_{N-K} \end{bmatrix} : Kp \times (N-1),$$

and

$$Y_+ = \begin{bmatrix} y_2 & y_3 & y_4 & \cdot & \cdot & y_N \\ y_3 & y_4 & y_5 & \cdot & \cdot & 0 \\ \cdot & \cdot & \cdot & \cdot & y_N & \cdot \\ \cdot & \cdot & \cdot & \cdot & 0 & \cdot \\ \cdot & \cdot & \cdot & \cdot & \cdot & \cdot \\ y_{J+1} & y_{J+2} & y_{J+3} & \cdot & 0 & 0 \end{bmatrix} : Jp \times (N-1).$$

Note that the matrix $Y_+ Y_-' / N$ is the same as the Hankel matrix (2), because $\hat{\Lambda}_k$ is calculated as $\Sigma_{t=1}^{N-k} y_{t+k} y_t' / N$.

Let the singular value decomposition of the matrix $Y = [Y_+' Y_-']'$ be given by $U \Sigma V'$. Then the matrix $Y \ Y'$ has $U \ \Sigma^2 U'$ as its singular value decomposition expression, from which we can extract by partitioning the matrix U comformably with the matrix Y

$$Y_+ Y_-' = [U_{11}\, U_{12}] \begin{bmatrix} \Sigma_1^2 & 0 \\ 0 & \Sigma_2^2 \end{bmatrix} \begin{bmatrix} U_{21}' \\ U_{22}' \end{bmatrix} .$$

Dividing the above by N, the singular value decomposition of (2) is obtained. This method is advantageous, because it avoids explicit sample covariance calculations, and hence has a lower condition number associated with it. This method is implemented in the TIMSAC-84 program package.

Stochastic Realization Estimator

We need three matrix equations to estimate three of the system matrices, A, C, and M. Just as the Hankel matrix is the covariance matrix between the future stacked vector y_t^+ and the stacked data vector y_{t-1}^-, we can get a relation involving the transition matrix A by calculating the covariance matrix between the shifted future vector y_{t+1}^+ and the stacked data vector, which is denoted by H^A. Shift up the matrix $H_{J,K}$ by one submatrix row and fill in the bottom rows by $\Lambda_{J+1}, \ldots, \Lambda_{J+K}$ to obtain it. Then note the relation that

$$H^A = \boldsymbol{O}_J A \Omega_K . \tag{2a}$$

This provides the first of the three relations. To obtain the other two, note that the first column of submatrices of $\boldsymbol{H}_{J,K}$ denoted by $H_{\cdot 1}$, is the product of $\boldsymbol{O}_J$ with M, i.e., it is a $Jp \times p$ matrix composed of the auto-covariance matrices Λ_1 through Λ_J. Finally we use a matrix which is the product of the matrix C with Ω_K, which is $p \times Kp$. It is the top row of the submatrices of $H_{J,K}$ denoted by $H_{1\cdot}$,

$$H_{\cdot 1} = \boldsymbol{O}_J M \quad \text{and} \quad H_{1\cdot} = C \Omega_K . \tag{2b}$$

The system matrices must be chosen to satisfy these three matrix relations. As we soon show, each of these equations can be regarded as the first-order-condition of a quadratic minimization problem, which is an overdetermined system of linear equations, that is, a system in which there are more equations than unknowns. For example, taking the transpose of the last equation in (2b) for the matrix C, we can regard it as a set of Kp equations for p column vectors of dimension n each. The number Kp is not less than n and usually much larger. The least squares solutions of such systems of equations are discussed in Golub and van Loan [1983, Chapter 6], for example, using the notion of the Moore-Penrose pseudo-inverse. It is defined in terms of the singular value decomposition of a matrix in the system of overdetermined equations.

We first calculate the singular value decomposition of the Hankel matrix as $U \Sigma V'$, where the non-zero singular values are arranged in non-increasing order along the main diagonal line of Σ. Then choose a coordinate system in which $\boldsymbol{O}_J = U \Sigma^{1/2}$ and $\Omega_K = \Sigma^{1/2} V'$. This coordinate system produces balanced models because $O_J' O_J = \Sigma = \Omega_K \Omega_K'$. For the time being, set aside the question of how many singular values to retain in Σ. We return to this question later in this chapter. The Moore-Penrose pseudoinverses of these two factors of the Hankel matrices are $\Sigma^{-1/2} V$ and $\Sigma^{-1/2} U'$ respectively. Using these, we propose the first type of estimators. We estimate the system matrices by

$$\hat{A} = \Sigma^{-1/2} U' H^A V \Sigma^{-1/2}, \tag{3a}$$

$$\hat{M} = \Sigma^{-1/2} U' H_{.1}, \tag{3b}$$

$$\hat{C} = H_{1.} V \Sigma^{-1/2}. \tag{3c}$$

These estimates are unique, because the Gramians are non-singular by the minimality of the model. We later show how these estimates are related to the instrumental variables (IV) estimators, and improve their asymptotic efficiency by modifying them.

Eq. (3a) shows that the eigenvalues of the estimated dynamic matrix are the roots of $0 = |\lambda\Sigma - U' H^A V|$; or, from the fact that $V' = \Sigma^{-1} U' H$, the roots of

$$0 = |\lambda\Sigma^2 - U' H^A H' U| .$$

When $n = Kp$, then the zeros are the same as those of $0 = |\lambda H - H^A|$.

To provide a bridge to the instrumental-variable estimators that we introduce next, and also to show that some of these estimators are also generalized least squares estimators, consider the matrix C as an example. Since this matrix is important in determining how the effects of state vector components are distributed among the data components, as we describe the notion of co-integration later in Chapter 11, this extra attention paid to the matrix C is well worth the effort. Write the observation equation for the data set as

$$Y = CZ + N ,$$

with

$$Y = [y_T \; y_{T-1} \cdots y_1] ,$$

and Z and N are defined analogously using the vectors z_t and e_t, where T is the number of data points. Taking the vectorized expression of the above produces

$$y_T^- = (Z' \otimes I)\theta + e_T^- ,$$

where θ is $vecC$. Because the innovation vectors are serially uncorrelated, the generalized least squares estimator chooses θ by

$$Min\, [y_T^- - (Z' \otimes I)\theta]' \, (I \otimes \Delta^{-1}) \, [y_T^- - (Z' \otimes I)\theta] ,$$

i.e., the vector θ is estimated by

$$\hat{\theta} = [(ZZ')^{-1} Z \otimes I] y_T^- .$$

In the original matrix form the matrix C is estimated by

$$\hat{C} = YZ'(ZZ')^{-1}.$$

Note that the estimate is independent of the weights used in the quadratic expression to be minimized, i.e., the ordinary and generalized least squares estimates are the same for this matrix.

The Instrumental Variables Estimator

By writing this estimator as

$$\hat{C} = (T^{-1}\sum y_t z_t')\,(T^{-1}\sum z_t z_t')^{-1}.$$

and recalling that the state vector is uncorrelated with the innovation vector, we see that this estimator uses the state vector z_t as instrumental variables. We set aside for the moment of the question how to calculate this state vector. Later we show that we use the expression in (8) of Chapter 6.

The preceding discussion suggests the use of any past data vector y_{t-j} as instruments, or for that matter, the stacked vector y_{t-1}^- as instruments. For example, the use of y_{t-2} would produce the equation

$$\hat{\Lambda}_2 = \hat{C}\hat{\Omega}_{K,2},$$

where $\Omega_{K,2}$ is the second submatrix in Ω_K, i.e., AM. We see that the use of the vector y_{t-1}^- as instruments yields exactly the relation used in the stochastic realization estimator, i.e., $\hat{H}_{1.} = \hat{C}\hat{\Omega}_K$.

Among all these possible vectors of instruments, which are the best? Following Sargan [1989] we note that $\Omega R^{-1} y_{t-1}^-$ is asymptotically the most efficient one for the matrix C. But this is the expression we derived in (8) of Chapter 6 for the state vector. We thus have an interesting characterization of the state vector as asymptotically the most efficient vector of instruments, at least for the matrix C. Recalling that state vectors are the "sufficient" summary or statistic of information carried by the past observation, this relation is not surprising at all. Using the state vector as instruments on the observation equation, the matrix C is estimated by solving

$$T^{-1}\sum y_t y_{t-1}^{-\prime} R^{-1}\hat{\Omega}' = \hat{C}\hat{\Pi},$$

where

$$\hat{\Pi} = T^{-1}\sum z_t z_t' = \hat{\Omega}R^{-1}\hat{\Omega}'.$$

This can be written succinctly as

$$H_{1.}R^{-1}\hat{\Omega}' = \hat{C}\hat{\Pi}. \tag{4}$$

We know the matrix $\hat{\Omega}$ as the second factor of the singular value decomposition of the Hankel matrix, and $\hat{\Pi}$ as the solution of the Riccati equation using the estimated matrices A, C, and M. In VAR we can bypass the solution of the Riccati equation because the expression $\Omega_K R^{-1} \Omega_K'$ for sufficiently large K is the correct solution of the Riccati equation. It merely restates the relation $\Pi_{t+1} = \Pi_t$. This fortunate situation exists because a finite truncation of the (infinite-dimensional) stacked data vector y_{t-1}^- at lag d proves to be sufficient for VAR(d). Several analytical examples are discussed to illustrate this point later in this chapter.

To interpret the estimator for the matrix M given by (3b) as an IV estimator, mutiply (1) of Chapter 6 from the right by y_{t-1}', noting that the vector x_t can be replaced by z_t

$$T^{-1} \sum y_t^+ y_{t-1}' = \hat{\boldsymbol{O}}_J \hat{M} ,$$

where $\hat{M}$ is equal to $T^{-1} \sum z_t y_{t-1}'$. The left side is exactly the sample version of $H_{.1}$. The observation equation of the backward innovation model of Section 7.1 can suggest another IV estimator. In that model the state vector, ζ_{t+1}, given in (10) of Section 7.1 is asymptotically the most efficient estimator for the matrix M'. Instead of (4) we obtain

$$H_{.1}' R_+^{-1} \boldsymbol{O}_J = \hat{M}' \hat{Z} , \tag{5}$$

where the covariance matrix Z of the backward innovation model is as given in Section 7.1, $\boldsymbol{O}_J' R_+^{-1} \boldsymbol{O}_J$. Having noticed this relation we can obtain the same equation for M from the first equation in (2b) by multiplying from the left by $\boldsymbol{O}_J' R_+^{-1}$.

The estimator for the matrix A of (3a) does not have a direct interpretation as the instrumental variables estimator. However, the product of $\boldsymbol{O} A$ can be so interpreted by using y_{t-1}^- as the vector of instruments on the stacked observation equation

$$y_{t+1}^+ = \boldsymbol{O}_J A z_t + \cdots ,$$

which was used earlier to derive H^A. Then the pseudo-inverse of the estimate of the observability matrix is used to recover an estimate of the matrix A.

Another estimator for A is suggested by the IV estimators for the matrices C and M we just described. Compare (2a) with the two equations in (2b). Multiply (2a) from the left by the same factor used in solving for M, and from the right by the factor used in solving for C, we define an estimator

$$\boldsymbol{O}_J' R_+^{-1} H^A R_-^{-1} \Omega_K = (\boldsymbol{O}_J' R_+^{-1} \boldsymbol{O}_J) A (\Omega_K R_-^{-1} \Omega_K') . \tag{6}$$

Eq. (4), (5), and (6) jointly constitute the instrumental-variable estimators of the system matrices.

To summarize, this section developed two sets of estimators using first the stochastic realization theory, and then improved their asymptotic efficiency by taking advantage of the close relation of the estimator with the IV estimators. See Rosenbrock [1970], Bosgra and van der Weiden [1980], Pernebo and Silverman [1982], and van Zee [1981] for related work on deterministic systems. Finally by using the state vector either in the forward or backward innovation models, as the case requires, the asymptotic efficiency of the estimators for the matrices A, C and M are improved.

9.2 Properties of Balanced Models

In the preceding section we set $O = U\Sigma^{1/2}$ and $\Omega = \Sigma^{1/2}V'$ to estimate system matrices for the innovation model in balanced representation. In this coordinate system the observability and Ω Gramians are equal to each other. Recall our discussions on the importance of proper choices of coordinate systems in Section 6.3. The class of estimators in which the matrix Σ is variously factored and associated with the observability and Ω matrices all share an important property called ''nesting of estimators''. However, only the balanced ones inherit stability property, as we soon demonstrate. Unfortunately, the instrumental-variable estimators are not nested, nor do they inherit the stability property of the transition matrix.

Nesting of System Matrix Estimates and Π

The balanced-model representation has several useful properties. One such property is the "nesting" of system matrices A, M, and C. These system matrices in any lower-dimensional model in balanced representation can be read off from a given model in balanced representation by taking the leading principal submatrix of the matrix A and the corresponding submatrices of M and C. For example, once a balanced model of dimension 7 is constructed from a given set of data, say, the system matrices A,C, and M of balanced models of dimension one through 6 can be read off without further calculations. Only the Riccati equation needs to be solved anew to obtain the covariance matrix for the innovation vector of a lower-dimensional innovation model, which is then used to re-estimate the matrix B and Δ. Even here the nesting property holds exactly for the state vector covariance matrix for VAR data-generating processes and approximately for other processes, as we soon show. For VAR processes, at least, building a model of certain size implies that all smaller models are at once specified.

To establish this nesting property, partition the matrices in the singular value decomposition of the Hankel matrix comformably,

$$O_J = [O_1 \;\; O_2] \text{ and } \Omega_K = \begin{bmatrix} \Omega_1 \\ \Omega_2 \end{bmatrix},$$

where U_1, Σ_1, and V_1 are the matrices corresponding to the first n_1 largest singular values, i.e., $O_1 = U_1\Sigma^{1/2}$ and $\Omega_1 = \Sigma_1^{1/2}V_1'$. Because U and V are orthogonal, we have

$$O_1^+ H^A \Omega_1^+ = \Sigma_1^{-1/2} U_1' [\, U_1\Sigma_1^{1/2} \; U_2\Sigma^{1/2} \,] A \begin{bmatrix} \Sigma_1^{1/2}V_1' \\ \Sigma_2^{1/2}V_2' \end{bmatrix} V_1\Sigma_1^{1/2}$$

$$= [\, I_{n_1} \;\; 0 \,] A \begin{bmatrix} I_{n_1} \\ 0 \end{bmatrix} = \hat{A}_{11}.$$

The estimate $\hat{A}_{11}$ is the upper-left-hand corner of $\hat{A}$, and is invariant when n_1 is increased, and similarly for $\hat{C}$. For example

$$\hat{C} = H^C \, [\, V_1\Sigma_1^{-1/2} \; V_2\Sigma_2^{-1/2} \,]$$

$$= [\, \hat{C}_1, \;\; \hat{C}_2 \,] \,.$$

The nesting property guarantees that a model constructed under misspecified model size is not a total loss. If the dimension is estimated to be smaller than the true one, then the estimated matrices A ,C, and M remain correct, i.e., the model using these system matrices provides a correct lower-dimensional approximation of the true model. If the model size is specified to be too large, then certain submatrices (the upper-left-hand corner in the case of the matrix A, upper submatrix for the matrix M, and the left-hand-side submatrix in C) are the correct system matrices; hence they need not be recalculated.

One could therefore say that the estimators are orthogonal in the same sense as an orthogonal polynomial expansion approximation to some continuous functions on an interval, say. To improve the accuracy, one need not recalculate the coefficients already calculated, and to drop a few terms the remaining coefficients are the correct ones to use.

Akaike's canonical variable approach, Akaike [1975], introduces additional state variables which are uncorrelated with the state variables already in place in increasing the model size. However, the system matrices are not nested. In our approach, the newly-introduced state variables are in general correlated with the existing ones. This is easy to see when we refer back to the definition of the state vector in (8) of Chapter 6. Retaining only the first n_1 largest singular values, the state vector is

$$z_t = \Omega_1 R^{-1} y_{t-1}^- \,.$$

When the rest of the singular values are also included, the state vector will add $\Omega_2 R^{-1} y_{t-1}^-$ as the second subvector. These two vectors have non-zero covariances, unless Σ_2 is zero.

Although the Gramians are affected by the ways the matrix Σ is factored, the eigenvalues of the transition matrix A are the same for this class of estimators. To see this, consider a special case in which the observability matrix is set to the matrix U and the Ω matrix to $\Sigma V'$. Then, instead of (3a), the estimator is given by $\hat{A} = U'H^A V\Sigma^{-1}$. However, the eigenvalues of both matrices are the roots of $|\lambda\Sigma - U'H^A V| = 0$.

Another useful fact of nesting in our estimation procedure is that these lower-dimensional nested models inherit the stability properties of the original model, as is shown in the next section.

Stability

Pernebo and Silverman [1982] have shown that all the lower-dimensional balanced models contained in an original balanced model are asymptotically stable if the original one is. This section follows their derivation of this useful fact.

Recall that the observability and reachability Gramians are the same diagonal matrix in a balanced representation. Denote this common matrix by Σ. When the equation for the reachability Gramian, given in (9) of Chapter 5, is multiplied from the left by A', and by A from the right, and added to the equation for the observability Gramian, given in (7) of the same chapter, the matrix Σ satisfies an algebraic matrix equation

$$A'A\Sigma A'A - \Sigma = -(C'C + A'MM'A) .$$

Let v be an eigenvector of $A'A$ with its corresponding eigenvalue, λ. Then the above equation yields, on multiplication by v' from the left, and by v from the right,

$$(\lambda^2 - 1)v'\Sigma v = -v'(C'C + A'MM'A)v \le 0 ,$$

establishing that the eigenvalues of $A'A$ are less than or equal to one in modulus. A further refinement of the argument can show that $|\lambda| < 1$ if the eigenvalues of Σ are all distinct (Pernebo and Silverman [1982]).

Consider a partition of a balanced model into two subsystems:

$$z_{t+1} = \begin{bmatrix} A_{11} & A_{12} \\ A_{21} & A_{22} \end{bmatrix} z_t + \begin{bmatrix} B_1 \\ B_2 \end{bmatrix} e_t$$

and

$$y_t = [C_1 \;\; C_2]\, z_t + e_t .$$

Assume that the system is reachable and observable so that $\Sigma > 0$. The two associated Gramians both become block diagonal, $\Sigma = \mathrm{diag}\,(\Sigma_1 , \Sigma_2)$. Suppose that the total system is asymptotically stable, i.e., $|A| < 1$. From the construction of a balanced representation, we know that $G_0 = G_\Omega = \Sigma$. To establish that every subsystem of an asymptotically stable balanced model is asymptotically stable, write the defining relation for the Ω Gramian for the subsystem 1

$$A_{11}\Sigma_1 A'_{11} + A_{12}\Sigma_2 A'_{12} - \Sigma_1 = -M_1 M'_1 ,$$

and multiply it by v' and v from left and right, respectively, to derive

$$(|\lambda|^2 - 1)v'\Sigma_1 v = -(v'A_{12}\Sigma_2 A'_{12}v + v'M_1 M'_1 v) \le 0 ,$$

where v is now redefined to satisfy $A_{11}v = \lambda v$, $v'v = 1$. Because $v'\Sigma_1 v > 0$, it easily follows that $|\lambda| \le 1$. The possibility that $|\lambda| = 1$ is eliminated, because then $v'A_{12} = 0$, and $v'M_1 = 0$ must follow from the fact that Σ_2 is positive definite. But this implies by the Hautus (characterization of reachability) of the pair (A, M),

$$(v', 0)\begin{bmatrix} A_{11} & A_{12} \\ A_{21} & A_{22} \end{bmatrix} = \lambda(v', 0) \quad \text{and} \quad (v', 0)\begin{bmatrix} M_1 \\ M_2 \end{bmatrix} = 0 ,$$

hence the system is not reachable, contrary to the assumption. Thus, the fact that $|\lambda| < 1$ is established, and subsystem 1 is asymptotically stable. Since subsystem 1 is any subsystem, subsystem 2 is also asymptotically stable.

Suppose the subsystems are obtained by partitioning the total system according to the criterion $\sigma_{min}(\Sigma_1) > \sigma_{max}(\Sigma_2)$ in the balanced representation. For subsystem 1, Σ_1 satisfies

$$A'_{11}\Sigma_1 A_{11} + A'_{21}\Sigma_2 A_{21} - \Sigma_1 = -C'_1 C_1 .$$

If subsystem 1 is not observable, there is a normalized eigenvector of $A_{11}, v, v'v = 1$ satisfying $A_{11}v = \lambda v$, and $C_1 v = 0$. Multiplying the above equation by v' and v from the left and right, respectively, it follows that

$$(1 - |\lambda|^2)v'\Sigma_1 v = v'A'_{21}\Sigma_2 A_{21}v .$$

Note that $v'\Sigma_1 v \ge \sigma_{min}(\Sigma_1)$. The right-hand side can be bounded by

$$v'A'_{21}\Sigma'_2 A_{21}v \le |A_{21}v|^2 \sigma_{max}(\Sigma_2) .$$

Balanced, stable models are such that $|A| \le 1$. This implies, in particular, that

$$\left|\begin{bmatrix} A_{11} \\ A_{21} \end{bmatrix} v\right| \le 1 \quad \text{or} \quad |A_{11}v|^2 + |A_{21}v|^2 \le 1 ,$$

hence $|A_{21}v|^2 \le 1 - |\lambda|^2$. Substituting these into (1), the inequality

$$(1 - |\lambda|^2)\sigma_{min}(\Gamma_1) \le (1 - |\lambda|^2)\sigma_{max}(\Gamma_2)$$

is established. Previous results show that $|\lambda| < 1$, hence $\sigma_{\min}(\Sigma_1) \leq \sigma_{\max}(\Sigma_2)$. This contradicts the assumed criterion for partitioning subsystems. Hence, subsystem 1 is observable. Proceeding analogously, subsystem 1 is also shown to be reachable. See Harshavarahana et al. [1984].

These estimates were constructed by a specific factorization of a Hankel matrix to produce the balanced-model representation as discussed in Chapter 5. Other factorizations lead to model representations related to the balanced one through some nonsingular transformations of state vectors. For example, let $O_J = U$ and $\Omega_K = \Sigma V'$. From our discussions in Section 6.3, the state vector of the estimated model by this new factorization is related to the balanced one by the matrix $S = \Sigma^{1/2}$, i.e., the state vector of this model is S times that for the balanced model, so that $\hat{C}_1$ is calculated independently of the calculation for $\hat{C}_2$.

One of the advantages of this estimation procedure is that there is no need to recalculate the estimates of these three matrices when n_1 is reduced to n^*. The upper-left-hand $n^* \times n^*$ submatrix of $\hat{A}$ and the first n^* row and column vectors in $\hat{M}$ and $\hat{C}$, respectively, are the correct system matrices for the balanced model of dimension n^*. The situation is analogous to that in calculating an approximate expansion of some function by a system of orthogonal functions. The coefficients already calculated need not be recalculated when the number of terms in the expansion is changed. In the canonical correlation approach of Akaike the orthogonality is imposed on the state vector components but not on the coefficient matrices. Changing the dimension of the state vector requires re-estimation of the coefficent matrices. In our algorithm only the matrices B and Δ need be recalculated when n is reduced.

The dimension of the estimated matrix $\hat{A}$ is the same as the number of singular values retained in Σ. The nature of the approximation and the errors of such constructions are discussed in Chapter 10 together with simple analytical examples.

9.3 Examples with Exact Covariance Matrices

Using exact expressions for covariance matrices, we present several analytical examples to illustrate the preceding developments in this section. Use of exact covariance matrix expressions eliminates statistical errors as a source of modeling error and assists us in our efforts for assessing the effectiveness of the proposed algorithm. Scalar-valued time series are discussed first. For a univariate series, the Hankel matrix of (.) is symmetric, and the orthogonal matrices U and V in its singular value decomposition are related by $U = DV$, where D is a diagonal matrix with entries 1 or -1. This special feature of the singular value decomposition implies some special features of system matrices in the balanced models. From (3) we see that $C' = DM$. The transition matrix A also has some special characters. To see this, let the model be given by

$$x_{t+1} = Ax_t + be_t, \quad y_t = cx_t + e_t.$$

Here we use lower-case letters b and c to denote column vectors. The two Lyapunov equations for the Gramian are

$$A\Sigma A' - \Sigma = -mm', \text{ and } A'\Sigma A - \Sigma = -c'c.$$

These equations are not independent, because each is of rank one. They differ only in the signs of the components. Then from the second equation we obtain $DA'\Sigma AD - D\Sigma D = -Dc'cD$, in which $D\Sigma D = \Sigma$ and hence $DA'D = A$, because of $m = Dc'$. For example, when the model is two-dimensional, then the matrix A is either

$$A = \begin{bmatrix} \alpha & \beta \\ \beta & \delta \end{bmatrix}$$

or the (2, 1) element of A is replaced by -β. In other words, the matrix A must either be symmetric or anti-symmetric, and the components of m are those of c or their negatives in the two-dimensional balanced models for univariate series. In general an off-diagonal element of the dynamic matrix a_{ij} is either equal to a_{ji} or its negative for univariate time series.

Next, a balanced state space model is constructed for a univariate AR(1) model, $y_t = ay_{t-1} + u_t$ with $cov\ u_t = q$. The sequence of covariances is easily seen to be $\lambda_j = qa^j/(1-a^2)$, $j = 0, 1, \ldots$. Suppose we use the data truncation with $K = 2$, even though in this simple example $K = 1$ suffices. Suppose we choose $v = u$, where the lower-case letters indicate that they are column vectors rather than matrices. The 2×2 Hankel matrix has the singular value decomposition $H = \sigma uu'$, with $\sigma = \lambda_1(1+a^2)$, and u' is proportional to $[1 \ \ a]$. Calculate c by (3c). It is equal to $\sqrt{\lambda_1}$. The parameter a is recovered as the estimate of the dynamic parameter from (3a). The Riccati equation is a quadratic polynomial in π, the variance of the state variable,

$$\pi = a^2\pi + (m - ac\,\pi)^2(\lambda_0 - c^2\pi)^{-1} = 0\,,$$

or, after some rearranging

$$\pi^2 - [2am/c + (1-a^2)\lambda_0/c^2]\pi + 1 = 0\,.$$

This equation has a real positive solution if and only if the next inequality,

$$[(1-a^2)\lambda_0/2c^2 + am/c]^2 \geq 1\,,$$

holds. By our choice of the vector v the ratio m/c is one. Note that λ_0/c^2 is a^{-1}, and this inequality is always satisfied if the exact values of the covariances are used. It is interesting to note that the same condition also appears in establishing that the solution of an associated Riccati differential equation approaches its equilibrium value as time goes to infinity.

When the estimated system parameters and the sample autovariance $\hat{\lambda}_0$ are used, this inequality may be violated due to statistical errors in them. Then the state space model cannot be constructed, because the Riccati equation has no positive solution. Such an example is found in the time series for the German index of industrial production, when the Hankel matrix is computed using $J=K=10$. The sample variance is $\lambda_0 = 0.002483$. The scalar model's system parameters are a=.9377 and m=c=.0494. With these values, the left side of the inequality becomes .9991, and the Riccati equation does not possess a positive solution. Either by changing K or by increasing the dimension of the model, this source of difficulty disappears.

Example 1: AR(2) Model. To convey the nature of balanced models simply, we continue to consider examples for which the balanced representation can be obtained analytically. Consider a special scalar AR (2) model

$$(1 - \phi L^2) y_t = u_t ,$$

where ϕ is less than one in magnitude. Denote the variance of u_t by q. Following our procedure given in Chapter 4, a space representation of this model is obtained by setting x_{1t} to $y_t - u_t$, and x_{2t} to ϕy_{t-1}. This choice of the state vector components produces a state space model in the observable canonical form with the system matrices given by the matrix

$$A = \begin{bmatrix} 0 & 1 \\ \phi & 0 \end{bmatrix} , \quad B' = [0 \ \ \phi] \text{ and } C = [1 \ \ 0].$$

The sequence of the auto-covariances of the series can be immediately obtained from the Yule-Walker equation, or from (6) of Chapter 3, since the matrix M is a two-dimensional column vector with the first component zero and the second $\phi\lambda_0$, with $\lambda_0 = q/(1-\phi^2)$. Its Riccati equation can easily be solved analytically. It is a diagonal matrix, $diag\,(\phi\lambda_0 \ \lambda_2)$. The autocovariance at lag k is $\phi^k \lambda_0$, $k = 1, 2, \ldots$ All covariances with odd lags are zero, because the first component of M is zero, using (3) of Chapter 3. The singular value decomposition of the 2×2 Hankel matrix has the matrix U, which has 1 along the counter diagonal line; its V is the 2-dimensional identity matrix, and the two singular values are equal to λ_2. The balanced representation results by factoring the above so that the observability matrix is the square root of λ_2 times the matrix U, and the Ω matrix is the same constant times the 2-dimensional identity matrix. The system matrices in balanced representation are

$$A = \begin{bmatrix} 0 & \phi \\ 1 & 0 \end{bmatrix}$$

and

$$C = [0 \ \ \lambda_2^{1/2}] , \text{ and } M = [\, \lambda_2^{1/2} \ \ 0] .$$

Note that ϕ appears as $A_{1,2}$ and zeros appear differently in the matrices C and M in the balanced representation. We now proceed to discuss vector-valued time series.

Models for VAR Processes

State space models for VAR processes are particularly simple to construct, a basic reason being that only a finite number of past data vectors is needed to express the current innovation vector. A few examples illustrate this point. An important consequence of this fact is that we can give a closed form expression for the solution of the Riccati equation for VAR processes. We also show how the parameters in VAR can be recovered from the system matrices.

Example 2: AR (1) Model. First, take an AR (1) model $y_{t+1} = \phi y_t + u_{t+1}$ with mean zero and serially uncorrelated disturbance term u_t. Let Q be the covariance matrix of u_t which is p-dimensional. Suppose that $|\phi| \neq 0$. (We drop this assumption later in this section.) Let Q be the covariance matrix of u_t. The Yule-Walker equations are

$$\Lambda_0 = \phi \Lambda_0 \phi' + Q \text{ , and, } \Lambda_1 = \phi \Lambda_0 .$$

When we set x_t equal to $y_t - u_t$, then the resulting state equation has ϕ both as the transition matrix A, and the matrix B which multiplies the innovation vector u_t. The matrix C in the observation equation is I_p. Therefore the covariance matrix of this state vector is equal to $\Lambda_0 - Q$--i.e., $\phi \Lambda_0 \phi'$ by the Yule-Walker relation above. We can directly verify that this expression satisfies the Riccati equation $\Pi = \phi \Pi \phi' + \phi Q \phi'$. Note that the matrix B is directly given and it is not necessary to solve the Riccati equation to determine it. As a matter of fact, if x_t is set equal to y_t, the Yule-Walker equation is the Riccati equation.

Suppose we now put the model in balanced form

$$x_{t+1} = Ax_t + Be_t \text{ and, } y_t = Cx_t + e_t \quad . \tag{7}$$

Advancing t by one unit in (7),

$$\begin{aligned} y_{t+1} &= C(Ax_t + Be_t) + e_{t+1} \\ &= CAx_t + CBe_t + e_{t+1} , \end{aligned}$$

where x_{t+1} is substituted out. Alternatively, y_{t+1} can also be expressed as

$$y_{t+1} = \phi(Cx_t + e_t) + u_{t+1} \quad . \tag{6}$$

The dynamic matrix ϕ has p eigenvalues, i.e., p dynamic modes. Such a process requires a p-dimensional state vector to describe dynamic behavior. Therefore the matrix C is square. Comparing the two alternative expressions above, we deduce a set of relations,

$$CB = \phi\,, \text{ and, } CA = \phi C.$$

The first is not surprising, since ϕ is the term in the Wold decomposition of this VAR process, and CB is the term in the impulse response sequence of the state space model. These two relations can be restated as $CA = CBC$, or $C(A - BC) = 0$. Since C is nonsingular, because the state space model is observable, we obtain a further relation that $A = BC$, or

$$B = AC^{-1}\,. \tag{8}$$

This result means that for this VAR (1) time series, the matrix B of the balanced minimal dimensional state space model can be obtained as (8), without solving the Riccati equation explicitly. In this special model $\boldsymbol{O} = C$ and $\Omega = M$, i.e.,

$$C = U\Sigma^{1/2} \text{ and } M = \Sigma^{1/2}V'$$

in the balanced representation, i.e,

$$C'C = MM' = \Sigma\,.$$

Therefore both C and M are nonsingular when Σ is.

The expression for Π, $\Omega\,\Lambda_0^{-1}\,\Omega'$, is simply $M\Lambda_0^{-1}M'$. This is a special case of the general formula for the solution of the Riccati equation of the state space model for VAR processes which is derived later. To see this more directly, we can proceed as follows: from the relations

$$\Pi - A\Pi A' = B\Delta B' \text{ and }, \; B\Delta = M - A\Pi C'\,,$$

we note that $B\Delta B'$ is

$$(M - A\Pi C')B' = MB' - A\Pi A'\,,$$

when $A = BC$ is used. We deduce that $\Pi = MB'$. Alternatively, substitute Δ out by $\Lambda_0 - C\Pi C'$ to obtain

$$B\Delta B' = B(\Lambda_0 - C\Pi C')B'$$

$$= B\Lambda_0 B' - A\Pi A'$$

or

$$\Pi = B\Lambda_0 B'\,.$$

From $MB' = B\Lambda_0 B'$, we obtain $M = B\Lambda_0$ if $\mathcal{N}(B) = 0$, or if B' is full rank. Then

$$B = M\Lambda_0^{-1},$$

and

$$\Pi = B\Lambda_0 B' = M\,\Lambda_0^{-1}M'.$$

Substitute this relation into the Riccati equation for the state vector covariance matrix $\Pi = cov\ z_t$, to express it as

$$\begin{aligned}\Pi &= A\Pi A' + B\Delta B' \\ &= A\Pi A' + AC^{-1}(\Lambda_0 - C\Pi C')C^{-T}A' \\ &= AC^{-1}\Lambda_0 C^{-T}A' \\ &= B\Lambda_0 B' = M\Lambda_0^{-1}M' \ .\end{aligned}$$

The theoretical rank of (a sufficiently large) Hankel matrix need not be an integer multiple of the dimension of the data vector. In the preceding example, the dimension of the state vector is the same as that of the data series, and also is the same as the rank of the Hankel matrix, because the matrix ϕ in the VAR is of full rank. If it is not, the rank of the Hankel matrix and consequently the dimension of the state vector may be less than the dimension of the time series. Recall the examples in Chapter 5 in which minimal and non-minimal state space models are presented. The next example also illustrates this point. See also the examples later in this section.

Deficient rank Nothing in the demonstration that $\phi\Lambda_0\phi'$ satisfies the Riccati equation in the AR(1) example depends on the rank of ϕ. Even when the matrix ϕ in the *AR*(1) model is rank deficient, the covariance matrix of a state vector is given by $\Omega R^{-1}\Omega'$ which specializes to $M\Lambda_0^{-1}M'$ with $M = \Lambda_1$.

To illustrate, consider

$$\phi = \alpha\beta'$$

with p-dimensional vectors α and β. The rank of ϕ is one rather than p. Let $\nu = \beta'\Lambda_0\beta$. From the Yule-Walker equation $\Lambda_0 = \phi\Lambda_0\phi' + Q$, we see that $\nu = (\beta' Q\beta)/[1-(\beta'\alpha)^2]$. The state vector defined by

$$z_t = \mathrm{E}(x_t \mid y_{t-1}) = \Lambda_1 \Lambda_0^{-1} y_{t-1}$$

has the covariance matrix

$$\begin{aligned} Z &= \Lambda_1 \Lambda_0^{-1} \Lambda_1' \\ &= \phi \Lambda_0 \phi' = \nu \alpha \alpha' , \end{aligned}$$

where we use the relation $\Lambda_1 = \phi \Lambda_0$. This Z satisfies the Riccati equation

$$\begin{aligned} Z &= \phi Z \phi' + \phi Q \phi' \\ &= \phi(\phi \Lambda_0 \phi' + Q) \phi' \\ &= \phi \Lambda_0 \phi' . \end{aligned}$$

More directly, we derive the expression

$$\phi Z \phi' + \phi Q \phi' = \frac{\beta' Q \beta}{1-(\beta'\alpha)^2} \alpha\alpha' = \nu \alpha \alpha' ,$$

which is equal to Z.

Actually if we know that $\phi = \alpha\beta'$ and if β is known, then we can deal directly with the scalar data-generating process

$$\beta' y_t = (\beta'\alpha)\,(\beta' y_{t-1}) + \beta' u_t .$$

In this example the state vector covariance matrix has rank one, which indicates that the dimension of the state vector should also be one. Use of dimensions larger than the true one does not imply that the extra elements introduced in the balanced representation of the matrices A and C are automatically small. To illustrate, take $p = 2$. The singular value decomposition of Λ_1 is

$$\Lambda_1 = u_1 \sigma_1 v_1' ,$$

with

$$u_1 = (\alpha'\alpha)^{-1/2} \alpha, \quad v_1 = (\beta' \Lambda_0^2 \beta)^{-1/2} \Lambda_0 \beta$$

and

$$\sigma_1 = [(\alpha'\alpha)(\beta' \Lambda_0^2 \beta)]^{1/2} .$$

Suppose that there is a small $\sigma_2 > 0$ because of finite sample size, i.e., $\hat{\Lambda}_1$ is calculated to be rank 2

$$\Lambda_1 = \sigma_1 u_1 v_1' + \sigma_2 u_2 v_2'$$

with

$$u_1' u_2 = \text{and } v_2' v_1 = 0.$$

Then the balanced model estimates a 2×2 matrix

$$\hat{A} = \Sigma^{-1/2} U' \phi \Lambda_1 V \Sigma^{1/2}$$

$$= \begin{bmatrix} (u_1' \alpha)\,(\beta' u_1) & \sqrt{\sigma_2/\sigma_1}\;(u_1' \alpha)(\beta' u_2) \\ \sqrt{\sigma_1/\sigma_2}(u_2' \alpha)\;(\beta' u_1) & (u_2' \alpha)(\beta' u_2) \end{bmatrix}$$

$$= \begin{bmatrix} (u_1' \alpha)(\beta' u_1) & \sqrt{\sigma_2/\sigma_1}\;(u_1' \alpha)(\beta' u_2) \\ 0 & 0 \end{bmatrix},$$

because $u_2' \alpha = 0$. The (1, 2) element of $\hat{A}$ is small only if σ_2/σ_1 is sufficiently small, whereas rank $\hat{A}$ is one even if σ_2/σ_1 is not sufficiently small. This example illustrates a general point: that a test of rank is preferable to testing whether edge elements are zero in some system matrices, such as the one above.

Example 3: AR(2) Model. The next example has a state vector with a dimension which is an integral multiple of that of y. This example shows that sequences of covariance matrices may be calculated from the Riccati equation in certain coordinates rather than from the Yule-Walker equation, and that a finite truncation of the (infinite-dimensional) data vector y_{t-1}^- suffices for VAR processes. Consider a VAR(2) process given by

$$y_t = \phi_1 y_{t-1} + \phi_2 y_{t-2} + u_t, \text{ with } cov\ u_t = \Delta .$$

By stacking lagged data vectors, we construct a state space representation for (8) as

$$\begin{bmatrix} y_t \\ y_{t-1} \end{bmatrix} = F \begin{bmatrix} y_{t-1} \\ y_{t-2} \end{bmatrix} + \begin{bmatrix} u_t \\ 0 \end{bmatrix} \text{ and , } \quad y_t = [I \ 0] \begin{bmatrix} y_t \\ y_{t-1} \end{bmatrix}, \tag{9}$$

with the transition matrix

$$F = \begin{bmatrix} \phi_1 & \phi_2 \\ I & 0 \end{bmatrix} .$$

Here the stacked vector $(y_t' \, y_{t-1}')'$ serves as a state vector. Consequently, the matrix M is equal to $[\Lambda_1' \Lambda_0']'$ in this coordinate system.

Because u_t is uncorrelated with the data vector y_{t-1} and earlier ones, the Riccati equation in this coordinate system is

$$R = FRF' + \begin{bmatrix} \Delta & 0 \\ 0 & 0 \end{bmatrix} \tag{10}$$

with

$$R = \begin{bmatrix} \Lambda_0 & \Lambda_1 \\ \Lambda_1' & \Lambda_0 \end{bmatrix} .$$

When the matrix F is known, this equation can be solved for R because it is a Lyapunov equation. This is an alternative to solving the Yule-Walker equation to obtain the exact expressions for the first two of the sequences of the auto-covariance matrices. Model (9) is an innovation model, because its state vector is uncorrelated with u_t. Its state vector is already of the form $z_t = \hat{E}(x_t \mid y_{t-1}, y_{t-2}) = x_t$. In this coordinate system $\Pi = R$, and we already know that it solves the Riccati equation (10).

In another representation two p-dimensional vectors, $x_{1t} = y_t - u_t$, and $x_{2t} = x_{t+1} - \phi_1 y_t = \phi_2 y_{t-1}$, are introduced as two substate vectors. Rearranging the definitional relation, the model representation becomes

$$\begin{bmatrix} x_1 \\ x_2 \end{bmatrix}_{t+1} = A \begin{bmatrix} x_1 \\ x_2 \end{bmatrix}_t + Bu_t \quad , \text{ and } \quad y_t = Cx_t + u_t$$

with

$$A = \begin{bmatrix} \phi_1 & I \\ \phi_2 & 0 \end{bmatrix}, \quad B = \begin{bmatrix} \phi_1 \\ \phi_2 \end{bmatrix}$$

and

$$C = [I \ 0].$$

The Kalman filter for x_t of this new representation $z_t = \hat{E}(x_t \mid y_{t-1}^-)$ obeys the same equation

$$z_{t+1} = Az_t + Be_t, \text{ and } e_t = y_t - Cz_t . \tag{11}$$

Although y_{t-1}^- is in general a vector which stacks all past observations, i.e., infinite-dimensional, we can truncate y_{t-1}^- as

$$y_{t-1}^- = \begin{bmatrix} y_{t-1} \\ y_{t-2} \end{bmatrix} \tag{12}$$

when the data generating process is (8). One way of seeing this is to note that all the lagged data auto-covariance matrices, $\Lambda_j, j \geq 2$, are uniquely determined by Λ_0, Λ_1, and Δ. The Yule-Walker Eqaution for (8) directly shows this:

$$\Lambda_0 = \phi_1 \Lambda_0 \phi_1' + \phi_2 \Lambda_0 \phi_2' + \phi_1 \Lambda_1 \phi_2' + \phi_2 \Lambda_1' \phi_1 + \Delta,$$

$$\Lambda_1 = \phi_1 \Lambda_0 + \phi_2 \Lambda_1',$$

and

$$\Lambda_j = \phi_1 \Lambda_{j-1} + \phi_2 \Lambda_{j-2}, \qquad j \geq 2.$$

We have already remarked that we can use the state space model introduced in (9) to solve for the covariance matrices. This is so because the covariance matrices are "external" descriptions of the data series and are independent of the coordinate systems used to describe it. Using (9) we have $\Lambda_{j+1} = [I \ 0]F^j \ [\Lambda_1' \Lambda_0]'$ from (3) of Chapter 3, so that we can write $[\Lambda_j \ \Lambda_{j+1}] = [I \ 0]F^j R$, $j = 1, \cdots$ since F times $[\Lambda_0' \Lambda_1]'$ is equal to $[\Lambda_1' \Lambda_0]$.

Consequently, the Hankel matrices of size min $(J, K) \geq 2$ depend only on R, F and Δ. For example, the Hankel matrix with $J = K = 2$ is expressible as

$$\begin{bmatrix} \Lambda_1 & \Lambda_2 \\ \Lambda_2 & \Lambda_3 \end{bmatrix} = \begin{bmatrix} \phi_1 & \phi_2 \\ \phi_1^2 + \phi_2 & \phi_1 \phi_2 \end{bmatrix} R.$$

From (12), the state vector depends only on y_{t-1} and y_{t-2} through

$$z_t = \Omega R^{-1} \begin{bmatrix} y_{t-1} \\ y_{t-2} \end{bmatrix} \tag{13}$$

with

$$\Omega = \mathrm{E}\,(x_t \ y_{t-1}^{-}{}')$$

$$= \mathrm{E}\,(z_t \ y_{t-1}^{-}{}').$$

In the coordinate system of (10)

$$x_t = \begin{bmatrix} y_t - e_t \\ \phi_2 y_{t-1} \end{bmatrix} = \begin{bmatrix} \phi_1 & \phi_2 \\ \phi_2 & 0 \end{bmatrix} \begin{bmatrix} y_{t-1} \\ y_{t-2} \end{bmatrix},$$

and hence $\Omega = \overline{\Omega} R$, with

$$\overline{\Omega} = \begin{bmatrix} \phi_1 & \phi_2 \\ \phi_2 & 0 \end{bmatrix} R.$$

These are particular cases of the relations valid in general when the data generating processes are VAR of some finite order. From (13) we know that in this process

$$z_{t+1} = \Omega R^{-1} \begin{bmatrix} y_t \\ y_{t-1} \end{bmatrix},$$

with R and Ω defined above. (In general y_t^- must be used.) Then, using z_t as instrument

$$E(z_{t+1} z_t') = A \Pi$$

with

$$\Pi = E(z_t z_t') = \Omega R^{-1} \Omega', \tag{14}$$

i.e.,

$$\begin{aligned} A &= E(z_{t+1} z_t') \Pi^{-1} \\ &= \Omega R^{-1} F R R^{-1} \Omega \Pi^{-1} \\ &= (\Omega R^{-1} F \Omega')(\Omega R^{-1} \Omega')^{-1}, \end{aligned}$$

because $\overline{\Omega}F = A\overline{\Omega}$. These are exact relations. Since $e_t = y_t - Cz_t$, $B\Delta = E(z_{t+1} e_t') = M - A\Pi C'$,
i.e.,

$$B = (M - A\Pi C')\Delta^{-1}, \tag{15}$$

the Riccati equation for Π is nothing but an equality of the covariance matrices of the two vectors $z_{t+1} - Az_t$ and Be_t, $cov(z_{t+1} - Az_t) = cov(Be_t)$, because the left-hand side is $\Pi - A\Pi A'$, by the fact that $E(z_{t+1} z_t') = A\Pi$, and the right-hand side is $B\Delta B'$. This concludes the demonstration that the covariance matrix of the state vector (14) satisfies the Riccati equation

$$\Pi = A\Pi BA' + B\Delta B',$$

with B defined in (15). If we know Λ_0 and Λ_1 exactly, and if we know Ω, then we can calculate Π by (14) without solving the Riccati equation.

Comparing the Wold decomposition and the impulse response sequence, the original coefficient matrices are recovered by $\phi_1 = CB$, and $\phi_2 = CAB - \phi_1^2$.

Choice of *K*

The preceding example shows that the K in determining the truncation of the data vector should be at least as large as p if VAR(p) is generating time series. Too small a value for K poses the danger of missing useful confirmation in some distant past observations. An example of this is provided by Findley [1985], in which the covariances behave as if they are generated by $AR(1)$ processes up to a certain lag. In this example the process, which is AR(4), will be mistakenly identified to be $AR(1)$ if K is too small.

Consider a scalar valued *AR* (4) process

$$y_t = \rho\, y_{t-1} - \alpha\rho\, y_{t-3} + \alpha\, y_{t-4} + e_t$$

with $|\rho| < 1$ and $|\alpha|$ small. To simplify algebra, we deal with correlation coefficients rather than covariances. The set of Yule-Walker equations can be rearranged to show that $r_i = \rho^i$; $i = 1, 2, 3$; and $r_4 = \rho^4 + \alpha(1-\rho^2)$, $r_5 = \rho^5 + \alpha\rho(1-\rho^2)$, and so on. We note that, up to lag 3, correlation coefficients are identical with those of *AR* (1) processes. The Hankel matrix with $K = 2$ involves only r_i , $i = 1 \sim 3$, and miss the departure of r_4 from ρ^4. The choice of $K = 3$ will correct this mistake.

On the other hand, too large a choice of K (relative to the number of data points) is not desirable, because auto-covariance matrices at large lags are calculated with less precision compared with those with smaller lags. Also the possibility increases that some spurious observations or outliers will corrupt the covariance matrix calculations. Recall the example in Section 5.5.

Models for MA Processes

Example 4: MA(1) Model. For MA data-generating processes, unlike VAR processea, the expression $\Omega R^{-1} \Omega'$ is not a solution of the associated Riccati equation. This simple example shows the effects of truncating the past history y_{t-1}^- at y_{t-1}. In this example, a sufficient finite summary of the past data set is e_{t-1}, which is not equal to y_{t-1}.

Consider the the vector-valued *MA* (1)

$$y_t = e_t + \theta\, e_{t-1}.$$

The covariances satisfy $\Lambda_0 = \Delta + \theta\, \Delta\, \theta'$, $\Lambda_1 = \theta\, \Delta$, and $\Lambda_j = 0$, $j \geq 2$.

A state space representation of this process is obtained by defining $x_t = y_t - e_t$, i.e., $y_t = x_t + e_t$. The dynamic equation reduces to

$$x_{t+1} = \theta\, e_t\ .$$

Note that the dynamic matrix A is zero, $B = \theta$, and $C = I$ in this coordinate system. When we project this model into the subspace spanned by y_{t-1},

$$z_t = \hat{E}\,(x_t \mid y_{t-1})$$

$$= M\, \Lambda_0^{-1}\, y_{t-1}$$

with

$$M = E\, x_t\, y_{t-1}'$$

$$= \theta\, \Delta.$$

In this representation the vectors z_t and e_t are uncorrelated--as they should be. The vector $z_t = M\ \Lambda_0^{-1}\ y_{t-1}$ evolves dynamically as

$$z_{t+1} = F\ z_t + G\ e_t$$

with

$$F = E\ (z_{t+1}\ z_t{}') \cdot \Pi^{-1}$$

$$= M\ \Lambda_0^{-1}\ \Lambda_1\ \Lambda_0^{-1}\ M'\ (M\ \Lambda_0^{-1}\ M')^{-1}$$

and

$$G = E\ (z_{t+1}\ e_t{}')\Delta^{-1} = M\ \Lambda_0^{-1}.$$

Unlike the VAR models we discussed earlier, the covariance matrix of z_t, $M \Lambda_0^{-1} M'$, does not satisfy the Riccati equation. The left-hand side is $M \Lambda_0^{-1}\ M'$ and the right-hand side is $M\ \Lambda_0^{-1}\ (\Delta + \Lambda_1\ \Lambda_0^{-1}\ \Lambda_1{}')\Lambda_0^{-1}\ M'$. These two are not equal to each other, because the expression in the parenthesis is not Λ_0. This implies that z_t is not a correct state vector. Suppose that e_{t-1} is directly available and redefine a state vector by

$$\zeta_t = \hat{E}\ (x_t \mid e_{t-1})$$

$$= \theta\ e_{t-1}.$$

Then its dynamics are

$$\zeta_{t+1} = \theta\ e_t\ ,$$

because $E\,(\zeta_{t+1}\ \zeta_t{}') = 0$.

The covariance matrix of this state vector, given by

$$\Pi_\zeta = \theta\Delta\theta'\ ,$$

satisfies the Riccati equation trivially, because the dynamic matrix A is zero and the matrix B is given by θ. Suppose $\| \theta \| < 1$. Then we can invert the data-generating process

$$e_t = y_t - \theta\ y_{t-1} + \theta^2 y_{t-2} - \cdots$$

Clearly, one can't reconstruct e_{t-1} from a finite truncation of y_{t-1}^-.

The next is a simple numerical example with

$$\theta = \text{diag. } (.1\ \ .5),\ \ \Delta = \text{diag. } (2,\ 1).$$

The singular value decomposition of Λ_1 is

$$\Lambda_1 = \text{diag. } (.2 \;\; .5) = U \Sigma V'$$

with

$$U = \begin{pmatrix} 0 & 1 \\ 1 & 0 \end{pmatrix}, \quad V = \begin{pmatrix} 0 & 1 \\ 1 & 0 \end{pmatrix} \quad \text{and } \Sigma^{1/2} = \text{diag.}(\sqrt{.5} \;\; \sqrt{.2}).$$

The balanced state space model is

$$z_{t+1} = B \; e_t \; ,$$

$$y_t = C z_t + e_t \; ,$$

with

$$C = \begin{bmatrix} 0 & \sqrt{.2} \\ \sqrt{.5} & 0 \end{bmatrix}, \quad M = \begin{bmatrix} 0 & \sqrt{.5} \\ \sqrt{.2} & 0 \end{bmatrix},$$

$$B = M \; \Delta^{-1} \, .$$

Therefore,

$$\Pi = B \; \Delta B' + \text{diag. } (.5 \;\; .1)$$

is the solution of the Riccati equation.

If we set

$$z_t = \hat{E}(x_t \mid y_{t-1})$$

$$= M \; \Lambda_0^{-1} \, y_{t-1},$$

then its covariance matrix is

$$cov \; z_t = M \; \Lambda_0^{-1} \, M' = \text{diag. } (.4 \;\; .099) \neq \text{diag. } (\sqrt{.5} \;\; .\sqrt{2}\,).$$

We can complicate the example further. The general conclusion is that the data-generating process $MA\,(q)$ requires $e_{t-1}, \; e_{t-2}, \ldots, e_{t-q}$ as the summary of past information, and that $y_{t-1}, \ldots y_{t-q}$ is not sufficient in defining the state vector. When e_{t-j} is not directly available, we need the entire past history y_{t-1}^-, in principle, to define the state vector.

Example 5: MA(2) Model. This simple example shows the effects of truncating the observation history y_{t-1}^- at some finite past. In this example a finite summary of the listing is available as e_{t-1} and e_{t-2} but not as y_{t-1} and y_{t-2}. The data-generating process is

$$y_t = e_t + \theta e_{t-2}.$$

The covariance matrices are given by $\Lambda_0 = \Lambda + \theta \, \Delta \, \theta'$, $\Lambda_1 = 0$, $\Lambda_2 = \theta \, \Delta$, and

$\Lambda_j = 0,\ j \geq 3$. A state space representation is obtained by introducing state variables $x_{1t} = y_t - e_t$ and $x_{2t-1} = x_{1t}$. Then, noting the relation that

$$x_{2t} = x_{1t+1}$$

$$= y_{t+1} - e_{t+1}$$

$$= \theta e_{t-1},$$

the state space model is given by

$$\begin{bmatrix} x_1 \\ x_2 \end{bmatrix}_t = \begin{bmatrix} 0 & I \\ 0 & 0 \end{bmatrix} \begin{bmatrix} x_1 \\ x_2 \end{bmatrix}_{t-1} + \begin{bmatrix} 0 \\ \theta \end{bmatrix} e_{t-1} ,$$

and

$$y_t = (I \ \ 0)\, x_t + e_t .$$

Let

$$e_{t-1}^- = \begin{bmatrix} e_{t-1} \\ e_{t-2} \end{bmatrix}$$

be the finite summary of the data set for this model.

Define a new state vector by

$$\zeta_t = \hat{E}\, (x_t \mid e_{t-1}^-)$$

$$= \Theta e_{t-1}^-$$

where

$$\Theta = E\,(x_t\ e_{t-1}^{-\prime})\,(cov\ e_{t-1}^-)^{-1}$$

$$= \begin{bmatrix} 0 & \theta \\ \theta & 0 \end{bmatrix} .$$

We can verify that the matrices A and B in the dynamic equation for this new state vector, given by

$$\zeta_{t+1} = A\,\zeta_t + B\ e_t ,$$

are equal to

$$A = \begin{bmatrix} 0 & I \\ 0 & 0 \end{bmatrix} \text{ and, } B = \begin{bmatrix} 0 \\ \theta \end{bmatrix}$$

by noting that e_t is uncorrelated with z_t, and that the auto-covariance matrices are

$$E(\zeta_{t+1}\ \zeta_t{}') = \Theta E(e_{t-1}^{-}\ e_{t-2}^{-}{}')\ \Theta' = \begin{bmatrix} 0 & \theta\Delta\theta' \\ 0 & 0 \end{bmatrix},$$

that the state covariance matrix is equal to

$$\Pi = E\zeta_t\zeta_t{}' = \Theta(I \otimes \Delta)\Theta' = \begin{bmatrix} \theta\Delta\theta' & 0 \\ 0 & \theta\Delta\theta' \end{bmatrix},$$

and that

$$B\Delta = E\zeta_{t+1}\, e_t{}'$$

$$= E\zeta_{t+1}(y_t - c\ z_t)'$$

$$= M - A\Pi C' = \begin{bmatrix} 0 \\ \theta \end{bmatrix} \Delta.$$

Here, the matrix M is given by

$$M = E\ \zeta_{t+1}\, y_t{}'$$

$$= \Theta E(e_t - y_t{}')$$

$$= \Theta \begin{bmatrix} \Delta \\ 0 \end{bmatrix}$$

$$= \begin{bmatrix} 0 \\ \theta\Delta \end{bmatrix}.$$

That this Π satisfies the Riccati equation can be seen by direct substitution. For example, the Riccati equation in submatrices can be directly solved:

$$\Pi_{11} = \Pi_{22} = \theta\Delta\theta',$$

and

$$\Pi_{22} = \Pi_{21} = 0.$$

Although use of the stacked vector $\eta_{t-1} = (y_{t-1}{}',\ y_{t-2}{}')$ correctly reproduces the system matrices, it does not summarize the past history as e_{t-1}^{-} does. One way for seeing this is to note the relation

$$\begin{bmatrix} y_{t-1} \\ y_{t-2} \end{bmatrix} = \begin{bmatrix} e_{t-1} \\ y_{e-2} \end{bmatrix} - \theta \begin{bmatrix} e_{t-3} \\ 3_{t-4} \end{bmatrix},$$

i.e.,

$$\begin{bmatrix} e_{t-1} \\ e_{t-2} \end{bmatrix} = \begin{bmatrix} y_{t-1} \\ y_{t-2} \end{bmatrix} + \theta \begin{bmatrix} y_{t-3} \\ y_{t-4} \end{bmatrix} + \theta^2 \begin{bmatrix} y_{t-5} \\ y_{t-6} \end{bmatrix} + \cdots,$$

assuming that $\| \theta \| < 1$.

Let

$$z_t = \hat{E}\,(x_t \mid y_{t-1}, y_{t-2})$$

$$= \Omega R^{-1} \begin{bmatrix} y_{t-1} \\ y_{t-2} \end{bmatrix},$$

where

$$\Omega = [M, AM] = \begin{bmatrix} 0 & \theta\Delta \\ \theta\Delta & 0 \end{bmatrix}.$$

Verify that the vector z_t is uncorrelated with e_t and its use as an instrument correctly produces

$$\Pi_\zeta = \Omega R^{-1} \Omega'.$$

The error in the Riccati equation is not zero,

$$\Pi_\zeta - A\,\Pi_\zeta A' - B\Delta B' = \begin{bmatrix} 0 & 0 \\ 0 & x \end{bmatrix} \neq 0,$$

with

$$x = -\theta^2 (\Delta^{-1} + \theta'\Delta^{-1}\theta)^{-1} \theta^{2\prime}.$$

Models for Vector-Valued ARMA Processes

The next few examples are designed to show that the expressions for the state vector covariance matrices with truncated data vector (14) no longer solve the Riccati equation when the data-generating processes are ARMA processes. The Riccati equation must be solved numerically.

Effects of finite truncation of the data history y_{t-1}^- to $y_{t-1,K}^- = (y_{t-1}', \ldots, y_{t-K}')'$ are at least two-fold. One type of effect is on statistical errors in estimating theoretical covariance matrices by sample ones. Even when this type of error is eliminated by using theoretical covariance matrices, there is another kind of effect which are informational, i.e., not using enough past history to summarize correctly the data history. This is the type of error in a state vector which is only approximately a sufficient statistic. This subsection illustrates the latter type of error by a series of simple examples in which theoretical covariance matrices are easily calculated and used to avoid introduction of the former type of errors.

Example 6: ARMA(1,1) Model. Unlike VAR model, the solution of the Riccati equation is no longer given by $\Omega R^{-1}\Omega'$ when y_{t-1}^- is truncated at some finite past when the data-generating process is described by a vector ARMA model. By varying the point at which the data are truncated, however, it may provide a good approximation to Π. An example is

$$y_t = \phi y_{t-1} + u_t + \theta\, u_{t-1},$$

where $\{u_t\}$ is a sequence of mean-zero serially-uncorrelated exogenous vectors with $cov\ u_t = \Delta$. Its Wold decomposition is

$$y_t = u_t + \psi u_{t-1} + \phi\, \psi\, u_{t-2} + \phi^2\, \psi\, u_{t-3} + \cdots$$

with $\psi = \phi + \theta$.

The auto-covariance matrices are:

$$\Lambda_0 = \phi\Lambda_0\phi' + \Delta + \theta\Delta\phi' + \phi\Delta\theta' + \theta\Delta\theta',$$

$$\Lambda_1 = \phi\Lambda_0 + \theta\Delta,$$

and

$$\Lambda_j = \phi\Lambda_{j-2},\ j \geq 2.$$

Its state space model is obtained by defining $x_t = y_t - e_t$. Then

$$x_{t+1} = \phi\, x_t + \psi\, e_t \ , \text{ and } y_t = x_t + e_t .$$

The corresponding representation of the model with

$$z_t = \hat{\mathrm{E}}\,(x_t \mid y_{t-1}^{-})$$

as a state vector is $z_{t+1} = \phi\, z_t + \psi e_t$, and $y_t = z_t + e_t$. The Riccati equation is $\Pi = \phi\Pi\phi' + \psi\Delta\psi'$ or by taking the vectorized version

$$vec\ \Pi = (I - \phi\otimes\phi)^{-1}\ (\psi\otimes\psi) vec\ \Delta.$$

Take K to be one. Then, an approximate state vector is defined by $\zeta_t = \hat{\mathrm{E}}(x_t \mid y_{t-1})$, which is equal to $\Lambda_1\Lambda_0^{-1}\, y_{t-1}$. Its covariance matrix $\Pi_\zeta = \Lambda_1\Lambda_0^{-1}\Lambda_1'$ does not satisfy the Riccati equation.

To see this simply suppose that

$$\theta = \begin{bmatrix} 0.1 & 0 \\ 0 & .2 \end{bmatrix},\ \phi = \begin{bmatrix} .3 & 0 \\ 0 & .5 \end{bmatrix},\ \text{and } \Delta = \begin{bmatrix} 7 & 0 \\ 0 & 11 \end{bmatrix}.$$

Then, the correct Π is

$$\Pi = \begin{bmatrix} 1.2308 & 0 \\ 0 & 7.1867 \end{bmatrix},$$

and

$$\Pi_\zeta = \begin{bmatrix} 1.2203 & 0 \\ 0 & 7.0128 \end{bmatrix}.$$

Next, increase K to 2, and calculate the approximate state vector to be

$$\zeta_t = \hat{\mathrm{E}}(x_t \mid y_{t-1}, y_{t-2})$$

$$= (\Lambda_1 \;\; \phi\Lambda_1)\, R_2^{-1} \begin{bmatrix} y_{t-1} \\ y_{t-2} \end{bmatrix},$$

where

$$R_2 = \begin{bmatrix} \Lambda_0 & \Lambda_1 \\ \Lambda_1' & \Lambda_0 \end{bmatrix}.$$

Then

$$\Pi_\zeta = \begin{bmatrix} 1.2307 & 0 \\ 0 & 7.1798 \end{bmatrix}.$$

When K is increased to three the approximate values become correct to three decimal digits:

$$\zeta_t = [\Lambda_1 \;\; \phi\Lambda_1 \;\; \phi^2\Lambda_1]\, R_3^{-1} \begin{bmatrix} y_{t-1} \\ y_{t-2} \\ y_{t-3} \end{bmatrix}$$

with

$$R_3 = \begin{bmatrix} \Lambda_0 & \Lambda_1 & \Lambda_2 \\ \Lambda_1' & \Lambda_0 & \Lambda_1 \\ \Lambda_2' & \Lambda_1' & \Lambda_0 \end{bmatrix},$$

and the approximation to the true state vector covariance matrix becomes

$$\Pi_\zeta = \begin{bmatrix} 1.2308 & 0 \\ 0 & 7.1864 \end{bmatrix}.$$

Example 7: ARMA(2,1) Model. This example illustrates the points made by the ARMA (1,1) example for a slightly more complex case. Let

$$y_t = \phi\, y_{t-2} + e_t + \theta e_{t-1}.$$

Its Wold decomposition is

$$y_t = (e_t + \phi e_{t-2} + \phi^2 e_{t-4} + \cdots)$$

$$+ (\theta e_{t-1} + \phi\theta e_{t-1} + \phi^2\theta e_{t-5} + \cdots).$$

With the state vector given by

$$x_t = \begin{bmatrix} y_t - e_t \\ \phi y_{t-1} \end{bmatrix},$$

its dynamic model is

$$x_t = \begin{bmatrix} 0 & I \\ \phi & 0 \end{bmatrix} x_{t-1} + \begin{bmatrix} \theta \\ \phi \end{bmatrix} e_{t-1}$$

$$y_t = [I \quad 0]\, x_t + e_t .$$

The auto-convariance matrices are:

$$\Lambda_0 = \phi\Lambda_0\phi' + \Delta + \theta\Delta\theta'$$

$$\Lambda_1 = \phi\Lambda_1{}' + \theta\Delta,$$

$$= \phi\Delta,\ \phi' + \phi\Delta\ \theta' + \theta\Delta,$$

$$\Lambda_j = \phi\Lambda_{j-2},\ j \geq 2$$

and the Riccati equation has the solution

$$\Pi_{11} = \theta\Delta\theta' + \phi\Delta\phi' + \phi\Pi_{11}\phi'$$

$$\Pi_{22} = \phi\Delta\phi' + \phi\theta\Delta(\phi\theta)' + \phi\Pi_{22}\phi'$$

$$\Pi_{12} = \theta\Delta\phi' + \phi\Delta(\phi\theta)' + \phi\Pi_{12}\phi'$$

$$\Pi_{21} = \Pi_{12}{}'.$$

The covariance matrix of the orthogonally-projected vector

$$z_t = \hat{\mathrm{E}}\,(x_t \mid y_{t-1}^-)$$

satisfies this Riccati equation.

With $K = 1$, an approximate state vector is

$$\zeta_t = \hat{\mathrm{E}}\,(x_t \mid y_{t-1})$$

$$= M\ \Lambda_0^{-1}\ y_{t-1}$$

with

$$M = \begin{bmatrix} \Lambda_1 \\ \phi\Lambda_0 \end{bmatrix},$$

and, the convariance matrix

$$\Pi_\zeta = M\ \Lambda_0^{-1}\ M' = \begin{bmatrix} \Lambda_{-1}\Lambda_0^{-1} & \Lambda_1\phi' \\ \phi\ \Lambda_1{}' & \phi\Lambda_0\phi' \end{bmatrix}.$$

From the relations for the components of the matrix Π, Π_{22} is given by

$$vec\Pi_{22} = (I = \phi\otimes\phi)^{-1}(\phi\theta\otimes\phi\theta + \phi\otimes\phi)vec\Delta \, .$$

Now $(\Pi_\zeta)_{22} = \phi\Lambda_0\phi'$ is such that

$$\begin{aligned} vec(\Pi_\zeta)_{22} &= \phi\otimes\phi vec\Lambda_0 \\ &= (\phi\otimes\phi)[(I - \phi\otimes\phi)^{-1}(I + \theta\otimes\theta)vec\Delta] \\ &= (I - \phi\otimes\phi)^{-1}(\phi\otimes\phi)(I + \theta\otimes\theta)vec\Delta \\ &= vec\Pi_{22}. \end{aligned}$$

Furthermore,

$$\begin{aligned} vec(\Pi_\zeta)_{12} &= (\phi\otimes I)vec\Lambda_1 \\ &= (\phi\otimes I)[(I - \phi\otimes\phi)^{-1}(\theta\otimes\phi + I\otimes\Theta)vec\Delta] \\ &= (I - \phi\otimes\phi)^{-1} (\phi\otimes I)(\theta\otimes\phi + I\otimes\Theta)vec\Delta \end{aligned}$$

which is the same as $vec\Pi_{12}$.

Only the expression $(\Pi_\zeta)_{11}$ does not coincide with the correct solution Π_{11}. For example, when the system matrices are specified by

$$\Theta = \begin{bmatrix} .1 & 0 \\ 0 & .2 \end{bmatrix}, \quad \phi = \begin{bmatrix} .3 & 0 \\ 0 & .5 \end{bmatrix}, \quad \text{and} \quad \Delta = \begin{bmatrix} 7 & 0 \\ 0 & 11 \end{bmatrix},$$

the values of the state vector covariance matrix are

$$\Pi_{11} = \text{diag}\,(.7652, 4.2533)$$

$$\Pi_{12} = \text{diag}\,(.3 \quad 2.2)$$

$$\Pi_{22} = \text{diag}\,(.6992 \quad 3.8133).$$

The matrix

$$\Lambda_1\Lambda_0^{-1}\Lambda_1' = \text{diag}(.1287 \quad 1.2692).$$

With $K = 2$, the approximate state vector is

$$\begin{aligned} \zeta_t &= \hat{E}(x_t \mid y_{t-1}, y_{t-2}) \\ &= \Omega_2 R_2^{-1} \begin{bmatrix} y_{t-1} \\ y_{t-1} \end{bmatrix} \end{aligned}$$

with

$$\Omega_2 = [M \quad AM]$$

$$= \begin{bmatrix} \Lambda_1 & \Lambda_2 \\ \Lambda_2 & \Lambda_3 \end{bmatrix} .$$

Then

$$(\Omega_2 R_2^{-1} \Omega_2')_{11} = \text{diag } (.7634 \quad 4.1592) ,$$

which is much closer to Π_{11} than $\Lambda_2 \Lambda_0^{-1} \Lambda_1'$. The other submatrices remain correct. Increasing K to 3 and 4 results in

$$(\Omega_3 R_3^{-1} \Omega_3')_{11} = \text{diag } (.7692 \quad 4.2496) ,$$

$$(\Omega_4 R_4^{-1} \Omega_4')_{11} = \text{diag } (.7692 \quad 4.2532) ,$$

and

$$(\Omega_5 R_5^{-1} \Omega_5')_{11} = \text{diag } (.7692 \quad 4.2533) .$$

Note that for the values $K = 4$ and 5, the value of Π_{11} is correct to four decimal places.

In the next two examples the effects of truncating the information set on the state vector covariance matrices are illustrated while keeping the correct model dimension intact. We illustrate the effects of using incorrect model size (lower dimensional state vectors than the correct ones) in the next section.

Example 8: ARMA (2, 1) Model. The state vector dimension of this example is not an integral multiple of the dimension of the data vector. The data is generated by

$$y_t = \phi_1 y_{t-1} + \phi_2 y_{t-2} + e_t + \theta_1 e_{t-1}$$

with

$$\phi_1 = diag\ (.5 \quad .6) \quad , \quad \phi_2 = diag\ (0 \quad -.05)$$

and

$$\theta_1 = \begin{bmatrix} 0 & 0 \\ -8 & -4.1 \end{bmatrix} .$$

The noise variance matrix is taken to be .3 times the identity matrix. Define the state variables in the usual way:

$$x_{1t} = y_{1t} - e_{1t} \ , \ x_{2t} = y_{2t} - e_{2t} \ , \text{ and } x_{3t} = -.05 y_{2t-1} \ .$$

The state space model is then given by

$$x_t = Ax_{t-1} + Be_{t-1}\ , \text{ and } y_t = Cx_t + e_t$$

with the system matrices given by

$$A = \begin{bmatrix} \phi_1 & \alpha \\ \beta & 0 \end{bmatrix}\ , \quad B = \begin{bmatrix} .5 & 0 \\ -8 & -3.5 \\ 0 & -.05 \end{bmatrix}\ , \quad C = [I_2\ 0]\ ,$$

$$\alpha = (0\ \ 1)'\ , \quad \beta = (0\ \ -.05)\ .$$

Note that the vector β is the second row of the matrix ϕ_2. These matrices are not in balanced form, however.

The filtered version of this model satisfies the same equation as (1) with x_t replaced by

$$z_t = \hat{E}(x_t \mid y_{t-1}^-)\ .$$

From the definition and the way we introduced the state vector, the matrix M is given by

$$M = \begin{bmatrix} \Lambda_1 \\ \beta\Lambda_0 \end{bmatrix}\ .$$

The covariance matrices in the matrix M are calculated by solving the Yule-Walker equation

$$\Lambda_0 = \phi_1\Lambda_1{}' + \phi_2\Lambda_2{}' + \Delta + \theta_1\Delta(\theta_1 + \phi_1)'\ ,$$

$$\Lambda_1 = \phi_1\Lambda_0 + \phi_2\Lambda_1{}' + \theta_1\Delta\ ,$$

$$\Lambda_2 = \phi_1\Lambda_1 + \phi_2\Lambda_0\ ,$$

$$\Lambda_j = \phi_1\Lambda_{j-1} + \phi_2\Lambda_{j-2}\ , \quad j \geq 3\ .$$

Because we know the system matrices in this example, the covariance matrix $\Pi = cov\ z_t$ can be obtained as the solution of the Lyapunov equation

$$(I - A \otimes A)^{-1}\,(B \otimes B)\ vec\ \Delta\ ,$$

with y_{t-1}^- truncated at $K = 2$--i.e, replaced by $(y_{t-1}{}', y_{t-2}{}')'$. With this change in the information set, the approximate state vector becomes

$$z_t = E(x_t \mid y_{t-1}, y_{t-2})$$
$$= \Omega_2 R_2^{-1} \begin{bmatrix} y_{t-1} \\ y_{t-2} \end{bmatrix},$$

with

$$\Omega_2 = [M, AM]$$

and

$$R_2 = cov \begin{bmatrix} y_{t-1} \\ y_{t-2} \end{bmatrix} .$$

To put the model in balanced form we need the singular value decomposition of the Hankel matrix. With K set at 2, note that only three singular values are nonzero, i.e., the rank of the matrix is three. Here we proceed in the original coordinate in which the numerical values of the system matrices are given, without putting the model into balanced form. An approximation of Π is then $\Omega_2 R_2^{-1} \Omega_2'$.

Numerically, the correct values are

$$\Pi = \begin{bmatrix} .100 & -1.6842 & .0421 \\ * & 34.1415 & -.9255 \\ * & * & .0861 \end{bmatrix},$$

and approximate values are

$$\Omega_2 R_2^{-1} \Omega_2' = \begin{bmatrix} .100 & -1.6842 & .0421 \\ * & 29.3791 & -.9255 \\ * & * & .0861 \end{bmatrix} .$$

Note that only the (2, 2) element is not correctly calculated. When $y_{t-1}, \cdots, y_{t-4}$ is used to approximate y_{t-1}^{-}, the (2, 2) element increases to 29.3984 but is still far from the correct value of 34.1415.

Now change the matrix θ_1 to

$$\theta_1 = \begin{bmatrix} 0 & 0 \\ .2 & .7 \end{bmatrix} .$$

This change affects only the matrix B. The new matrix is

$$B = \begin{bmatrix} .5 & 0 \\ -.2 & -.1 \\ 0 & -.05 \end{bmatrix} .$$

The exact state vector covariance matrix is now

$$\Pi = \begin{bmatrix} .1000 & -.0421 & .0011 \\ * & .0260 & .0007 \\ * & * & .0008 \end{bmatrix}$$

and some approximations are

$$\Omega_2 R_2^{-1} \Omega_2' = \begin{bmatrix} .1272 & -.0534 & .0004 \\ * & .0227 & -.0003 \\ * & * & .0001 \end{bmatrix},$$

when only the past two data vectors are retained, and with three past data vectors

$$\Omega_3 R_3^{-1} \Omega_3' = \begin{bmatrix} .1273 & -.0535 & .0004 \\ * & .0228 & -.0003 \\ * & * & .0001 \end{bmatrix}.$$

See Example in Section 10.3, in which this model is put into balanced form.

Example 9: ARMA (2, 2) model. Now take the data generating process to be

$$y_t = \phi\, y_{t-2} + e_t + \theta\, e_{t-2}.$$

Its odd auto-covariance matrices vanish, and even ones are given by

$$\Lambda_0 = \phi\, \Lambda_0\, \phi' + \Delta + \theta\Delta\theta' + \theta\Delta\phi' + \phi\Delta\theta'$$

$$\Lambda_2 = \phi\, \Lambda_0 + \theta\Delta$$

$$\Lambda_{2j} = \phi\, \Lambda_{2(j-1)},\ j \geq 1.$$

With the state vector

$$x_t = \begin{bmatrix} y_t - e_t \\ \phi y_{t-1} + \theta e_{t-1} \end{bmatrix}$$

the model is

$$x_t = \begin{pmatrix} 0 & I \\ \phi & 0 \end{pmatrix} x_{t-1} + \begin{pmatrix} 0 \\ \psi \end{pmatrix} e_{t-1}$$

with

$$\psi = \theta + \phi.$$

The covariance matrix of the correct state vector

$$\zeta_t = \hat{\mathrm{E}}\,(x_t \mid y_{t-1}^{-})$$

has the solution of the Riccati equation for the state vector covariance matrix of the form

$$\Pi_{22} = \Pi_{11} = \phi\, \Pi_{11}\, \phi' + \psi \Delta \psi'$$

and

$$\Pi_{12} = \Pi_{21} = 0.$$

An approximate state vector

$$\zeta_t = \hat{\mathrm{E}}\,(x_t \mid y_{t-1}, y_{t-2})$$

$$= \Omega_\zeta R_2^{-1} \begin{bmatrix} y_{t-1} \\ y_{t-2} \end{bmatrix}$$

with

$$\Omega_\zeta = \begin{bmatrix} 0 & \Lambda_2 \\ \Lambda_2 & 0 \end{bmatrix}, \quad R_2 = \mathrm{diag}\ (\Lambda_0\, \Lambda_0)$$

has the covariance matrix such that

$$(\Pi_\zeta)_{11} = (\Pi_\zeta)_{12} = \Lambda_2\, \Lambda_0^{-1}\, \Lambda_2'$$

$$(\Pi_\zeta)_{12} = (\Pi_\zeta)_{12} = 0.$$

With the same numerical example

$$\theta = (\mathrm{diag}\ \ (.1\ \ .2),\ \ \phi = (.3\ \ .5) \text{and}\ \ \Delta = \mathrm{diag}\ (7,\ 11),$$

the numberical values are

$$\Pi_{11} = \ \mathrm{diag}\ (1.2308\ \ 7.1807),$$

$$(\Pi_\zeta)_{11} = \ \mathrm{diag}\ (1.2203\ \ 7.0225).$$

Increasing K to 4 results in

$$(\Pi_\zeta)_{11} = \ \mathrm{diag}\ (1.2307\ \ 7.1798).$$

And with $K = 7$ we get four correct decimal places:

$$(\Pi_\zeta)_{11} = \ \mathrm{diag}\ (1.2308\ \ 7.1864).$$

9.4 Numerical Examples

Using exact covariance matrix information, the next few examples illustrate changes in system behavior when high-dimensional models are approximated by lower-dimensional ones.

Example 10: AR(2) Model. A two-dimensional model is

$$y_t = A_1 y_{t-1} + A_2 y_{t-2} + u_t \ ,$$

where

$$A_1 = \begin{bmatrix} .5 & 0 \\ 0 & .65 \end{bmatrix} \ , \ A_2 = \begin{bmatrix} 0 & .1 \\ .1 & .2 \end{bmatrix}$$

and

$$Q = cov\ u_t = \begin{bmatrix} 1.5 & 2 \\ 2 & 5 \end{bmatrix} \ .$$

This AR (2) can be put into a state space form by defining

$$z_t = (y_t{}', y_{t-1}{}')',$$

with

$$z_{t+1} = \Phi z_t + \begin{bmatrix} I \\ 0 \end{bmatrix} u_t \ ; \ y_t = (I \ \ 0) z_t \ ,$$

where

$$\Phi = \begin{bmatrix} A_1 & A_2 \\ I & 0 \end{bmatrix} : 4 \times 4 \ .$$

Since Φ is nonsingular, and

$$rank \begin{bmatrix} I & 0 \\ A_1 & A_2 \end{bmatrix} = 4 \ ,$$

this state space model is minimal dimensional with $dim\ z_t = 4$.

The covariance matrices Λ_0 and Λ_1 are determined by solving either the Yule-Walker equation or the corresponding Riccati equation as shown in Example 3, for example. The numerical values are

$$\Lambda_0 = \begin{bmatrix} 2.853 & 5.9289 \\ * & 18.6053 \end{bmatrix} \ , \ \Lambda_1 = \begin{bmatrix} 1.8804 & 4.5380 \\ 4.9494 & 15.7355 \end{bmatrix} \ .$$

Other covariance matrices are determined iteratively by

$$\Lambda_{k+1} = A_1 \Lambda_k + A_2 \Lambda_{k+1} \ , \ k = 1, 2, \cdots$$

The four singular values of the Hankel matrix are 31.697, .5181, .2255, and .0018. These theoretically correct Λ's are used to build the Hankel matrix from which the matrices in the balanced representation are obtained:

$$A = \begin{bmatrix} .9061 & -.0214 & -.0614 & .0009 \\ .0538 & .4211 & -.0311 & .0594 \\ .0525 & -.2152 & -.2658 & -.0190 \\ -.0032 & -.0635 & -.0006 & .0866 \end{bmatrix},$$

$$C = \begin{bmatrix} 1.1694 & -.5849 & .1966 & .0150 \\ 3.9935 & .2765 & .2659 & -.0080 \end{bmatrix},$$

and

$$M' = \begin{bmatrix} 1.2827 & -.6442 & .0194 & -.0162 \\ 3.9586 & .0474 & -.3236 & .0082 \end{bmatrix}.$$

The covariance matrix of the state vector of the balanced model is

$$\Pi = cov\ z_t = \Omega R^{-1}\Omega',$$

where

$$\Omega = \Sigma^{1/2} V'$$

is the second factor in the singular value decomposition of the (4×4) Hankel matrix and

$$R = cov \begin{bmatrix} y_t \\ y_{t-1} \end{bmatrix} = \begin{bmatrix} \Lambda_0 & \Lambda_1 \\ \Lambda_1' & \Lambda_0 \end{bmatrix}.$$

The eigenvalues of Π are .5887, .4819, .0700, and .0035. The smallest eigenvalue of Π is very small, which suggests that a three-dimensional or perhaps a two-dimensional state space model may be nearly as good as the four-dimensional model. The eigenvalues of the matrix A are .9022, .4197,–.2698, and .0979. Those of the (3×3) upper left-hand corner of A are .9022, .4310, and –.2698.

From the definition of the state vector $z_t = \Omega R^{-1} y_{t-1}^-$, the three-dimensional approximate state vector corresponding to the three largest singular values is

$$z_t^a = \Omega_1 R^{-1} y_{t-1}^-,$$

where

$$\Omega_1 = \Sigma_1^{1/2} V_1'$$

with

$$\Sigma = diag\ (\Sigma_1, \Sigma_2)$$

and

$$V' = \begin{bmatrix} V_1' \\ V_2' \end{bmatrix}.$$

Thus $\Pi^a = \Omega_1 R^{-1} \Omega_1'$ is the 3×3 upper left-hand corner of $\Pi = \Omega R^{-1} \Omega'$. The eigenvalues of Π^a are .8587, .4813, and .0070. The covariance matrix of the innovation vector is the same as Q, $\Delta = \Lambda_0 - C \Pi C'$, and the matrix B is determined, as usual, by $B = (M - A \Pi C')\Delta^{-1}$.

This matrix Π solves the Riccati equation exactly because $\Pi - A \Pi A' - B \Delta B'$ is of the order 10^{-13}.

The impulse responses are

$$CB = \begin{bmatrix} .5 & 0 \\ 0 & .65 \end{bmatrix}$$

$$CAB = \begin{bmatrix} .25 & .1 \\ .1 & .6225 \end{bmatrix}$$

$$CA^2B = \begin{bmatrix} .125 & .115 \\ .115 & .5346 \end{bmatrix}$$

$$\vdots$$

$$CA^9B = \begin{bmatrix} .0197 & .0708 \\ .0708 & .2572 \end{bmatrix}$$

$$\vdots$$

$$CA^{19}B = \begin{bmatrix} .007 & .0253 \\ .0253 & .0919 \end{bmatrix} .$$

Since the ratio of the smallest over the largest singular values is of the order 10^{-5}, an appropriate model with $n = 3$ will have impulse-response characteristics which are nearly indistinguishable from the true ones.

The B matrix becomes

$$B_a = \begin{bmatrix} .0381 & .1568 \\ -.7224 & .2559 \\ .1756 & -.1753 \end{bmatrix} ,$$

and the impulse response matrices are

$$C_a B_a = \begin{bmatrix} .5016 & -.0008 \\ -.0008 & .6504 \end{bmatrix}$$

$$C_a A_a B_a = \begin{bmatrix} .2476 & .1011 \\ .1011 & .6221 \end{bmatrix}$$

$$C_a A_a^2 B_a^2 = \begin{bmatrix} .1254 & .1150 \\ .1149 & .5346 \end{bmatrix}$$

$$C_a A_a^9 B_a = \begin{bmatrix} .0197 & .0708 \\ .0707 & .2593 \end{bmatrix}$$

$$C_a A_a^{19} B_a = \begin{bmatrix} .0070 & .0253 \\ .0253 & .0920 \end{bmatrix}$$

With $n = 2$, the approximation get worbe:

$$CB = \begin{bmatrix} .446 & .046 \\ -.104 & .731 \end{bmatrix}$$

$$CAB = \begin{bmatrix} .218 & .102 \\ .063 & .609 \end{bmatrix}$$

$$CA^2B = \begin{bmatrix} .119 & .117 \\ .123 & .529 \end{bmatrix}$$

$$CA^9B = \begin{bmatrix} .025 & .069 \\ .090 & .251 \end{bmatrix}$$

$$CA^{19}B = \begin{bmatrix} .009 & .025 \\ .003 & .091 \end{bmatrix} .$$

Example 11: AR(2) Model. This AR(2) model is specified analogously to Example 1, with the matrices

$$\Phi_1 = \begin{bmatrix} .50 & -.30 \\ .20 & .65 \end{bmatrix} , \quad \Phi_2 = \begin{bmatrix} -.50 & .30 \\ 0 & -.40 \end{bmatrix} , \quad Q = \begin{bmatrix} 1 & -.08 \\ -.08 & 1 \end{bmatrix} .$$

Unlike the model of Example 1 this example shows that the smallness of the ratio σ_n / σ_1 is not always a reliable guide in reducing the dimension to $n - 1$. This is because of the fact that an approximate model is forced to replace one pair of complex roots by a real one, when the state vector dimension is reduced by one. The dynamic matrix has two pairs of complex values when the dimension is 4.

The covariance matrices are

$$\Lambda_0 = \begin{bmatrix} 2.5662 & -1.4948 \\ * & 1.5021 \end{bmatrix}, \quad \Lambda_1 = \begin{bmatrix} .9484 & -1.0296 \\ -.0465 & .4839 \end{bmatrix},$$

$$\Lambda_2 = \begin{bmatrix} -1.2434 & .5381 \\ .7574 & -.4922 \end{bmatrix}, \quad \Lambda_3 = \begin{bmatrix} -1.3371 & 1.0767 \\ .2622 & -.4059 \end{bmatrix},$$

$$\Lambda_4 = \begin{bmatrix} .1017 & .2434 \\ -3.999 & .1484 \end{bmatrix}.$$

The singular values of the 4×4 Hankel matrix are $\sigma_1 = 2.4803$, $\sigma_2 = 2.1195$, $\sigma_3 = 1.764$, and $\sigma_4 = .0622$.

The eigenvalues of this 4×4 matrix A are $.2429 \pm j.7744$ and $.3321 \pm j.4397$. When the dimension is dropped to three, they become $.2486 \pm j.7614$ and $.1491$.

An indication that a model with $n = 3$ is not a good approximation is given by a large error in solving the Riccati equation by the upper left-hand 3×3 submatrix of Π. The symplectic matrix Φ of (30) in Aoki (1987, p. 104) is 6×6 and has three pairs of complex eigenvalues, one of which has magnitude exactly one, which violates one of the assumptions under which the non-recursive solution procedure is developed there.

*Example 12.** This model is AR(2) with a three-dimensional y vector:

$$y_t = A_1 y_{t-1} + A_2 y_{t-2} + u_t$$

with

$$A_1 = \begin{bmatrix} .5 & 0 & .2 \\ 0 & .65 & 0 \\ .2 & 0 & .3 \end{bmatrix}, \quad A_2 = \begin{bmatrix} 0 & .1 & 0 \\ .1 & .2 & 0 \\ 0 & 0 & .45 \end{bmatrix}$$

$$\mathrm{cov}\ u_t = \begin{bmatrix} 1.5 & 2 & 1.5 \\ 2 & 5 & 4.5 \\ 1.5 & 4.5 & 16 \end{bmatrix}.$$

The dimension of the state vector is 6. The eigenvalues of the dynamic matrix are all real: .9330, .8616, .3766, .1002, −.2683, and −.5531.

The 6×6 Hankel matrix has six singular values: 104.5004, 13.4154, 4.0049, .5432, .2035, and .0024. Because the last singular value is much smaller than the other five, an approximate model with state vector dimension of 5 or even 4 appears to be a good one in terms of approximating the impulse responses of the original six-dimensional model. Another indication of this is given by the eigenvalues of Π. The smallest one is .0041, and the next smallest is .0698. The eigenvalues of the 5-th dimensional dynamic matrix A are

* This example is supplied by D. Sartore

.9330, .8617, .3863, −.5531, and −.2683. The 5×5 or 4×4 upper left-hand corner of Π serves as the first approximation of the state vector covariance matrices of dimension 5 and 4 respectively.

Using these approximate Π expressions to calculate approximate expressions for Δ and B, the error matrix $E(\Pi)$

$$E(\Pi) = \Pi - A\Pi A' - B\Delta B'$$

is of the order 10^{-3} for $n = 5$, and 10^{-2} for $n = 4$. One application of the Newton-Raphson step $vec\, \Pi_{new} = vec\, \Pi_{old} - (I - F \otimes F)^{-1}\, vec\, E(\Pi_{old})$ (where $F = A - BC$ is evaluated using Π_{old}), improves these to be of the order 10^{-9} and 10^{-5} respectively. After a few periods, the differences in the multiplier profiles of $n = 6$ and $n = 4$ are not visible in their graphical plots.

9.5 Monte Carlo Experiments

This section reports on Monte Carlo simulation results for vector-valued AR(1) and AR(2) examples. Some simulation experiments reported in the literature investigate small-sample properties of estimators such as maximum likelihood, least squares, or their variants for scalar-valued time series model parameters. Cooper and Thompson [1977] use 200 replications for a first-order moving-average model, $y_t = \mu + n_t - \theta n_{t-1}$, with a non-zero μ with 59 samples. Ansley and Newbold [1980] use up to 1000 replications with sample size 50 and 100 to test three estimators for ARMA models of order up to two. Cryer and Ledolter [1981] obtain the exact distribution of the estimate of the moving-average model of order one for sample size two. They all report on estimates of θ in the first-order moving-average model having a considerable concentration at or very near the boundaries of 1 and -1 for values of θ close to these boundaries. All of the tested estimators exhibit this pileup effect, and hence a positive probability of estimated models being unstable.

When the autoregressive polynomials and the moving-average ones have near common factors in ARMA models, then estimates tend to be poor, in particular if the zeros of the common factors are close to one in magnitude, as reported by Schwert [1988], for example.

We have tried sample sizes up to 600 and the number of replications of 3600 for some selected examples.

AR(1) Models

The data are generated by

$$y_t = \phi y_{t-1} + u_t \ , \tag{16}$$

where u_t is mean zero, serially uncorrelated and

$$cov\, u_t = Q > 0 \ .$$

The covariance matrices satisfy the Yule-Walker equation

$$\Lambda_0 = \phi\Lambda_1' + Q \quad ,$$

$$\Lambda_1 = \phi\Lambda_0$$

or

$$\Lambda_0 = \phi\Lambda_0\phi' + Q \ . \tag{17}$$

A state space representation of (1) is

$$x_t = \phi x_{t-1} + \phi u_{t-1} \tag{18}$$

$$y_t = x_t + u_t \quad .$$

In this example $e_t = y_t - y_{t|t-1} = u_t$, i.e., $cov\ e_t = \Delta = Q$. Since $dim\ y_t = p$, this state vector is also p-dimensional. Unless ϕ is of rank $(p-1)$ or less, p is the correct state vector dimension and the Hankel matrix $H = \Lambda_1$ has rank p. The orthogonal projected image of x_t onto the subspace spanned by y_{t-1}^- is $s_t = E(x_t \mid y_{t-1}^-)$. The vector of s_t satisfies the same equation as (18). This represenation is not in balanced form. The matrix M is $E(x_t y_{t-1}') = \Lambda_1$. For the process given in (1), y_{t-1}^- can be replaced by y_{t-1} alone. Let its singular value decomposition be

$$\Lambda_1 = U\Sigma V' \quad .$$

Using subscript b to denote matrices in balanced form, we obtain the innovation model in balanced representation

$$z_{t+1} = A_b z_t + B_b e_t \text{ ,and } y_t = C_b z_t + e_t \quad , \tag{19}$$

where

$$C_b = \Lambda_1 V \Sigma^{-1/2}$$

$$= U\Sigma^{1/2} ,$$

$$M_b = \Sigma^{-1/2} U' \Lambda_1$$

$$= \Sigma^{1/2} V' ;$$

and

$$A_b = \Sigma^{-1/2} U' \Lambda_2 V \Sigma^{-1/2} = \Sigma^{-1/2} U' \phi U \Sigma^{1/2} .$$

To find B_b in the above, we note that the state vectors z_t and s_t are related by

$z_t = Ss_t$, with $S = \Sigma^{-1/2}U'$, because $A_b = S\phi S^{-1}$ and $C_b = S^{-1}$. Therefore $B_b = S\phi$ must hold.

Note that $C_b B_b = \phi$ and $C_b A_b B_b = \phi^2$. This is as it should be, because the impulse response sequence for y_t is $y_t = e_t + \phi e_{t-1} + \phi^2 e_{t-2} + \cdots$, which is invariant with respect to the coordinates used.

From (19), recalling that z_t is uncorrelated with e_t,

$$\Lambda_0 = C_b \Pi_b C_b' + Q$$

$$= C_b \Pi_b C_b' + \Lambda_0 - \phi\Lambda_0\phi',$$

or

$$C_b \Pi_b C_b' = \phi\Lambda_0\phi',$$

where (2) is used to eliminate Q. Since C_b is invertible in this example, we obtain, without solving the Riccati equation, the covariance matrix of the state vector as

$$\Pi_b = \Sigma^{-1/2}U'\phi\Lambda_0\Lambda_o^{-1}(\Lambda_0\phi')U\Sigma^{-1/2}$$

$$= \Sigma^{1/2}V'\Lambda_0^{-1}V\Sigma^{1/2},$$

because $\phi\Lambda_0 = \Lambda_1 = U\Sigma V'$.

We note that $\Pi_b = \Omega R^{-1}\Omega'$, because Ω is equal to M_b, i.e., $\Omega = \Sigma^{1/2}V'$ and $R = \Lambda_0$, since y_{t-1}^- is replaced by y_{t-1} alone.

We have run four experiments with $p = 2$:

$$\textit{model 1} \quad \phi = \begin{pmatrix} .7 & 1 \\ -.4 & .7 \end{pmatrix}, \quad Q = \textit{diag}\,(.1 \;\; .1) \qquad \textit{S-N ratio} = 11.0 \;;$$

$$\textit{model 2} \quad \phi = \begin{pmatrix} .7 & .8 \\ -.4 & .6 \end{pmatrix}, \quad Q = \begin{pmatrix} .1 & .05 \\ .05 & .1 \end{pmatrix} \qquad \textit{S-N ratio} = 4.5 \;;$$

$$\textit{model 3} \quad \phi = \begin{pmatrix} .7 & 0 \\ -.3 & .6 \end{pmatrix}, \quad Q = \begin{pmatrix} .1 & .08 \\ .08 & .1 \end{pmatrix} \qquad \textit{S-N ratio} = 1.7 \;;$$

and

$$\textit{model 4} \quad \phi = \begin{pmatrix} .7 & .8 \\ -.1 & .6 \end{pmatrix}, \quad Q = \begin{pmatrix} .1 & .05 \\ .05 & .1 \end{pmatrix} \qquad \textit{S-N ratio} = 3.9 \,.$$

Each model, except model 3, has a pair of complex conjugate eigenvalues. Model 1 has the largest eigenvalue in magnitude. The signal-to-noise ratios as well as the sample sizes affect performances of the algorithm significantly. The S-N ratio is measured here by the ratio $tr\,\Lambda_0 / tr\,Q$.

The precision with which the individual element is estimated may be inferred by calculating the asymptotic variance matrices of the estimators. A large element in the error variance matrices signals poor performances in estimating it. As we describe in Section 10.3, the asymptotic error variance matrix for the matrix C is given by the expression

$$T\,cov\,(vec\;\delta C) \rightarrow \Pi^{-1} \otimes \Delta$$

and for the matrix A by

$$T cov\;\;(vec\delta A) \rightarrow \Pi^{-1} \otimes B\,\Delta B' + \Pi^{-1} W (I \otimes \Delta) W' ,$$

where the matrix W is as given in (13) of Section 10.3.

In model 1 the error covariance matrices are

$$\Pi^{-1} \otimes \Delta = \begin{bmatrix} \underline{.107} & 0 & -.032 & 0 \\ 0 & \underline{.107} & 0 & -.032 \\ -.032 & 0 & \underline{.114} & 0 \\ 0 & -.032 & 0 & \underline{.114} \end{bmatrix}$$

and with J and K set to one,

$$\Pi^{-1} \otimes (B\,\Delta B' + W \Delta W') = \begin{bmatrix} \underline{.216} & -.007 & -.064 & .002 \\ * & .301 & .002 & -.089 \\ * & * & \underline{.230} & -.008 \\ * & * & * & \underline{.321} \end{bmatrix} ,$$

where large entries are underlined.

In model 2, they are

$$\Pi^{-1} \otimes \Delta = \begin{bmatrix} \underline{.111} & .055 & -.025 & -.012 \\ .055 & \underline{.111} & -.012 & -.025 \\ -.025 & -.012 & \underline{.128} & .064 \\ -.012 & -.025 & .064 & \underline{.128} \end{bmatrix}$$

and

$$\Pi^{-1} \otimes (B\,\Delta B' + W \Delta W') = \begin{bmatrix} \underline{.449} & .310 & -.100 & -.069 \\ .310 & \underline{.810} & -.069 & -.181 \\ .100 & -.069 & \underline{.517} & .357 \\ -.069 & -.181 & .357 & \underline{.934} \end{bmatrix} .$$

For model 3, the asymptotic error covariance matrices are

$$\Pi^{-1} \otimes \Delta = \begin{bmatrix} \underline{.161} & .129 & .035 & .028 \\ .129 & \underline{.161} & .028 & .035 \\ .035 & .028 & \underline{.156} & .125 \\ .028 & .035 & .125 & \underline{.156} \end{bmatrix}$$

and

$$\Pi^{-1} \otimes (B\Delta B' + W\Delta W'') = \begin{bmatrix} \underline{1.726} & \underline{1.452} & .378 & .318 \\ \underline{1.452} & \underline{2.509} & .318 & .550 \\ .378 & .318 & \underline{1.675} & \underline{1.410} \\ .318 & .550 & \underline{1.410} & \underline{2.435} \end{bmatrix}.$$

The error covariance matrices in model 4 are

$$\Pi^{-1} \otimes \Delta = \begin{bmatrix} .114 & .057 & -.037 & -.019 \\ -.057 & .114 & -.019 & -.037 \\ -.037 & -.019 & \underline{.188} & .094 \\ -.019 & -.037 & .094 & \underline{.188} \end{bmatrix}$$

and

$$\Pi^{-1} \otimes (B\Delta B' + W\Delta W'') = \begin{bmatrix} .795 & .811 & .174 & .178 \\ .811 & \underline{2.793} & .178 & .612 \\ .174 & .178 & .772 & .787 \\ .174 & .612 & .787 & \underline{2.711} \end{bmatrix}.$$

Model 3 has the largest error covariance matrices for the estimates of the matrices C and A. This is to be expected, in that Model 3 has the worst S-N ratio. System matrices of models with complex eigenvalues tend to be estimated better than those with real eigenvalues only, because the data signals tend to be 'persistently exciting' in the language of system theory. See Ljung [1987, p.362] for this concept.

Experimental Results

Small scale simulations are run with sample sizes 40, 80, 120, 250, 320, 500, and 600, mostly replications of the order of several hundred up to 10^3. The the Hankel matrices of 2×2 up to 4×4 have been run. Two by two Hankel matrices produce best estimates. Then for the sample size 500, 3600 replications are run for all four models with the (2×2) Hankel matrices. Sample standard errors decrease approximately at the rate of $T^{-1/2}$ as the sample size increases. For example in model 1, the standard errors of C_{11} with the theoretical value 1.109, which is the largest, behave as follows:

T	40	80	120	240	320	500	600
s.d.	.269	.224	.184	.124	.101	.103	.079

AR(2) Models

For a p-dimensional data-generating process

$$y_t = \phi_1 y_{t-1} + \phi_2 y_{t-2} + u_t$$

the covariance matrices satisfy

$$\Lambda_0 = \phi_1 \Lambda_1' + \phi_2 \Lambda_2' + Q \quad ,$$

$$\Lambda_1 = \phi_1 \Lambda_0 + \phi_2 \Lambda_1' \quad ,$$

$$\Lambda_2 = \phi_1 \Lambda_1 + \phi_2 \Lambda_0 \quad ,$$

or

$$\Lambda_0 = \phi_1 \Lambda_0 \phi_1' + \phi_2 \Lambda_0 \phi_2' + \phi_1 \Lambda_1 \phi_2' + \phi_2 \Lambda_1' \phi_1'$$

and

$$\Lambda_1' = \Lambda_0 \phi_1' + \phi_1 \Lambda_0 \phi_2' + \phi_2 \Lambda_1' \phi_2' \quad .$$

These are solved simultaneously using the Kronecker products. For example,

$$vec = (I - \phi_2 \otimes \phi_2)^{-1} (I \otimes \phi_1 + \phi_1 \otimes \phi_2) \, vec \, \Lambda_0$$

$$vec \, \Lambda_1' = (I - \phi_2 \otimes \phi_2)^{-1} (\phi_1 \otimes I + \phi_2 \otimes \phi_1) \, vec \, \Lambda_0$$

and

$$vec \, \Lambda_0 = (I - \phi_1 \otimes \phi_1 - \phi_2 \otimes \phi_2)^{-1} [(\phi_2 \otimes \phi_1) \, vec \, \Lambda_1 + (\phi_1 \otimes \phi_2) \, vec \, \Lambda_1'] \quad .$$

When the matrix ϕ_2 is of rank p, the $2p$-dimensional state vector is required.

A state space model representation is obtained for the state vector

$$x_t = \begin{bmatrix} y_t - u_t \\ \phi_2 y_{t-1} \end{bmatrix}$$

as

$$x_t = A x_{t-1} + B u_t$$

$$y_t = C x_t + u_t$$

with

$$A = \begin{bmatrix} \phi_1 & I \\ \phi_2 & 0 \end{bmatrix} \quad , \quad B = \begin{bmatrix} \phi_1 \\ \phi_2 \end{bmatrix} \quad \text{and} \quad C = [I_p \quad 0] \ .$$

By projecting x_t onto the subspace spanned by y_{t-1}^-, the filtered dynamic model is

$$z_t = Az_{t-1} + Be_t$$

$$y_t = Cz_t + e_t$$

with

$$e_t = u_t \ .$$

In Section 3, we have shown for $p = 2$ that the state covariance matrix is given by

$$\Pi = \Omega R^{-1} \Omega' \quad \text{with} \quad \Omega = [M \quad AM] \ ,$$

$$M = E(x_t y_{t-1}') \quad \text{and} \quad R = cov\,[y_{t-1}', y_{t-2}']' \ .$$

We summarize the results of simulation in the next five tables.

Model 1

	Theoretical Values	Sample Means	Sample Standard Devi.
σ_1	1.268	1.338	.25
σ_2	.702	.671	.13
A	$\begin{bmatrix} .5357 & .6737 \\ -.6638 & .8643 \end{bmatrix}$	$\begin{bmatrix} .5340 & .6716 \\ -.6310 & .8606 \end{bmatrix}$	$\begin{bmatrix} .013 & .012 \\ .016 & .017 \end{bmatrix}$
C	$\begin{bmatrix} 1.1089 & .2343 \\ -.3304 & .7856 \end{bmatrix}$	$\begin{bmatrix} 1.1032 & .2332 \\ -.3291 & .7813 \end{bmatrix}$	$\begin{bmatrix} .103 & .030 \\ .041 & .075 \end{bmatrix}$
M	$\begin{bmatrix} 1.068 & 1.062 \\ -.3158 & -.3142 \end{bmatrix}$	$\begin{bmatrix} 1.062 & .4436 \\ -.3142 & .7525 \end{bmatrix}$	$\begin{bmatrix} .102 & .042 \\ .031 & .075 \end{bmatrix}$

Model 2

	Theoretical Values	Sample Means	Sample Standard Devi.
σ_1	.528	.528	.066
σ_2	.250	.249	.036
A	$\begin{bmatrix} .5781 & .5768 \\ .7219 & -.5897 \end{bmatrix}$	$\begin{bmatrix} .5768 & .5432 \\ -.5897 & .7182 \end{bmatrix}$	$\begin{bmatrix} .016 & .023 \\ .022 & .034 \end{bmatrix}$
C	$\begin{bmatrix} .6943 & .1481 \\ -.2153 & .4774 \end{bmatrix}$	$\begin{bmatrix} .6925 & 1474 \\ .2144 & 4757 \end{bmatrix}$	$\begin{bmatrix} .045 & .016 \\ .021 & .034 \end{bmatrix}$
M	$\begin{bmatrix} .7690 & .2596 \\ -.1785 & .4669 \end{bmatrix}$	$\begin{bmatrix} .6770 & .2591 \\ -.1782 & .4650 \end{bmatrix}$	$\begin{bmatrix} .043 & .026 \\ .020 & .033 \end{bmatrix}$

Model 3

	Theoretical Values	Sample Means	Sample Standard Devi.
σ_1	5.490	5.477	.852
σ_2	.0669	.0674	.0102
A_b	$\begin{bmatrix} .8283 & -.3302 \\ .0887 & .4717 \end{bmatrix}$	$\begin{bmatrix} .8230 & -.3326 \\ .0907 & .4726 \end{bmatrix}$	$\begin{bmatrix} .025 & .031 \\ .034 & .098 \end{bmatrix}$
C_b	$\begin{bmatrix} .1784 & -.2580 \\ -2.3364 & -.0197 \end{bmatrix}$	$\begin{bmatrix} .1757 & -2.581 \\ -2.3266 & -.0193 \end{bmatrix}$	$\begin{bmatrix} .033 & .020 \\ .180 & .002 \end{bmatrix}$
M_b	$\begin{bmatrix} .4008 & -2.3086 \\ -2.549 & -.0043 \end{bmatrix}$	$\begin{bmatrix} .4000 & -2.2988 \\ -.2550 & -.0444 \end{bmatrix}$	$\begin{bmatrix} .023 & .181 \\ .019 & .003 \end{bmatrix}$

Model 4

	Theoretical Values	Sample Values	Standard Deviations
σ_1	.5672	.5656	.0723
σ_2	.0686	.0683	.0092
A	$\begin{bmatrix} .7390 & .2756 \\ -.3100 & .5610 \end{bmatrix}$	$\begin{bmatrix} .7364 & .2759 \\ -.3095 & .5581 \end{bmatrix}$	$\begin{bmatrix} .025 & .033 \\ .032 & .071 \end{bmatrix}$
C	$\begin{bmatrix} .7519 & -.0147 \\ .0423 & .2616 \end{bmatrix}$	$\begin{bmatrix} .7452 & -.0143 \\ .0418 & .2604 \end{bmatrix}$	$\begin{bmatrix} .048 & .005 \\ .017 & .018 \end{bmatrix}$
M	$\begin{bmatrix} .7049 & .2651 \\ -.0922 & .2452 \end{bmatrix}$	$\begin{bmatrix} .7024 & .2640 \\ -.0920 & .2440 \end{bmatrix}$	$\begin{bmatrix} .048 & .018 \\ .009 & .016 \end{bmatrix}$

AR(2)

	Theoretical Values	Sample Means	Sample Standard Devi.
σ_1	2.4803	2.4728	.400
σ_2	2.1195	2.1133	.261
σ_3	.1764	.1765	.020
σ_4	.0622	.0618	.025
L_0	$\begin{bmatrix} 2.5662 & -1.4948 \\ * & 1.5021 \end{bmatrix}$	$\begin{bmatrix} 2.5704 & -1.4961 \\ * & 1.5036 \end{bmatrix}$	$\begin{bmatrix} .267 & .155 \\ .155 & .125 \end{bmatrix}$
L_1	$\begin{bmatrix} .9484 & -1.0296 \\ -.0465 & .4839 \end{bmatrix}$	$\begin{bmatrix} .9461 & -1.0247 \\ -.0476 & .4822 \end{bmatrix}$	$\begin{bmatrix} .108 & .124 \\ .089 & .059 \end{bmatrix}$

$$Q \qquad \begin{bmatrix} 1.000 & -.8000 \\ & 1.000 \end{bmatrix} \qquad \begin{bmatrix} .9776 & -7.868 \\ & .9877 \end{bmatrix} \qquad \begin{bmatrix} .147 & .0628 \\ & .053 \end{bmatrix}$$

$$A \qquad \begin{bmatrix} .5437 & -.8730 & -.0770 & .0511 \\ .7971 & -.0258 & 0 & -.0482 \\ -.4243 & -.1194 & .1283 & .5812 \\ -.0669 & -.0594 & -.4187 & .5038 \end{bmatrix} \qquad \begin{bmatrix} .5330 & -.8492 & -.0698 & .0351 \\ .7496 & -.0151 & 10^{-4} & -.0388 \\ -.3826 & -.0923 & .1340 & .5506 \\ -.0572 & -.0480 & -.3986 & .3028 \end{bmatrix} \qquad \begin{bmatrix} .084 & .256 & .050 & .113 \\ .219 & .052 & .030 & .083 \\ .182 & .081 & .070 & .193 \\ .111 & .044 & .156 & 7.485 \end{bmatrix}$$

$$C \qquad \begin{bmatrix} -.5299 & -1.2082 & .1621 & -.0553 \\ .5089 & .4260 & .3779 & -.0048 \end{bmatrix} \qquad \begin{bmatrix} -.5557 & -1.1575 & .1467 & .0482 \\ .5076 & .4025 & .3419 & .0037 \end{bmatrix} \qquad \begin{bmatrix} .137 & .260 & .067 & .028 \\ .107 & .109 & .157 & .019 \end{bmatrix}$$

$$M \qquad \begin{bmatrix} .5869 & -.0568 \\ -1.000 & .9216 \\ .2227 & .3176 \\ -.0787 & -.0405 \end{bmatrix} \qquad \begin{bmatrix} .5365 & -.0251 \\ -.9897 & .8902 \\ .2024 & .2876 \\ -.0699 & -.0363 \end{bmatrix} \qquad \begin{bmatrix} .204 & .135 \\ .228 & .198 \\ .093 & .131 \\ .042 & .030 \end{bmatrix}$$

9.6 Model Selection

One of the crucial choices in selecting a model is that of the model size, i.e., the state vector dimension. Actually, in estimating models using this algorithm, we basically have only two decisions to make, one being the model size and the other the choice of K, which determines the size of the Hankel matrix to be used. Effective numbers of common factors in static factor models are discussed, for example, in Rao [1965, p. 496]. Akaike [1975, 1976] introduced state vectors as vectors of canonical variates between the data and future observations in building state space models for time series as outlined in Chapter 8. This may be regarded as a generalization of the similar notion for two vectors in factor models. It is a genuine generalization because, unlike static factor models, dynamic factors can't be a mixture of past and unobserved future realization of the data vectors. The direction of time is important. Recall our discussion in Chapter 2 about the minimal splitting subspace. There we noted that state vectors of state space models belong to a minimal splitting space and serve as common factors. The number of common factors is equal to the dimension of the state vector.

This section uses known distributional properties of the canonical correlation coefficients to provide a test for the dimension of the minimal state space models. Other approaches to model selection use measures such as the AIC of Akaike or Schwarz's criterion. We have already noted that the ratios of singular values naturally arise in measuring relative errors of transfer function approximation in the Hankel norm. We have also noted that sole reliance on the ratios of the singular values alone is sometimes unsatisfactory in proper selection of model size because a pair of complex-valued eigenvalues may be forced to be replaced by a single real-valued eigenvalue when the model size is reduced by one. Such a reduction in model size may produce a model with quite a different sequence of impulse responses from the original one, even when the ratios of singular values are very small.

A number of similarities exist between the singular values and the canonical correlation coefficients that are known in the statistical literature, as in Rao (1979). One such result, Theorem 3.1 of Rao [1979], states that

$$\lambda_{N+1-i}\ [cov\,(P'R_+^{-1/2}y_t^+ - \Gamma Q'R_-^{-1/2}y_{t-1}^-)\,]=1-\rho_i^2\,,$$

where the singular value decomposition of the scaled Hankel matrix, $R_+^{-1/2}HR_-^{-1/2}$, is given by $P\Gamma Q'$, where $N=Jp$, $i=1, 2,..., n$, and $J \le K$. This equality holds simultaneously--i.e., the determinant of the covariance matrix on the left is equal to that of the matrix $(I - \Gamma^2)$. Any other matrix S in defining a summary $S'y_{t-1}^-$ produces an inequality $\ge$ rather than the equality in the above. This also shows that the covariance matrix of the innovation is the smallest when the state vector is defined by (8) of Chapter 6.

We examine next the statistical consequences of alternative choices of the state vector dimension. Instead of developing distributional theory for singular values, we will relate the singular values to the canonical correlation coefficients, in view of the fact that the distributional theory for the latter is well-developed. We use the test proposed by Bartlett [1939], which is also described in Kshirsager [1972, p. 326] and used by Otter [1975] in a related context. Denote the singular value decomposition of H by $U\tilde{\Sigma}\tilde{V}'$, where $\tilde{\Sigma} = [diag\, \Sigma_{Jp} O]$ and $\tilde{V} = [V, \bar{V}]$, where V is $Jp \times Jp$ and $\bar{V}$ is $(K-J)p \times Kp$. Note that $U\tilde{\Sigma}\tilde{V}' = U\Sigma_{Jp}V'$. Some of the singular values in Σ_{Jp} may be zero, and $n = rank\, H = rank\, \Sigma_{Jp}$, that is we take $n \le Jp$. As before, the covariance matrices of y_t^+ and y_{t-1}^- are denoted, respectively, by R_+ and R_-.

Let the Cholesky decomposition of R_+ be L_+L_+', where L_+ is a lower triangular matrix. Define L_- likewise. Calculating the covariance matrix of $[y_t^{+\prime}\ y_{t-1}^{-\prime}]'$, the canonical correlation matrix Γ between the stacked future and past vectors appears in the singular value decomposition of the scaled Hankel matrix

$$L_+^{-1}HL_-^{-T} = P\Gamma Q' ,$$

with

$$P'P = I_{Jp}\ , Q'Q = I_{Kp}\ \ and\ \ \Gamma = [diag\,(\rho_1, \cdots, \rho_{Jp})0]\ ,$$

in which canonical correlation coefficients ρ's are arranged in decreasing order. We may take the matrix Q to be Jp by Kp and Γ to be Jp by Jp without any loss of generality. Note that

$$\begin{bmatrix} P'L_+^{-1} & 0 \\ 0 & Q'L_-^{-1} \end{bmatrix} \begin{bmatrix} R_+ & H \\ H' & R_- \end{bmatrix} \begin{bmatrix} L_+^{-T}P & 0 \\ 0 & L_-^{-T}Q \end{bmatrix} = \begin{bmatrix} I_{Jp} & \Gamma \\ \Gamma & I_{Kp} \end{bmatrix} .$$

On taking the determinants of both sides, we have

$$|R_+|^{-1}\ |R_+ - HR_-^{-1}H'| = |I - \Gamma^2| . \tag{8}$$

Partition the matrix Γ into two parts and let Γ_i, $i=1, 2$ be the corresponding matrix of canonical correlation coefficients. Partition the matrix Q conformably into $[Q_1\ Q_2]$. Then, following Bartlett [1939], a certain constant times the logarithm of the determinant shown below is distributed according to the chi-squares distribution, which can be used to sequentially test the significance of the choice of state vector dimension. Note that

$$\lambda = |I - \Gamma_2^2| = |I - \Gamma^2|\ /\ |I - \Gamma_1^2| .$$

Its alternative expression involving the Hankel matrix is given by

$$\lambda = |R_+ - HR_-^{-1}H'|\ /\ |L_+P_1P'_1L'_+ - HL_-^{-T}Q_1Q_1'L_-^{-1}H'| ,$$

because Γ_1 is equal to $P_1'L_+^{-1}HL_-^{-T}Q_1$. The numerator is the determinant of the one-step-ahead forecast error by the complete model, i.e., the determinant of the covariance of the innovation noise vector, raised to the J-th power, and the denominator has a similar interpretation. It is not exactly equal to the determinant of the one-step-ahead forecast error of the approximate model raised to the J-th power because $P_1'L_+^{-1}H_2L_-^{-T}Q_1$ is not zero, when the Hankel matrix is decomposed into a sum $H_1 + H_2$, where H_1 inherits the n largest singular values of H, but close to it when the singular values of H_2 are "much" smaller than those in H_1.

Bartlett uses an approximation

$$-\{N - n - ½(2Kp + 1)\} ln \prod_{j > n+1} (1 - \rho_j^2) , \tag{9}$$

where ρ is the estimate of the canonical correlation coefficient, which is a chi-squared variable with $(Jp - n)(Kp - n)$ degrees of freedom. The test is conducted for $n = 1, 2, \cdots$ sequentially. When the test for $n = n^*$ is not rejected while the test with $n = 1, \cdots, n^* - 1$ is rejected, the number $n = n^*$ is the dimension of the state vector. Lawley [1959] suggests using the factor $\{N - n - ½((J+K)p + 1) + \Sigma_{i=1}^{n} 1 / \rho_j^2\}$ rather than that in (8). A Monte Carlo study conducted by Mendoza et al. [1978] concludes that Lawley's approximation is most desirable, followed by the Bartlett's approximation. It is interesting to observe that this statistic is proportional to the mutual information between the two canonical variates calculated by Gel'fand and Yaglom [1959] mentioned in Chapter 8.

Examples

This section presents two examples, bivariate bilateral exchange rate series, and a five-dimensional wheat price series discussed in Pena and Box [1987]. First, a biweekly series of two bilateral exchange rates (Pound Sterling and Deutsch Mark vs. US dollar) from June 1973 to February 1986, a total of 306 data points, has been modeled. With $K = 2$, the sample values of the canonical correlation coefficients are $\rho_1 = .996$, $\rho_2 = .989$, $\rho_3 = .150$, and $\rho_4 = .004$. The singular values of H are $\sigma_1 = .27$, $\sigma_2 = .006$; and the first four digits of σ_3 and σ_4 are zero.

We denote the sample values of Bartlett's statistics by b_n, where the subscript denotes the number of singular values retained in the modeling. The results are: b_1=1157.4 , b_2=6.85 and b_3= 3.76. The critical values at the 5% level are $\chi_9^2 = 16.9$, $\chi_4^2 = 9.5$, and $\chi_1^2 = 3.8$. The choice of n of 2 or 3 is indicated. The "whiteness" of the residuals as measured by the Durbin-Watson statistic or the Ljung-Box statistic is quite satisfactory and nearly the same for n =2 and n =3.

Another class of models with $dim\ y_t^+ = 2$ and $dim\ y_t^- = 4$ has been constructed as a partial check on robustness. Here the values of sample canonical correlation coefficients are $\rho_1 = .994$, $\rho_2 = .989$, $\rho_3 = .232$, and $\rho_4 = .045$. The singular values are $\sigma_1 = .38$, $\sigma_2 = .008$, and the first four digits of σ_3 and σ_4 are zero. The critical values at the 5% level are $\chi_{12}^2 = 21.03$ for n=2 and $\chi_5^2 = 11.07$ for n=3. The values of the statistic are b_2=17.1 and b_3= .6. Consequently, the state vector dimension $n = 2$ seems to be indicated, since the "whiteness" of the residuals is about the same as in the case of J=K=2.

As a second example, the five-dimensional annual wheat price series used by Pena and Box [1987], a total of 126 data points, is presented. Forecasts of the wheat prices one and two years ahead have been modeled--i.e., J=1 and 2 have been tried. Before we discuss the full five-dimensional series, it is informative to discuss two cases of three-dimensional series: one composed of three southern regions (Avila, Segovia, and Valladolid), and the other combining data for two southern regions of Avila and Segovia with data for one northern region, Palencia. We detect two factors in the former. This finding agrees with that by Pena and Box. However, we find that four factors are needed to adequately forecast the mixture of southern and northern wheat prices together. The same holds for the five series. This example also confirms the observation that the singular values drop off towards zero more rapidly than the canonical correlation coefficients.

9.7 Incorporating Exogenous Variables

In forecasting time series, some exogenous variables are often included to improve forecasting precision. It is also of some interest to evaluate dynamic effects of changes in exogenous variables on observed data in a framework of policy or decision analysis. We are thus naturally led to the question of how to incorporate exogenous variables in state space time series models, and how to do this simply, that is, without too many modifications of algorithms for estimating state space models for time series sans exogenous variables. More specifically, this section shows how to incorporate exogenous variables in the framework of the innovation state space modeling algorithm described in this book.

When exogenous variables are appended to data vectors to form augmented data vectors, sample autocovariance matrices of the augmented data vectors possess some structural regularity, such as being block triangular. Therefore, it may appear that we must estimate model parameters while imposing this structural constraint. Rather than this tack, we adopt a procedure which avoids dealing directly with constrained covariances by subtracting out the effects of the exogenous variables from certain auto-covariance matrices. This approach allows us to use the modeling algorithm without any modification.

Regression Model

One of the simplest ways to include exogenous variables is to construct a multivariate regression model

$$y_t = \theta x_t + v_t \ ,$$

where y_t is a p-dimensional data vector, and the residual vector v_t will in general be (highly) correlated. The residuals are then modeled by a state space or ARMA model. The matrix θ may be estimated by the least squares method

$$\hat{\theta} = \Sigma y_t x_t'(\Sigma x_t x_t')^{-1} \ .$$

The residuals $\hat{v}_t = y_t - \hat{\theta} x_t$ are then given by

$$\hat{v}_t = [I - X_t(X_t'X_t)^{-1}X_t']y_t$$

where

$$X_t' = [x_t, x_{t-1}, \cdots].$$

Defining $\hat{V}_t' = [\hat{v}_t, \hat{v}_{t-1}, \cdots]$, we have $X_t'\hat{V}_t = 0$, thus, the residuals are uncorrelated with x_t, x_{t-1}. The state vector of the model for the residuals may then be taken to be uncorrelated with x's. This model is a natural one when exogenous variables are uncorrelated.

Dynamic Model

If the exogenous variables are (mean zero) stationary and serially correlated, the above approach may be improved further by incorporating the dynamics of exogenous signals as part of the model. Suppose that the exogenous variable is being generated by

$$x_{t+1} = Fx_t + \varepsilon_t \ , \tag{22}$$

and the residual vector $y_t - \theta x_t$ is regarded as a new data vector, and is modeled by a state space model. In other words, we assume the data generating structure to be

$$y_t = \theta x_t + Cz_t + e_t \ , \tag{23}$$

$$z_{t+1} = Az_t + Be_t \ ,$$

where

$$\Delta = cov = \begin{bmatrix} \Delta_e & 0 \\ 0 & \Delta_\varepsilon \end{bmatrix} . \tag{24}$$

We assume that the vector x_t is observed without error, and that the off-diagonal element of (24) is zero, because the fact that y_t does not Granger-cause x_t implies that $E\varepsilon_t x_t' = E\varepsilon_t y_t' = 0$, or $E\varepsilon_t e_t' = 0$. See Hosoya [1977]. The forecast of y_t, $y_{t|t}$, is $y_t = \theta x_t + Cz_t$ if x_t is available in forming the forecast, or $y_{t|t-1} = \theta x_{t|t-1} + Cz_t$ where $x_{t|t-1} = Fx_{t-1}$, if x_t is not available.

This model is different from the regression model because the dynamics of the exogenous vector are recognized. Thus, future data contain some components of the exogenous vector, which is forecastable if the parameters of (22) are known. To simplify the algebra, let $\Phi = diag\,(A, F)$, and $\Theta = (C, \theta)$, and write (22) and (23) jointly. Let w_t be the column vector which stacks z_t over x_t. Then the model is

$$w_{t+1} = \Phi w_t + \Psi n_t \,, \tag{25}$$

where n_t is the column vector composed of e_t and ε_t, and

$$y_t = \Theta w_t + e_t \,.$$

We can also allow the innovation of the exogenous vector to be present in the state equation (23). In other words, the matrix Ψ in (25) can have an off-diagonal submatrix

$$\Psi = \begin{bmatrix} B & \psi \\ 0 & I \end{bmatrix} .$$

Note that the covariance matrix of the data has a structure more complicated than any so far encountered in this book:

$$\Lambda_l = Ey_{t+l}y_t{}' = CA^{l-1}M_y + \theta F^l \Gamma_o{}' \;, \quad l \geq 1 \,, \tag{26}$$

where

$$M_y = Ez_{t+1}y_t{}' = A\,\Pi C' + A\,\Sigma\theta' + B\,\Delta_e$$

with

$$cov \begin{bmatrix} z_t \\ x_t \end{bmatrix} = \begin{bmatrix} \Pi & \Sigma \\ \Sigma' & X \end{bmatrix} = \Xi \;, \; and \;\; \Gamma_o = Ey_t x_t{}' = C\,\Sigma + \theta X \;.$$

They depend on the matrices in both (23) and (22). To separate out these estimates, the effects of exogenous variable in Λ_l will be isolated, and the matrices in (26) are appropriately modified in order to apply the algorithm developed in Section 1. To this end, define the cross-covariance matrix between y_{t+l} and x_t as

$$\begin{aligned} \Gamma_l &= Ey_{t+l}x_t{}' \\ &= (C \;\; \theta)\Phi^l N \;, \quad l \geq 1 \,, \end{aligned} \tag{27}$$

where

$$N = E \begin{bmatrix} z_t \\ x_t \end{bmatrix} x_t{}' = \begin{bmatrix} \Sigma \\ X \end{bmatrix} .$$

Construct a Hankel matrix with these Γ's and note that it has the factored form

$$H_x = \begin{bmatrix} \Gamma_0 & \Gamma_1 & \cdots \\ \Gamma_1 & \cdots & \\ \cdots & \cdots & \end{bmatrix} = \begin{bmatrix} \Theta \\ \Theta\Phi \\ \vdots \end{bmatrix} (N, \Phi N, \cdots) .$$

Identifying the two stacked matrices on the right with the appropriate factors in the singular value decompostion of H_x as described in Section 1, we estimate the matrices Θ, Φ, and N.*

From (25), the covariance matrix for the vector z_t and x_t satisfies

$$\Xi = \Phi \Xi \Phi' + \Psi \Delta \Psi' . \tag{28}$$

From (26) the matrix Σ is the same as

$$\Sigma = A \Sigma F' + \psi \Delta_\varepsilon . \tag{29}$$

Substituting this into the matrix Γ_o, it is rewrittten as

$$\bar{\Gamma}_0 = C \psi \Delta_\varepsilon ,$$

where

$$\bar{\Gamma}_0 = \Gamma_0 - CA \Sigma F' - \theta X .$$

Similarly, from (27) we define

$$\bar{\Gamma}_l = CA^{l-1} \psi \Delta_\varepsilon ,$$

by subtracting out the expressions we have estimated from

$$\begin{aligned} \Gamma_l &= CA^l \Sigma + \theta F^l X \\ &= CA^{l-1}(A \Sigma F' + \psi \Delta_\varepsilon) + \theta F^l X \quad , \; l \geq 1 . \end{aligned}$$

The matrix equation

$$\begin{bmatrix} \bar{\Gamma}_o \\ \bar{\Gamma}_1 \\ \vdots \end{bmatrix} = \begin{bmatrix} C \\ CA \\ \vdots \end{bmatrix} \psi \Delta_\varepsilon$$

is next solved to estimate $\psi\Delta_\varepsilon$, now that the first factor on the right has already been estimated.

The algebraic equation for Π changes into

$$\Pi = A \Pi A' + B \Delta_e B' + \psi \Delta_\varepsilon \psi' .$$

To solve this equation for the matrices Π, B, and Δ_e, subtract out the known expressions from Λ_0 and define

$$\bar{\Lambda}_0 = \Lambda_0 - C\,\Sigma\theta' - \theta\Sigma' C' - \theta X\,\theta' ,$$

which is equal to $C\Pi C' + \Delta_e$. Next, use (5) and define $\bar{M}_y$ by subtracting out $A\,\Sigma\theta'$ from M_y, which is equal to $A\,\Pi C' + B\,\Delta_e$. We see that with the matrices $\bar{\Lambda}_0$ and $\bar{M}_y$ replacing Λ_0 and M, the algorithm given in Section 1 can be modified to obtain Π, B, and Δ_e.

The same procedure can be applied to other situations as well. For example, suppose that dynamic interaction between z_t and x_t is introduced by redefining Φ to be

$$\Phi = \begin{bmatrix} A & \phi \\ 0 & F \end{bmatrix} .$$

The autocovariance matrices are such that

$$M_y = (A\,\Pi + \phi\Sigma')C' + A\,\Sigma\theta' + B\,\Delta_\varepsilon$$

and

$$M_x = Ex_{t+1}y_t{}' = F\,(C\,\Sigma + \theta X)' .$$

From the Hankel matrix Θ_x, we can still estimate C, θ, A, ϕ, and Σ. Then

$$\bar{\Lambda}_l = CA^{\,l-1}M_y ,$$

where

$$\bar{\Lambda}_l = \Lambda_l - (C\,\Phi_{l-1} + \theta F^{\,l-1})M_x ,$$

where

$$\Phi^l = \begin{bmatrix} A^l & \phi_l \\ 0 & F^l \end{bmatrix} \tag{30}$$

implicitly defines Φ_l. Since M_x is estimated, so are the $\bar{\Lambda}$'s which are used to estimate $\bar{M}_y$. Then redefining $\bar{M}_y$ as $M_y - \Phi\Sigma' C' - A\,\Sigma\theta'$, we still have the relation $B\,\Delta_e = \bar{M}_y - A\,\Pi C'$. Hence the Riccati equation for Π can be constructed and solved as before.

10 Approximate Models and Error Analysis

Exogenous disturbances affect time series variables in complex and varied ways. Their relationships are usually only approximately captured by models. In the frequency domain, rational transfer functions of the models are best viewed as approximations to more complex rational, or possibly irrational, transfer functions. In the time domain, finite-dimensional state space (innovation) models merely approximate dynamic phenomena of greater complexity that cannot be conveniently captured. Model builders can only hope to reproduce some salient features of actual dynamics by judicious choice of the dimension and the values of model system matrices, and by analyzing the consequences of adding (deleting) a time series from the data vector, or of retaining (dropping) correlations between some of the data components.

10.1 Structural Sensitivity

We examine the effects of correlations among the two components of data vectors to evaluate the consequences of modeling a bivariate series as a single time series. Suppose a two-dimensional, weakly stationary bivariate series $\{y_t\}$ has the covariace matrices specified by the two-dimensional matrices

$$\Lambda_l = \begin{bmatrix} \lambda_l & \omega_l \\ \nu_l & \mu_l \end{bmatrix}, \qquad l \geq 1,$$

where

$$\lambda_l = c^2 a^{l-1}, \quad |a| < 1, \qquad \omega_l = \rho\kappa_1' a^{l-1}, \quad |\rho| < 1,$$

$$\nu_l = \rho\kappa_2' f^{l-1}, \ \mu_l = h^2 f^{l-1}, \text{ and } |f| < 1,$$

for some $a, f, c, h, \rho, \kappa_1'$, and κ_2'.

When the parameter ρ is set at zero, Λ_l becomes diagonal and two separate dynamic models for the two components of the series result. We now formally derive this fact. Denote by H_0 the corresponding Hankel matrix containing these diagonal Λ's. It has a rank 2 singular value decomposition, given by

$$H_0 = U \Sigma U', \tag{1}$$

where

$$U = [u, v], \quad \Sigma = diag(\sigma_1, \sigma_2),$$

with

$$\sigma_1 = c^2/(1 - a^2), \text{ and } \sigma_2 = h^2/(1 - f^2).$$

The infinite-dimensional vectors in the matrix U are defined by

$$u' = \sqrt{1-a^2}\,(1\;0\;a\;0\;a^2 \cdots),$$

and

$$v' = \sqrt{1-f^2}\,(0\;1\;0\;f\;0\;f^2 \cdots).$$

Note that the vectors u and v are orthonormal, i.e., $u'u = v'v = 1$, and $u'v = 0$.

When H_0 is shifted up by 2 rows, $\overset{A}{H_0}$ results. From (1), this is expressible as

$$\overset{A}{H_0} = a\sigma_1 uu' + f\sigma_2 vv'.$$

Therefore, the balanced dynamic matrix A is given by

$$A_{bal} = \Sigma^{-1/2} U' \overset{A}{H_0} U \Sigma^{-1/2}$$

$$= diag\,(a, f).$$

The matrix C becomes diagonal

$$C = \overset{c}{H_0} V \Sigma^{-1/2}$$

$$= diag(c, h),$$

where we use the relation

$$\overset{c}{H_0} = [c\sqrt{\sigma_1}u' \;\; h\sqrt{\sigma_2}v']'.$$

From the symmetry of H_0, $\overset{M}{H_0} = (\overset{c}{H_0})'$. Thus the matrix M is the same as C.

The two-dimensional balanced model for this bivariate series thus becomes

$$z_{t+1} = \begin{bmatrix} a & 0 \\ 0 & f \end{bmatrix} z_t + Be_t, \tag{2}$$

$$y_t = \begin{bmatrix} c & 0 \\ 0 & h \end{bmatrix} z_t + e_t,$$

where the matrix B is determined by solving the Riccati equation and the covariance matrix of the innovation vector, Δ, can be also calculated as detailed in Chapter 9. They are also diagonal. To derive them, first recall that $\Delta = cov\; e_t = \Lambda_0 - C\Pi C'$. If Λ_0 is $diag\,(\lambda_1, \lambda_3)$, i.e., if the correlations between the two components of the data vector are ignored, then the covariance matrix of the state vector Π becomes diagonal, $\Pi = diag\,(\pi_1, \pi_3)$, where π_1 and π_3 are governed by a scalar Riccati equation of the type shown in (4) of Chapter 9, and are given by

$$\pi_i = \alpha_i - \sqrt{\alpha_i^2 - 1}, \qquad i = 1, 3, \tag{3}$$

where

$$\alpha_1 = a + (1 - a^2)\lambda_1 / 2c^2,$$

and α_3 is analogously defined by replacing a by f, c by h, and λ_1 by λ_2. Assume that $\alpha_i > 1$, $i = 1, 3$ to ensure positive π's. The covariance matrix of the innovation vector is thus equal to $\Delta = diag\,(\lambda_1 - c^2\pi_1, \lambda_3 - h^2\pi_3)$, and the matrix B is given by $B = diag\,(b, g)$, where $b = c(1 - a\pi_1)$,and $g = h(1 - f\pi_3)$.

Eq. (3) and the subsequent calculations of the matrices B and Δ establish that the original bivariate series becomes two separate series driven by two uncorrelated innovation sequences (i.e., by the two components of e_t) when all the correlation coefficients (both contemporaneous and lagged) are equal to zero. The balanced model is given by (2) where the innovation covariance matrix is diagonal.

Effects of static correlations show up as off-diagonal elements in Λ_0, which affects Δ and Π matrices. It can be shown that it is always better not to neglect small off-diagonal elements in Λ_0 by comparing the one-step-ahead forecast error covariance matrices. We next examine the effects of correlations among lagged components of the series.

Off-Diagonal Elements of Λ_0. *Before* turning to the examination of the effects of off-diagonal elements in lagged covariance matrices, we show that inclusion of the off-diagonal elements of Λ_0 generally increases the covariance matrix of the state vector of (2), and improves the accuracy of the one-step-ahead prediction of the model. The same result, however, does not always follow when off-diagonal elements of lagged covariances are incorporated, a topic to which we turn in the next subsection.

It is quite possible that the off-diagonal elements of contemporaneous and lagged correlations behave differently, so that the lagged covariance matrices are nearly diagonal while Λ_0 is not. To isolate the effects of off-diagonal elements of the contemporaneous covariance matrix from those of the lagged ones, this subsection assumes that the only off-diagonal elements in the covariance matrices are in

$$\Lambda_0 = \begin{bmatrix} \lambda_1 & \lambda_2 \\ \lambda_2 & \lambda_3 \end{bmatrix}.$$

The off-diagonal element, λ_2, causes the innovation covariance matrix to have non-zero off-diagonal elements, i.e., the earlier diagonal matrix denoted by Δ_0 becomes a new non-diagonal one denoted by Δ. For definiteness we compare the (1, 1) element computed in two ways: one with the value of λ_2 assumed to be zero and the other non-zero.

From the definition

$$\Delta = \Lambda_0 - C\pi C' = \begin{bmatrix} \lambda_1 - c^2\pi_1 & \lambda_2 - ch\pi_2 \\ \lambda_2 - ch\pi_2 & \lambda_3 - h^2\pi_3 \end{bmatrix},$$

to the first order in the magnitude of λ_2, the newly introduced off-diagonal element of Λ_0 affects the (1, 1) element of Δ by $-c^2\delta\pi_1$, where $\delta\pi_1$ stands for a (small) change in π_1, because the changes in the values of the elements c and h can be shown to be of the second order of smallness. (See the next subsection.)

Thus, the (1, 1) element of Δ is smaller than that of Δ_0 if and only if $\delta\pi_1 \geq 0$ is true. We next show that inclusion of the neglected correlation, i.e., use of more information, improves (one-step-ahead) forecasting accuracy. Denote the (1, 1) element of Π by π_1. From the Riccati equation for Π, (9) or (9c) of Chapter 5, π_1 satisfies

$$-(\pi_1^2 - 2\alpha_1\pi_1 + 1)c^2\Delta_3 = (1-a^2)\pi_1\Delta_2^2 + 2abh\,\pi_2\Delta_2 + (a\pi_2)^2h^2\Delta_1\,, \quad (4)$$

where Δ_1 denotes the (1, 1) element, Δ_3 the (2, 2) element, and Δ_2 the off-diagonal element of the covariance matrix of the innovation vector. When λ_2 is zero, the right-hand side of (4) becomes zero because both π_2 and Δ_2 become zero. (They can be shown to be of the order $|\lambda_2|$.) Then, (4) simplifies to (4) of chapter 9, and the solution is given by (3). The non-zero right-hand side modifies the Riccati equation for π_1 from that shown in (4) of Chapter 9 to

$$\pi_1^2 - 2(\alpha_1 + \varepsilon)\pi_1 + 1 + \theta = 0\,,$$

where

$$\varepsilon = [2a^2ch\,\pi_2\Delta_2 - (1-a^2)\Delta_2^2 + (ach\,\pi_2)^2]\,/\,2c^2\Delta_3\,,$$

and

$$\theta = [2ach\,\pi_2\Delta_2 + (a\pi_2)^2h^2\Delta_1]/c^2\Delta_3.$$

The new π_1 is not less than the original one if and only if the inequality

$$\theta/2 \geq \varepsilon(\pi_1)_0$$

is satisfied, where the subscript 0 denotes the original value of π_1 specified by (3). This inequality becomes a quadratic form in the ratio of Δ_2/π_2 denoted by x in

$$(ah)^2\Delta_1 + 2ach(1 - a\pi_1)x + \pi_1(1-a^2)x^2 \geq 0\,.$$

When Δ_1 is substituted out by $\lambda_1 - c^2\pi_1$, and (3) is used, this inequality always holds, because the discriminant of this quadratic form is zero. Therefore, we conclude that when the off-diagonal elements in Λ_0 are considered, while neglecting those in the lagged covariance matrices, Λ_l, $l \geq 1$, the predictive accuracy of the first series does not deteriorate.

Off-Diagonal Elements of Λ_l. Now suppose ρ is not zero. Off-diagonal elements are introduced to all lagged covariance matrices. The new Hankel matrix has some additional structure attributable to these elements. To (1) is added a matrix

$$H_1 = \kappa_1 u \theta' + \kappa_2 v \phi' , \tag{5}$$

where

$$\kappa_1 = \rho\kappa_1'/(1-a^2), \quad \kappa_2 = \rho\kappa_2'/(1-f^2),$$

$$\theta' = \sqrt{1-a^2}(\, 0\, 1\, 0\, a \;\; 0\, a^2 \cdots),$$

and

$$\phi' = \sqrt{1-f^2}(\, 1\, 0\, f \;\; 0\, f^2 \cdots).$$

Note that κ_1 and κ_2 are proportional to ρ . The vectors satisfy the relations

$$\theta'\theta = 1, \;\; \theta' u = 0, \;\; \theta' v = \frac{\sqrt{(1-a^2)(1-f^2)}}{1-af} = \omega,$$

$$\theta'\phi = 0, \;\; \phi' u = \omega, \;\; \phi' v = 0, \;\; \text{and} \;\; \phi'\phi = 1 ,$$

where the parameter ω is defined by the above equality.

To obtain the singular value decomposition of H, first form HH' and note that

$$HH'[u \;\; v] = [u \;\; v]\Gamma ,$$

where

$$\Gamma = \begin{bmatrix} \gamma_1 & \gamma_2 \\ \gamma_2 & \gamma_3 \end{bmatrix} ,$$

with

$$\gamma_1 = \sigma_1^2 + \kappa_1^2, \quad \gamma_2 = (\sigma_1\kappa_2 + \sigma_2\kappa_1)\omega,$$

and

$$\gamma_3 = \sigma_2^2 + \kappa_2^2.$$

Therefore, its matrix of the eigenvectors is given by $U = (u, v)\Theta$, where $\Gamma\Theta = \Theta\Xi^2$, $\Theta'\Theta = I$, and $\Xi^2 = diag\,(\xi_1 \;\, \xi_2)$, which yields the relation $HH'U = U\Xi^2$, where ξ's are the eigenvalues of Γ. They are given by $\xi_1 = \gamma_1 + \gamma_2^2/(\gamma_1 - \gamma_3) + o\,(\rho^2)$, and $\xi_2 = \gamma_3 - \gamma_2^2/(\gamma_1 - \gamma_3) + o\,(\rho^2)$. The elements of the matrix Θ are given by

$$\theta_{11} = 1 - \varepsilon^2 + o(\rho^2)\,, \text{ and } \theta_{21} = \varepsilon + o(\rho^2)\,,$$

and

$$\theta_{12} = -\varepsilon + o(\rho^2)\,, \text{ and } \theta_{22} = \theta_{11}\,,$$

where the parameter ε stands for $\gamma_2/(\gamma_1 - \gamma_3)$, which is of the order ρ.

The matrix V of the singular value decomposition of the new Hankel matrix is next obtained by the relation $V = H'U\Xi^{-1}$, which becomes

$$V = [\sqrt{\gamma_1}\zeta_1 \;\; \sqrt{\gamma_2}\zeta_2]\Theta\Xi^{-1}\,,$$

where

$$\zeta_1 = (\sigma_1 u + \kappa_1\theta)/\sqrt{\gamma_1}, \text{ and } \zeta_2 = (\sigma_2\upsilon + k_2\phi)/\sqrt{\gamma_3}.$$

Note that $\zeta_1'\zeta_1 = \zeta_2'\zeta_2 = 1$, and $\zeta_1'\zeta_2 = \gamma_2/\sqrt{\gamma_1\gamma_3}$.

The new Hankel matrix H, which is the sum of (1) and (5), has the rank two singular value decomposition

$$H = \sqrt{\gamma_1}\, u\zeta_1' + \sqrt{\gamma_3}v\zeta_2'\,. \tag{6}$$

Analogous to the sum expression for H, the matrices H^A, H^M, and H^C are written as the sums of two parts, the first parts corresponding to the case where ρ is set to zero, and the second due to the off-diagonal elements in the covariance matrices. From (6), they are expressible as

$$H^A = a\sqrt{\gamma_1}u\zeta_1' + f\sqrt{\gamma_2}v\zeta_2'\,,$$

$$H^M = \sqrt{1-a^2}\; u(\sigma_1 \;\kappa_1) + \sqrt{1-f^2}\; v(\kappa_2 \;\sigma_2)\,,$$

and

$$H^C = \sqrt{(1-a^2)\gamma_1}\begin{bmatrix}1\\0\end{bmatrix}\zeta_1' + \sqrt{(1-f^2)\gamma_3}\begin{bmatrix}0\\1\end{bmatrix}\zeta_2'\,.$$

The dynamic matrix of the balanced model with nonzero ρ is then obtained by the evaluating the following formula

$$A = \Xi^{-1/2}U'H^A V\Xi^{-1/2}\,.$$

We note that $U'H^A V$ becomes

$$\Theta'\begin{bmatrix} a\gamma_1 & a\gamma_2 \\ f\gamma_2 & f\gamma_3 \end{bmatrix}\Theta\,.$$

Utilizing the orthogonality of the matrix Θ, the elements of the matrix A are given by

$$a_{11} = a + \varepsilon^2(f - a) + o(\rho^2)\text{, and } a_{12} = \varepsilon\gamma_3(f - a)/[\gamma_1\gamma_3^3]^{1/4} + o(\rho^2)\,,$$

$$a_{21} = \varepsilon\gamma_1(f - a)/[\gamma_1^3\gamma_3]^{1/4} + o(\rho^2)\text{, and } a_{22} = f - \varepsilon^2(f - a) + o(\rho^2)\,.$$

Analogous calculations produce the remaining system matrices. The elements of the matrix C are:

$$c_{11} = [c - \varepsilon(c\varepsilon - h\kappa_2/\sqrt{\sigma_1\sigma_2})]/D_1\,,$$

$$c_{21} = [h\kappa_2/\sigma_2 - \varepsilon(c\sqrt{\sigma_1/\sigma_2} + \varepsilon h\kappa_2/\sigma_2)]/D_2\,,$$

$$c_{12} = [c\kappa_1/\sigma_1 + \varepsilon(h\sqrt{\sigma_2/\sigma_1} - \varepsilon c\kappa_1/\sigma_1]/D_1\,,$$

and

$$c_{22} = [h - \varepsilon(\varepsilon h + c\kappa_1/\sqrt{\sigma_1\sigma_2})]/D_2\,,$$

where

$$D_1 = [1 + (\kappa_1/\sigma_1)^2 + (\gamma_2/\sigma_1)^2/(\gamma_1 - \gamma_3)]^{1/4}\,,$$

and

$$D_2 = [1 + (\kappa_2/\sigma_2)^2 - (\gamma_2/\sigma_2)^2/(\gamma_1 - \gamma_3)]^{1/4}\,.$$

The matrix M is analogously calculated, and found to be identical to the matrix C. Observe that the diagonal elements are the same as in the system with ρ zero, up to the second order, and that the off-diagonal elements are of the order $O(\rho)$ if none of a and f is close to one in magnitude and the singular values are distinct.

10.2 Error Norms

Because the number of data points is usually small, any method of building time series models necessarily produces only approximate models. The method used in this book is no exception. The models are affected by alternative choices of the numerical rank of the Hankel matrix and of the integers J and K in defining the finite section of the Hankel matrix used in the actual singular value decomposition. Some measures are needed to gauge relative sizes of various sources of errors affecting the models.

To produce a lower-dimensional approximation to time series models also requires some measures of goodness of fit to indicate how large the errors of approximation may be. Comparing time paths generated by alternative models in fitting data points, or in predictions outside the data sets, is much more intricate than comparing mere numbers. Such comparisons are made by defining "norms" for assigning numbers, or sequences of numbers or vectors to time functions.

One way to construct approximate dynamic systems is via approximation of transfer functions. Assigning a norm to a transfer function can be done in at least two ways. One such norm has already been discussed in Section 5.7. To assign a norm in this way to a transfer function $C(zI-A)^{-1}M$, the covariance matrices associated with the transfer matrix $C(zI-A)^{-1}M$, which are the coefficient matrices when the transfer function is expanded in powers of z^{-1}, are used to construct a Hankel matrix, and its largest singular value is assigned as the norm of the corresponding transfer matrix. (A constant additive matrix of the transfer matrix has no role to play in this norm-assignment procedure, hence it is dropped in this discussion.) Another way of viewing this definition of a norm is to regard the transfer matrix $C(zI-A)^{-1}B$, say, as the z-transform of the Hankel matrix

$$G(z)=\sum_{i=1}^{\infty} H_i z^{-i},$$

and to regard the relation between the z-transform of the stacked future observation vectors and that of the stacked past realization of the noise vectors as

$$\eta(z)=G(z)\varepsilon(z),$$

where

$$\eta(z)=\sum_{j=1}^{\infty} y_{t+j|t} z^{-j},$$

and

$$\varepsilon(z)=\sum_{j=0}^{-\infty} e_{t+j} z^{-j}=\sum_{k=1}^{\infty} e_{t+1-k} z^{k-1}$$

in terms of the original time domain relation

$$y_{t+1|t}^{+}=\boldsymbol{H}e_t^{-}.$$

When the L^2-norms of the infinite vectors are finite,

$$|y_{t+1|t}|^2=\sum_{j=1}^{\infty} y_{t+j|t}'y_{t+j|t},$$

and

$$|e_t^{-}|^2=\sum_{k=1}^{\infty} e_{t-j}'e_{t-j},$$

the norm of the Hankel matrix, when regarded as a linear operator on L^2, is defined by

$$|\boldsymbol{H}|_H^2=sup\frac{|y_{t+1|t}|^2}{|e_t^{-}|^2},$$

where sup is taken over all non-zero e_t^{-}. In this way we see that the largest singular value of the Hankel matrix is equal to the square root of the spectral radius of the product of the transpose of the Hankel matrix with itself. The subscript H denotes the Hankel norm.

Another, perhaps more direct, way to assign a norm to a transfer function is to use the maximum of the magnitude $|G(e^{i\theta})|$ as its norm, where $z=e^{i\theta}$ describes the circumference of the unit disc in the complex plane as θ ranges from zero to 2π. These two norms are in general not the same. The former is denoted by subscript H, and the latter by ∞. In general $\|G(z)\|_H < \|G(z)\|_\infty$. If $G(z)$ is the transfer function of an all-pass filter, i.e., if $G(z)G(1/z)' = \sigma^2 I$ for some σ, then $\|G(z)\|_\infty$ is clearly equal to σ and both norms coincide. This is the only case of a rational transfer function with the same norms in these two senses. Glover [1984] has proved the following

Theorem. For any rational transfer function G(z) with poles inside the unit disc in the complex plane, the following equality holds

$$|G(z)|_H = inf|G(z) - F(z)|_\infty ,$$

where the infinum is taken over the set of rational transfer functions with poles outside the unit disc. (Such functions are transfer functions for anti-causal systems.)

By expanding the transfer function in a formal power series in z, Markov parameters can be interpreted as the Fourier coefficients. By setting z to $e^{i\theta}$ in the power series expansion of the system transfer function

$$G(z) = C(zI - A)^{-1}M = \sum_{l=1}^{\infty} \Lambda_l z^{-l} ,$$

or

$$C(zI - A)^{-1}B = \sum_{i=1}^{\infty} H_i z^{-i} ,$$

we observe that this series is a one-sided Fourier series because the dynamics are causal. The coefficients for positive powers of z, i.e., the coefficients associated with $e^{ik\theta}$ with positive k, do not appear.

Since functions analytic in $|z| < 1$ have positive power series expansion in z, with no coefficients with negative powers of z, their Hankel matrices are zero. Thus, Hankel matrices of two functions of z which differ from each other by some analytic function are the same, hence their Hankel norms are also the same. See Adamjan, Arov, and Krein [1978], and earlier works by Nehari [1957] and Clark [1968].

Scalar Example. We compute the two norms of a scalar dynamic system,

$$x_{t+1} = a\, x_t + b\, n_t, \quad y_t = cx_t .$$

Its transfer function $G(z) = bc/(z-a)$ has the norm

$$|G(z)|_{\infty}^{2} = \max_{|z|=1} |G(z)G'(1/z)|$$

$$= \max_{\theta} [\,(bc)^2/(1-2a\cos\theta + a^2)\,] = (bc)^2/(1-a)^2.$$

The observability Gramian is a scalar and is given by $c/(1-a^2)$. The reachability Gramian is the same expression when c is replaced by b. The Markov parameters are of the form $ca^{k-1}b$. The Hankel norm is given by $|bc|/(1-a^2)$, with the normalized eigenvector $u' = (1-a^2)^{1/2}(1, a, a^2, \cdots)'$. We see that $1/(1-a^2) \le 1/(1-a)$, i.e., the Hankel norm is less.

Two-Dimensional Example. In this example, the dynamic system is of second-order, and y and n are both scalar time series. The system parameters are constrained by the fact that the model is in balanced representation to explicitly bound the error of approximating a second-order transfer function by a first order one. Let the model be given by

$$x_{t+1} = Ax_t + bn_t,\quad y_t = cx_t + n_t.$$

The lower-case letters b and c denote two-dimensional vectors. The transfer function is $c(zI - A)^{-1}m$, which is the z-transform of the covariances for $\{y_t\}$, where m is also two-dimensional, where the matrix A is either

$$A = \begin{bmatrix} \alpha & \beta \\ \beta & \delta \end{bmatrix}$$

or the (2, 1) element of A is replaced by -β. This special feature of the balanced representation has already been pointed out. To be concrete, we take the matrix A to be symmetric and set m to c. The one-dimensional balanced model has the transfer function $c_1^2/(z-\alpha)$, where $c = (c_1\ c_2)$.

The difference between the actual and approximate transfer function, $e(z) = c(zI - A)^{-1}m - c_1^2/(z-\alpha)$, is expressible as

$$e(z) = \hat{c}(z)\Delta^{-1}(z)\hat{c}(z),$$

where $\hat{c}(z) = c_1\phi(z)\beta + c_2$, $\Delta(z) = z - \delta - \beta^2/(z-\alpha)$, and $\phi(z) = 1/(z-\alpha)$. This expression is the result of writing $(zI - A)^{-1}$ as

$$\begin{bmatrix} \phi(z) + \phi(z)\beta\Delta^{-1}\beta\phi(z) & \phi(z)\beta\Delta^{-1} \\ \Delta^{-1}\beta\phi(z) & \Delta^{-1} \end{bmatrix},$$

which is a 2×2 case of a general inversion formula for matrices involving two blocks of submatrices: see the next example. Next, in calculating the expression for $e(z)e(1/z)$, we eliminate the expressions for c_1^2, c_1c_2, and c_2^2 by the

Lyapunov equation for the Gramian, such that

$$c_1{}^2 = (1-\alpha)^e \sigma_1 - \beta^2 \sigma_2 , \text{ and } c_1 c_2 = -(\sigma_1 \alpha + \sigma_2 \delta)\beta ,$$

and so on. The coefficient of σ_1 vanishes. We obtain

$$\hat{b}(z)\hat{b}(1/z) = \sigma_2[1 - h(z)h(1/z)] ,$$

where $h(z) = \delta + \beta^2 \phi(z)$, which by definition is equal to $z - \Delta(z)$. Therefore, the norm of the error of approximating the transfer function is given by

$$|e(z)|_\infty / \sigma_2 = \max | E(z)E(1/z)|^{1/2} ,$$

where $E(z) = 1/z + zp(z) - \Delta(1/z)$, and where $p(z) = \Delta^{-1}(z)\Delta(1/z)$. On the unit circle the maximand can be bounded as

$$|\delta + \frac{\beta^2}{1/z - \alpha} + zp(z)| \le 1 + |\delta + \frac{\beta^2}{e^{-i\theta}}| ,$$

because the magnitude of $p(z)$ is one on the unit circle. Using the inequality $\beta^2 < (1-\alpha)(1-\delta)$ twice, we see that the second term above is bounded by 1. Hence we establish that

$$|e(z)|_\infty \le 2\,\sigma_2.$$

Since the Hankel norm of $e(z)$ is less than that of $|\cdot|_\infty$, and since the Hankel norm of the original transfer function is equal to σ_1 by definition, we have an upper bound on the relative error:

$$|e(z)|_H / |G(z)|_H \le 2\sigma_2/\sigma_1.$$

More generally, by successive reduction of dimension by one, an n-dimensional transfer function and its k-dimensional approximation is such that the relative error of approximation is bounded from above by

$$\sigma_{k+1} / \sigma_k \le |G(z) - \hat{G}(z)|_H / |G(z)|_H \le 2\sum_{i=k+1}^{n} \sigma_i / \sigma_1,$$

because the Hankel norm of the approximation error of the transfer function $G(z)$ by $\hat{G}(z)$ is bounded from below by σ_{k+1} .

Approximation of an n-Dimensional Transfer Function. This example discusses approximating an n-dimensional dynamic system with a system of dimension n-1, because this specialized approximation of a system with another of one less state vector dimension can be treated without a specialized assumption on the dynamic matrix, such as block diagonality. Partition all the system matrices into $(n-1)$-dimensional and one dimensional subsets:

$$\begin{bmatrix} A & \alpha \\ \beta' & a \end{bmatrix}, \quad \begin{bmatrix} M \\ \theta' \end{bmatrix}, \text{ and } \quad (C \;\; \delta).$$

Here, A is $(n-1)\times(n-1)$, M is $(n-1)\times p$, and C is $p\times(n-1)$. The vectors α and β are $(n-1)$-dimensional, and θ and δ are p-dimensional.

Now approximate the transfer function of this system, $W(z)$, by that constructed from the $(n-1)$-dimensional subsystem, $\hat{W}(z)$. The error of this transfer function approximation is denoted by $E(z)$. Noting that

$$\begin{bmatrix} zI-A & -\alpha \\ -\beta' & z-a \end{bmatrix}^{-1} = \begin{bmatrix} \phi(z)+\phi(z)\alpha\Delta^{-1}(z)\beta'\phi(z) & \phi(z)\alpha\Delta^{-1} \\ \Delta^{-1}(z)\beta'\phi(z) & \Delta^{-1} \end{bmatrix},$$

where

$$\phi(z)=(zI-A)^{-1}, \text{ and } \Delta(z)=z-a-\beta'\phi(z)\alpha,$$

the error term $E(z)$ can be rewritten as $\hat{C}(z)\,\Delta^{-1}(z)\,\hat{M}(z)$, where

$$\hat{C}(z)=C\phi(z)\alpha+\delta,$$

and

$$\hat{M}(z)=\beta'\phi(z)M+\theta'.$$

Next calculate the norm of this transfer function error matrix:

$$|E(z)E(1/z)'| = |\hat{C}(z)\Delta^{-1}(z)\hat{M}(z)\hat{M}(z)'\Delta^{-1}(1/z)\hat{C}(1/z)'|$$

$$= |\Delta^{-1}(z)\hat{M}(z)\hat{M}(z)'\Delta^{-T}(z)\hat{C}(1/z)'\hat{C}(z)|\,.$$

The product $\hat{C}'(1/\hat{z})C(z)$ becomes, after the Lyapunov equation for the observability Gramian is used to substitute out $C'C$ etc.,

$$\sigma[1-(a+\psi'(1/z))(a+\psi(z)],$$

where

$$\psi(z)=\beta'\phi(z)\alpha\,.$$

In the above, every term not containing σ cancels out when the relation $I+\phi(z)A=z\phi(z)$ is noted. Note that the product $\hat{M}(z)\hat{M}'(z)$ can be similarly reduced by noting the relations in its Lyapunov equation for the Gramian, (14) of Chapter 5,

$$A'\Sigma A + \sigma\beta\beta' - \Sigma = -CC',$$

$$\alpha'\Sigma A + \sigma a\beta' = -\delta'C',$$

and

$$\sigma a^2 + \alpha'\Sigma\alpha - \sigma = -\delta'\delta.$$

Here Σ and σ are the $(n-1)$- and one-dimensional elements of the diagonal matrix of the Gramian in the balanced representation. After the substitution, the two remaining factors become

$$\hat{M}(z)\hat{M}(1/z)' = [1 - (a + \psi(z))(a + \psi(1/z)')]\sigma.$$

Note that the expression in parentheses is equal to $z - \Delta(z)$. After these preliminary reductions the error transfer function is expressed as

$$\sigma^{-2}E(z)E(1/z)' = [1/z + zp(z) - \Delta(1/z)'][its\ \ transpose\ \ with\ \ z]$$

where $p(z) = \Delta^{-1}(z)\Delta(1/z)'$.

Therefore, the norm of this error matrix is proportional to σ, the neglected singular value of the original dynamic matrix A. For example, if either α or β is zero, so that the dynamics are recursive, then the value of the error matrix is equal to $\sigma(1+|a|)$, which is less than 2σ because a is less than one in magnitude for an asymptotically stable system. The final expression is

$$|E(z)|_H = |E(z)|_\infty \le 2\sigma.$$

10.3 Asymptotic Error Covariance Matrices of Estimators

In this section, asymptotic error covariance matrices of the IV estimators are derived. The basic framework is the innovation model (1) of Chapter 9, together with the assumption that the matrix A is asymptotically stable and (1) is minimal, i.e., the pair (A, B) is a reachable pair. Let $\mathcal{F}_t$ be an increasing sequence of σ-fields generated by z_t^- and e_t^-. Assume that $E(e_t \mid \mathcal{F}_{t-1}) = 0$, and $E(e_t e_t' \mid \mathcal{F}_{t-1}) = \Delta_t$. In addition we assume that as a goes to infinity

$$sup_t E[e_t'e_t I(e_t'e_t > a) \mid \mathcal{F}_{t-1}] \to 0 \text{ in probability}.$$

Throughout we use a version of the central limit theorem for martingale differences that the "vec" of $T^{-1/2}\sum_t z_t e_t'$ converges in distribution to a normal distribution function with mean zero and finite variance, under suitable conditions. Assuming some moment conditions such a theorem was proved in Billingsley [1961] . See also Lai and Wei [1985]. A more general version, obtained by Anderson and Kunimoto [1989], allows for time-varying conditional covariance matrices for the martingale differences $\{e_t\}$. We follow them in stating the next

Lemma (Anderson and Kunimoto). Under the set of assumptions stated above, $T^{-1}\sum e_t e_t'$ converges in probability to Δ, $T^{-1}\sum e_t z_t'$ converges to zero in probability, and $T^{-1}\sum z_t z_t'$ converges in probability to Π.

If we further assume that

$$T^{-1} \sum_{t-1=max(r,s)}^{T} (\Delta_t \otimes e_{t-r} e_{t-s}') \to \delta_{r,s}(\Delta \otimes \Delta) \text{ , in probability ,}$$

then we have the next

Theorem (Anderson and Kunimoto). The expression $T^{-1/2} vec(\sum_{t=1}^{T} z_t e_t' B')$ converges in distribution to a normal one, with mean zero and covariance matrix $B \Delta B' \otimes \Pi$ as a goes to infinity.

We also have

Lemma. The expression $T^{-1/2} e_t z_{t-i}'$ is a martingale difference array and $T^{-1/2} \Sigma\, vec\, e_t z_{t-i}'$ converges in distribution to $N(0, \Pi \otimes \Delta)$ as T goes to infinity, $i = 1, 2, \cdots$, where $\Delta = cov\ e_t$ and $\Pi = var\ z_t$. Similarly $T^{-1/2} vec\ \Sigma e_t y_{t-1}'$ converges in distribution to $N(0, \Lambda_0 \otimes \Delta)$, where $\Lambda_0 = var\ y_t$.

Variances of $\hat{\Delta}$ and $\hat{Z}$

We next calculate the variances of these estimates when e_t and z_t are scalar and e_t is a normal random variable.

When e s are scalar and $NID(0, \sigma^2)$, the variance of $\hat{\Delta} = T^{-1}\Sigma e_t^2$ is easy to calculate. Using σ^2 rather than Δ^2, we obtain

$$var\,(T^{-1}\Sigma e_t^2) = T^{-2} E\,(\Sigma_{i \neq j} e_i^2 e_j^2 + \Sigma_i e_i^4) - \sigma^4$$

$$= \sigma^4 [\,T\,(T-1) + 3T - T^2\,] / T^2 = 2\sigma^4 / T$$

where we use a known fact about normal random variables that $E(e_t^4) = 3\sigma^4$.

When the dimension of e_t and z_t is one, we can explicitly calculate the variance of $\hat{\Pi} = T^{-1}\Sigma z_t^2$ as follows.

Let

$$z_{t+1} = a z_t + b e_t,$$

where e_t is $NID(O, \sigma^2)$. Define a $(T \times T)$ matrix J by

$$\begin{bmatrix} z_T \\ z_{T-1} \\ \vdots \\ z_1 \end{bmatrix} = bJ \begin{bmatrix} e_{T-1} \\ e_{T-2} \\ \vdots \\ e_0 \end{bmatrix} \text{,i.e., } J = \begin{bmatrix} 1 & a & \cdots & a^{T-1} \\ 0 & 1 & & a^{T-2} \\ \vdots & & \ddots & \vdots \\ \vdots & & & a \\ 0 & & & 1 \end{bmatrix}.$$

Then diagonalize $J'J$ by a coordinate change

$$\Sigma_1^T z_t^2 = b^2 e'J'Je = b^2 \Sigma_{i=1}^T \delta_i \xi_i^2 ,$$

where e is the stacked e_ts, $Je = D\xi$, and ξ stacks ξ_is, where ξ_i is $NID(0, \sigma^2)$, and hence $E\,\Sigma_1^T z_t^2 = (\Sigma\delta_i)\sigma^2$, with

$$\Sigma\delta_i = tr\, J'J = \frac{T}{1-a^2}\left[1 - \frac{a^2(1-a^{2T})}{(1-a^2)T}\right] b^2.$$

We have

$$E(T^{-1}\Sigma z_t^2) = Z\left[1 - \frac{a^2(1-a^{2T})}{(1-a^2)T}\right],$$

where $Z = \sigma^2 b^2/(1-a^2)$ is the solution of the algebraic Riccati equation $Z = a^2 Z + b^2\sigma^2$. To calculate the variance of Z we note that

$$var\, T^{-1}\Sigma z_t^2 = E(T^{-1}\Sigma z_t^2)^2 - [\, E(T^{-1}\Sigma z_t^2)\,]^2 ,$$

and use the relation

$$(\Sigma z_t^2)^2 = (\Sigma\delta_i\xi_i^2)^2 = \sum_{i\neq j}\delta_i\delta_j\xi_i^2\xi_j^2 + \Sigma\delta_i^2\xi_i^4$$

and hence

$$E(\Sigma z_t^2)^2 = (\sum_{i\neq j}\delta_i\delta_j + 3\Sigma\delta_i^2)\sigma^4.$$

Thus

$$var\,(T^{-1}\Sigma z_t^2) = 2b^2\sigma^4\Sigma\delta_j^2/T.$$

We need the trace of $(J'J)^2$. To calculate this, we first note that $J'J$ is a Toeplitz matrix. Let $\lambda_1 \cdots \lambda_T$ be its eigenvalues. Then $\Sigma\delta_i^2 = tr\,(J'J)^2 = \Sigma\lambda_i^2$. Grenander and Szegö [1958, p. 71] derived the expression for the characteristic polynomial for $J'J$, and have obtained the eigenvalues for large T as

$$\lambda_i = f\left(\pi + \frac{i\pi + \varepsilon}{T+1}\right) \quad \text{and} \quad \lambda_{n+1-i} = f\left(\frac{i\pi+\varepsilon}{T+1}\right),$$

where

$$f(x) = \frac{1 - a^2}{1 - 2a \cos x + a^2} \le 1$$

and $\varepsilon \to 0$ as $T \to \infty$.
An upper bound on the trace is then simply T and

$$E(\hat{\Pi} - Z)^2 \le 2b^2\sigma^4 / T.$$

Errors of System Matrix Estimates

We next turn to analysis of the system matrices estimated by the instrumental variables method described in Chapter 9.

Matrix C. Using $z_t = \Omega R_-^{-1} y_{t-1}^-$ as instruments on the observation equation

$$y_t = C z_t + e_t,$$

estimate the matrix C by

$$T^{-1} \Sigma y_t z_t' = \hat{C} T^{-1} \Sigma z_t z_t' \quad ,$$

i.e.,

$$\hat{C} = (T^{-1} \Sigma y_t z_t') \hat{\Pi}^{-1} \quad , \tag{7}$$

where

$$\hat{\Pi} = T^{-1} \Sigma z_t z_t' \quad .$$

The estimation error matrix $\delta C = \hat{C} - C$ is given by

$$\delta C \hat{\Pi} = T^{-1} \Sigma e_t z_t'. \tag{8}$$

Using its vectorized expression

$$T^{-1/2}(\hat{\Pi} \otimes I) vec\, \delta C = T^{-1/2} \Sigma z_t \otimes e_t$$

and the theorem, we deduce the asymptotic error covariance matrix of δC to be

$$T(\hat{\Pi} \otimes I) cov\,(vec\, \delta C)\,(\hat{\Pi} \otimes I) \to \Pi \otimes \Delta\,, \tag{9}$$

i.e.,

$$T\, cov\,(vec \delta C) \to \Pi^{-1} \otimes \Delta \;, \tag{10}$$

because as T goes to infinity, $\hat{\Pi} \to \Pi$.

In the case of the stochastic realization estimator, given in (3c) of Chapter 9, the expression in (8) is replaced by $\hat{C} \hat{\Omega} = T^{-1} \sum e_t y_{t-1}^-{}'$. Instead of (9), we arrive at the expression given in

$$T(\hat{\Omega}'\otimes I)cov\ (vec\delta C)(\hat{\Omega}\otimes I)\rightarrow R_{-}\otimes\Delta\ , \qquad (11)$$

or by taking the pseudo-inverse of Ω

$$Tcov\ (vec\delta C)\rightarrow \Xi\otimes\Delta\ ,$$

where $\Xi=\Sigma^{-1/2}V'R_{-}V\Sigma^{-1/2}$. By comparing these two expressions we arrive at

Proposition. In VAR models in which the minimal dimension of the model is an integer multiple of that of the data vector, the asymptotic covariance matrices of the stochastic realization estimator and the IV estimator with the state vector as instruments are the same.

This important property of the stochastic realization estimator follows by comparing (9) and (11), and noting that the state vector covariance matrix is given by $\Omega R_{-}^{-1}\Omega'$ in such situations and that the matrix V in the singular value decomposition of the Hankel matrix is such that VV' is the identity matrix, so that $V(V'R_{-}^{-1}V)^{-1}V'$ is equal to R_{-}. Only when the state vector covariance matrices are not equal to $\Omega R_{-}^{-1}\Omega'$, the asymptotic covariance matrix given in (9) is smaller than that given in (11) in the partial ordering of the symmetric positive definite matrices.

Example. To show that the limiting covariance matrix in (3) is smaller than that in (4) in non-VAR models, consider Example 8 in Section 9.3. The model in this example is not a VAR model and, furthermore, the state vector dimension is not an integral multiple of the data vector dimension. When we put this model in balanced form, its state vector covariance matrix becomes

$$\begin{bmatrix} 1.913 & * & * \\ -1.009 & .745 & * \\ -.807 & .496 & 2.257 \end{bmatrix} .$$

The matrices R_{-} and $\Omega'\Pi^{-1}\Omega$ are identical in the upper-left (3×3) corner, but differs in the last row (and column because the matrix is symmetric). It is $[-.842\ \ 18.509\ \ -1.184\ \ 34.415]$ in the former. The last two elements of this row are $-.1805$ and 9.961 in the latter. The difference has two positive and two zero eigenvalues.

Matrix M. In Section 9.1 we derived the expression for the IV estimator for the matrix M as

$$O_J'R_{+}^{-1}\hat{H}_{\cdot 1}=\hat{Z}\hat{M}\ ,$$

where Z is the state vector covariance matrix for the backward innovation model as given in Section 7.1, where

$$\hat{H}_{\cdot 1} = T^{-1}\sum y_{t+1}^{+} y_t{}' .$$

Substituting for y_{t+1}^{+} the expression in terms of the observability matrix and z_{t+1}, similar to (1) of Chapter 6, the estimator error is given by

$$\delta\hat{M} = \hat{Z}^{-1}\boldsymbol{O}_J{}'R_{+}^{-1}GT^{-1}\sum e_{t+1}^{+} y_t{}' .$$

Vectorizing this, the asymptotic error covariance matrix as T goes to infinity is given by

$$T cov\ (vec\delta M) \to \Lambda_0 \otimes W(I \otimes \Delta)W' ,$$

where

$$W = Z^{-1}\boldsymbol{O}_J{}'R_{+}^{-1}G ,$$

where the matrix G is a lower-triangular Toeplitz matrix with the identity matrix I_p on the main diagonal and the impulse response matrices H_1, H_2,... on the succeeding subdiagonal lines.

Matrix A. The IV estimator for the matrix A has been given in (6) of Section 9.1. The left side of that equation can be interpreted as the covariance matrix between the state vector for the backward innovation model as given in (10) of Section 7.1 and that for the forward innovation model, given in (8) of Chapter 6. Accordingly, using z_t as instruments in the stacked future observation equation

$$\begin{aligned} y_{t+1}^{+} &= \boldsymbol{O}\, z_{t+1} + Ge_{t+1}^{+} \\ &= \boldsymbol{O}Az_t + \boldsymbol{O}Be_t + Ge_{t+1}^{+} , \end{aligned}$$

multiply it from the left by $\boldsymbol{O}_J{}'R_{+}^{-1}$ to obtain

$$\boldsymbol{O}_J{}'R_{+}^{-1}\hat{H}^A R_{-}^{-1}\hat{\Omega}_K = \hat{Z}\hat{A}\hat{\Pi} ,$$

where Z and Π are the state vector covariance matrices of the backward and forward innovation models, respectively, as given in Section 7.1. Therefore the error of the estimator is given by

$$\hat{Z}\,\delta A\,\hat{\Pi} = \hat{Z}BT^{-1}\sum e_t z_t{}' + \boldsymbol{O}_J{}'R_{+}^{-1}GT^{-1}\sum e_{t+1}^{+} z_t{}' .$$

Vectorizing this and letting T approach infinity, we derive the limiting expression for the covariance matrix as

$$T cov\ (vec\delta A) \to \Pi^{-1} \otimes B\,\Delta B' + \Pi^{-1} \otimes W(I \otimes \Delta)W' ,$$

where the matrix W is as defined in the subsection on the matrix M estimator error analysis.

In the case of the stochastic realization estimator, given in (3a) of Chapter 9, estimator errors are governed by

$$O_J \delta A \hat{\Omega} = T^{-1} \sum (O_J B e_t + G e_{t+1}^{+}) y_{t-1}^{-\prime},$$

from which the error covariance matrix is seen to converge to $\Xi \otimes (B \Delta B' + O_J^{\dagger} G (I \otimes \Delta) G' O_J')$. We have propositions for matrices M and A, similar to the one for the matrix C

Matrix Δ. The residuals are calculated by

$$\hat{e}_t = y_t - \hat{C} z_t = e_t - \delta C z_t \; . \tag{12}$$

Then

$$\hat{\Delta} = T^{-1} \Sigma \hat{e}_t \hat{e}_t'$$

$$= T^{-1} \Sigma y_t y_t' - \hat{C} (T^{-1} \Sigma z_t z_t') \hat{C}' = \hat{\Lambda}_0 - \hat{C} \hat{\Pi} \hat{C}' \; .$$

From (12), the error matrix is

$$\sqrt{T}\, \delta\Delta = \sqrt{T}\, (\hat{\Delta} - \Delta)$$

$$= -\delta C (T^{-1/2} \Sigma z_t e_t') - (T^{-1/2} e_t z_t') \delta C' + \delta C (T^{-1/2} \Sigma z_t z_t') \delta C' .$$

We know that $\delta C = O_p(T^{-1/2})$. Consequently

$$T\, cov\, (vec\, \Delta) \to 0 \; , \tag{13}$$

i.e., $\hat{\Delta} \to \Delta$ in *probability* , as T goes to infinity.

10.4 Other Statistical Aspects

This section briefly describes a test for residuals, and the variability of sample correlation coefficients and sample covariances for scalar series.

Test for Residuals

For one thing, the residuals should behave as innovations, i.e., should be uncorrelated. One should at least test for the "whiteness" of the residuals in comparing alternative models.

Denote scalar-valued sample residuals by $\{r_t, t = 1, \ldots, N\}$. The Durbin-Watson statistic, d, is defined by

$$d = \sum_{t=2}^{N} (r_t - r_{t-1})^2 / \sum_{1}^{N} r_t^2.$$

This statistic is approximately equal to $2(1 - \rho_1)$, where ρ_1 is the first-order sample correlation coefficient. Theoretically ρ_1 should be zero. Hence, d should be close to 2.

The Box-Pierce statistic tests for correlations of l residuals together to see if they are too large to be generated by white noise, Box and Pierce [1970]. It is defined by

$$Q = N \sum_{k=1}^{l} \hat{\rho}_k^2 ,$$

where N is the number of samples, and $\hat{\rho}_k$ is a sample correlation of order k. The number l should be taken reasonably large--say, 20. A variant of this to correct for small sample size is the Ljung-Box statistic, defined by

$$Q' = N(N+2) \sum_{1}^{l} \hat{\rho}_k^2/(n - k).$$

See Ljung and Box [1978] for an asymptotically-more-efficient version of the χ^2 test. The statistic approaches asymptotically the chi-square distribution with degrees of freedom equal to l minus the number of parameters. Tests of model adequacy based on these statistics are generally called portmanteau tests. Some of these tests may have low power with small sample size, and especially when models are mis-specified. See Davies and Newbold [1979], for example.

Akaike avoids formulating model adequacy as hypothesis testing. Rather he proposes FPE (final prediction error) and later AIC (an information criterion , or Akaike's information criterion) which is to be minimized with respect to model size.

Variability of Sample Correlation Coefficients

Kendall and Stuart [1966, Chapter 48] discuss statistical variability of sample correlation coefficients of scalar-valued, mean-zero, jointly normal random variables $\{u_t\}$. Sample correlations are computed as

$$\hat{\rho}_t = \sum_i u_i u_{i+t} / N .$$

Their covariances are expressible as

$$cov\ (\hat{\rho}_t , \hat{\rho}_{t+s}) = E\ (\sum_i \sum_j u_i u_{i+t} u_j u_{j+t+s}) - \rho_t \rho_{t+s} ,$$

where ρs are the true correlations. Using the expression for the mean of a product of four normal random variables (see Parzen [1962, p. 93], for example),

$$E\ (u_i u_{i+t} u_j u_{j+t+s}) = \rho_t \rho_{t+s} + \rho_{j-i} \rho_{s+j-i} + \rho_{j-i+t+s} \rho_{j-i-t} .$$

The expression for the covariance can therefore be written as

$$\sum_k [\, \rho_k \rho_{k+s} + \rho_{k+t+s} \rho_{k-s} \,] / N \, .$$

Specializing the above formula by setting t and s to zero, the variance of $\hat{\rho}_0$ is obtained as

$$var\ (\hat{\rho}_0) = 2 \sum_k \rho_k^2 / N \, .$$

By setting s and t separately to zero, one obtains

$$var\ (\hat{\rho}_t) = \sum_k (\rho_k^2 + \rho_{k+t}\ \rho_k) / N \, ,$$

and

$$cov\ (\hat{\rho}_0 \hat{\rho}_s) = 2 \sum_k \rho_k \rho_{k+s} / N \, .$$

The sample correlation coefficient, defined by

$$r_j = \hat{\rho}_j / \hat{\rho}_0 \, ,$$

varies according to

$$\partial r_j = \partial \hat{\rho}_j / \hat{\rho}_0 - \hat{\rho}_j \partial \hat{\rho}_0 / \hat{\rho}_0^2 \, .$$

Therefore, its variance is equal to

$$var\ (r_j) = var\ (\hat{\rho}_j) / \hat{\rho}_0^2 - 2\, \hat{\rho}_j\, cov\ (\hat{\rho}_0, \hat{\rho}_j) / \hat{\rho}_0^3 + \hat{\rho}_j^2\, var\ (\hat{\rho}_0) / \hat{\rho}_0^4 \, .$$

Taking $\hat{\rho}_0$ to be 1, this expression simplifies, when u's are uncorrelated, to

$$var\ (r_j) = 1 / N \quad j \geq 1 \, .$$

The standard deviation of the sample correlation coefficient is $1 / \sqrt{N}$.

Variances of Sample Covariances

Asymptotic means and variances of the sample mean and covariance have been derived in Anderson [1971, Sec. 8.3], Fuller [1976, Sec. 6.2], and Porat and Friedlander [1985]. Here we give the formula for the variances of the autocovariances of univariate series in state space form.

Then the matrices C and M become a n-dim row vector c' and n-dim column vector m in

$$\lambda_k = c'A^{k-1}m \quad , \quad k = 1, 2, \cdots$$

In our class of data-generating processes, Anderson's assumption

$$\Sigma_{-\infty}^{\infty}\lambda_k^2 < \infty$$

is satisfied because the matrix A is asymptotically stable. Anderson (1971 , p. 446) derives

$$lim\, T\, var\, (\hat{\lambda}_k) = \Sigma_{-\infty}^{\infty}\lambda_l^2 + \Sigma\lambda_{l-k}\lambda_{l+k}$$

$$= 2\lambda_k^2 + \Sigma_0^{\infty}\lambda_{p+k+1}^2 + \Sigma_0^{\infty}\lambda_{l-k+1}^2 + 2\Sigma_0^{\infty}\lambda_{l-k+1}\lambda_{l+k+1} \quad . \quad (*)$$

Noting that

$$\lambda_k^2 = c'A^{k-1}mm'(A')^{k-1}c \ ,$$

we can write

$$\Sigma_0^{\infty}\lambda_{l+k+1}^2 = c'A^kXA'c$$

with

$$X = \Sigma_{l=0}^{\infty}A^l mm'A'l \ .$$

We know that this is the unique solution of the Lyapunov equation

$$X - AXA' = mm' \quad .$$

The cross-product term in (*) is also expressible, using this matrix X, as

$$\Sigma_{l=0}^{\infty}\lambda_{l-k+1}\lambda_{l+k+1} = c'A^{-k}XA^{k\prime}c \quad .$$

Collecting these we have a convenient expression for the covariance of $\hat{\lambda}_k$ as

$$lim\, T\, var\, (\hat{\lambda}_k) = L \ ,$$

where

$$L = c'(2A^{k-1}mm'(A')^{k-1} + A^kXA^{k\prime} + A^{-k}X(A')^{-k} + 2A^{-k}X(A')^k)c \quad .$$

11. Integrated Time Series

Earlier in Chapter 3 we touched on random trends and described two aggregation schemes to decompose series into trends and other, weakly stationary, components. Series with random walk components are called integrated (of order one) because they are the sums (integrals) of weakly-stationary components. When some linear combinations of components of an integrated vector-valued series become weakly stationary rather than being integrated, we say that these components are cointegrated. This notion is proposed by Granger--see Granger [1981], Granger and Weiss [1983], and Engle and Granger [1987], for example. In state space modeling of cointegrated series, some components of time series at different time points may enter as state variables. We therefore interpret the notion of cointegration broadly and allow for the possibility of linear combination of not-necessarily-contemporaneous components of time series being weakly stationary, as in Aoki [1990], for example. We devote this chapter to modeling of integrated time series and related series such as cointegrated or nearly integrated series.

Integrated series are characterized by the presence of ''unit roots,'' i.e., the dynamic matrices of their state space models have eigenvalues with unit magnitude. Such series require processing in addition to that for weakly-stationary series, such as those described in Chapter 9. For example, the autocovariance matrices used in Chapter 9 in estimating model matrices become unbounded as the sample size becomes unbounded. Often this additional processing takes the form of decomposing series into an integrated part and a weakly stationary part. In the absence of some additional theory or assumption on the data-generating processes, such a decomposition is inherently arbitrary and not unique. Two decompositons, one well-known and the other based on a state-space approach, are described in Sections 1 and 2. In the literature, the dimension of the data vector, p, and the number of unit roots are sometimes taken to be the same. In the state space representation of this book, the number of unit roots may be smaller then p, and the effects of this vector-valued random walk are distributed among the p components of data through the matrix C, which acts as a disaggregation matrix. One of the two aggregation schemes described in Section 3.3 is used to extend the basic algorithm described in Chapter 9 to integrated series, called a two-step procedure in Section 5. We also describe how to find such linear combinations of cointegrated series. The error bounds we produced in Section 10.3 are not applicable to integrated series. Error analysis is redone by taking explicit accounts of these unit roots in Section 7. When some eigenvalues are less than but close to one in magnitude, the series may be more profitably analyzed as a slight departure from integrated series rather than as a weakly-stationary one. Phillips [1987], Chan and Wei [1988] , and Chan [1988] have developed analytical tools for such approximations, some of which are described in Section 10.

Instead of the usual model for a random walk

$$y_t = y_{t-1} + u_t \,, \tag{1}$$

we specify for integrated series a model of the form

$$\tau_{t+1} = \tau_t + Bw_t \tag{2}$$

$$y_t = C\tau_t + w_t \,,$$

where w_t is weakly stationary and has a rational spectral density. This assumption allows us to use the algorithm discussed in Chapter 9 to model $\{w_t\}$ immediately without any further development. Similarly for a vector valued y_t. The dimension of the state vector is discussed in Section 5.

The representation (2) can be turned into (1) by substituting τ_t out. Since

$$y_t = C(\tau_{t-1} + Bw_{t-1}) + w_t = y_{t-1} + (CB - I_p)w_{t-1} + w_t \,,$$

we see that the noises are related by

$$u_t = w_t + (CB - I_p)w_{t-1} \,.$$

Therefore, if $\|CB - I_p\| < 1$, then the above can be inverted to express w_t in terms of u's. The representation (1) has p unit roots at 1. We later show that the state space representation can accommodate a smaller number of unit roots as well as complex roots on the unit circle or the root at -1. These roots may arise from seasonal components as described in Section 6.

11.1 The Beveridge and Nelson Decomposition

Suppose that the first difference of a time series is stationary, and we express it as a sum of two components, $\Delta y_t = \Delta y_{1t} + \Delta y_{0t}$. We call such a decomposition orthogonal if the spectral density of Δy_t is the sum of two spectral densities of the component series, namely the cross-spectral density of the two component series is identically zero, or the cross covariances of the two component series are zero at all leads and lags. If y_{0t} is weakly stationary, then the spectral density of its first difference is zero at zero frequency. This follows from the spectral density expression, given in (9) and (10) of Chapter 3, or (5) of Appendix 10. The transfer function for this difference is $(1 - q^{-1})$ times the transfer function for y_{0t}, where $q = e^{i\omega}$, which is one at $\omega = 0$.

A well-known decomposition of time series into permanent (random walk) and transitory (weakly stationary) components is proposed by Beveridge and Nelson [1981] and has been adopted by Stock and Watson [1988] and others. To convey their proposed scheme simply, consider a univariate series $\{y_t\}$, and assume that the first difference $y_t - y_{t-1}$ is weakly stationary, and has a representation as a Wold decomposition

$$y_t - y_{t-1} = \mu + A(L)u_t \ , \tag{3}$$

where μ is a drift term, and

$$A(L) = 1 + a_1 L + a_2 L^2 + \ldots$$

such that $\Sigma a_i^2 < \infty$, and u_t is mean zero, i.i.d. process. Note that $var\ (y_t - y_{t-1})$ is given by $(\Sigma_0^\infty a_i^2)\sigma^2$, where $a_0 = 1$, is finite by assumption. Using L as the lag operator we rewrite (3) as

$$(1-L)y_t = \mu + A(1)u_t + [\ A\ (L) - A(1)\ ]\ u_t,$$

and decompose y_t into $y_{1t} + y_{2t}$ where

$$(1-L)y_{1t} = \mu + A(1)\ u_t, \tag{4}$$

and

$$(1-L)y_{2t} = [\ A(L) - A(1)\ ]\ u_t \ . \tag{5}$$

The component (4) is called a random walk with drift, if $\mu \neq 0$. Its variance is $A(1)^2\sigma^2$. Recall our discussion of the method of Gilbert in Chapter 4. The residue matrix of the unit root of the partial fraction expansion of (3) is precisely equal to A(1).

Suppose that the information set $I_t = \{u_t, u_{t-1} \ldots\}$ is available. Then, (4) implies that, aside from μ, $A(1)u_t$ is the one-step ahead forecast error of y_{1t}, using the information set I_t. Rewrite (5) as

$$y_{2t} = \frac{A(L) - A(1)}{1-L} = A^*(L)\ u_t \ , \tag{6}$$

where $A^*(L)$ is defined by this equality, and has no singularity at $L = 1$, if the series has a single unit root, and $A^*(1) \neq 0$. When the process y_t has a rational spectral density function, we can calculate $A^*(L)$ explicitly in terms of its state space model as we do shortly later. Expand $A^*(L)$ into an infinite series as $h_0 + h_1 L + \cdots$. Then in the time domain (6) is the same as

$$y_{2t} = h_0 u_t + h_1 u_{t-1} + \cdots \ ,$$

and hence

$$y_{2t} = y_{2t|t-1} + h_0 u_t .$$

Together, (6) and (7) yield

$$y_t = y_{t|t-1} + u_t \ ,$$

namely, the innovation is the same as u_t, since $A(1) + h_0$ sums to one. Here the subscript on y means the conditional expectation with respect to the information set I_{t-1}.

One-step-ahead forecast error of y_{1t} is $A(1)u_t$ and one of y_{2t} is $h_0 u_t$. In words, the innovation u_t is allocated to y_{1t} and y_{2t} as $A(1)u_t$ and $h_0 u_t$, so that the conditional variances are

$$var\,(y_{1t} \mid \mathcal{F}_{t-1}) = A(1)^2\sigma^2$$

and

$$var\,(y_{2t} \mid \mathcal{F}_{t-1}) = h_0^2\sigma^2$$

and hence the ratio is given by

$$\frac{var\,(y_{1t} \mid \mathcal{F}_{t-1})}{var\,(y_{1t} \mid \mathcal{F}_{t-1}) + var\,(y_{2t} \mid \mathcal{F}_{t-1})} = \frac{A(1)^2}{A(1)^2 + h_0^2}.$$

As pointed out by Quah [1988], among others, orthogonal decompositions of time series are not unique. To see this simply, let $\Theta(L)$ be a transfer function of a causal dynamics, i.e., non-anticipatory filter, normalized to be one at frequency zero. Then for any time series such as (3), we can decompose it using $\mu + \Theta(l)A(1)u_t$ as the first term and $([1 - \Theta(L)]A(1) + A(L) - A(1))u_t$ as the second term. The second term is the first difference of a weakly-stationary series because its spectral density vanishes at zero frequency. Since the spectral density is non-negative, we must have the spectral density of the first part not greater than that of the first difference of the original series. This condition is stated as $\Theta(\omega)$ not to exceed the spectral density of the first differenced series normalized by its value at zero frequency. Subject to this condition, any Θ will do in the decomposition.

In multivariate series, the Beveridge-Nelson modeling of integrated series would posit $\Delta y_t = \mu + A(L)e_t$ where $A(L)$ is now a $(p \times p)$ matrix. This formulation implies the existence of p unit roots at 1. Stock and Watson [1988] use a common trend formulation whereby they allow for the number of unit roots to be less than that of the component series in data. They use

$$y_t = C\tau_t + v_t$$

$$\tau_t = \tau_{t-1} + [\,I_k,\ 0\,]\,u_\tau$$

and $v_t = F(L)u_t$, where C is $p \times k$, where $k \le p$, and u_t consists of two subvectors of dimension k and $p - k$, respectively, u_t^τ, and u_t^c . This formulation specifies the transfer function between u_t and y_t as

$$\hat{y} = \left[\frac{C}{1 - L}(I_k,\ 0) + F(L) \right] u_t .$$

In Sections 4 and 5 we propose a two-step modeling procedure in which we use a different decomposition of the series into a common trend part defined by

$$y_t = C\tau_t + w_t ,$$

where C is $p \times k, p < k$, and a weakly-stationary part which is governed, aside from a possible non-zero mean, by

$$\tau_t = \tau_{t-1} + Bw_{t-1},$$

$$w_t = Hz_t + e_t$$

and the dynamics of the state vector for the weakly stationary part is given by

$$z_t = Fz_{t-1} + Ge_{t-1}.$$

In this representation a k-dimensional random walk series is shared by the p components of the series in which the matrix C acts as a disaggregator of random trends.

11.2 State Space Decomposition

As briefly indicated at the end of the previous section state space modeling provides a way of constructing models for integrated processes such that models have unit roots fewer in number than the dimension of the data series. Suppose that the state space model is built in two stages. This construction is more fully discussed in Section 5. For simplicity assume that the number of unit roots is one. Then, the first step estimates a model for "random trend" part of the series by

$$\tau_{t+1} = \rho\tau_t + bw_t$$
$$y_t = c\,\tau_t + w_t \quad ,$$

where w_t is a weakly-stationary residual series with sample mean $\bar{w}$, and the parameter ρ is some number not greater than one. The second step estimates the residuals by

$$z_{t+1} = Fz_t + ge_t$$
$$w_t - \bar{w} = h'z_t + e_t \quad ,$$

where e_t is the residual series of the second step, which is serially uncorrelated , and any of the eigenvalues of F is smaller in magnitude than ρ. Then the transfer function from the innovation sequence to the data series is given, using q as the lead operator; and, taking the transfer functions of these two sets of equations as

$$y_t = G(q)\, e_t ,$$

it can be written as a product of the two transfer functions

$$G(q) = G_1(q)\, G_2(q)\,,$$

where

$$G_1(q) = 1 + \frac{bc}{q - \rho}$$

and

$$G_2(q) = 1 + h'(qI - F)^{-1} g.$$

It is the same as the transfer function for the joint vector composed of τ_t and z_t,

$$G(q) = 1 + (c \quad h') \begin{bmatrix} q - \rho & -bh' \\ 0 & qI - F \end{bmatrix}^{-1} \begin{bmatrix} b \\ g \end{bmatrix}.$$

We can discuss several decompositions of the transfer function $G(q)$. For example, define

$$y_{1t} = \frac{bc}{q - \rho} G_2(q)\, e_t = \left[\frac{bc}{q - \rho} G_2(\rho) + bcG_2^*(q) \right] e_t\,,$$

where

$$G_2^*(q) = [\, G_2(q) - G_2(\rho)\,] \,/\, (q - \rho)$$

and

$$y_{2t} = G_2(q)\, e_t\,.$$

Another is to define

$$y_{1t} = \frac{bcG_2(\rho)}{q - \rho} e_t\,,$$

and

$$y_{2t} = [\, G_2(q) + bcG_2^*(q)\,]\, e_t\,.$$

Suppose $\rho = 1$. Then the first decomposition yields

$$(1 - q)\, y_{1t} = qbcG_2(q)\, e_t\,,$$

or

$$y_{1t} - y_{1t-1} = bc\, [\, e_{t-1} + h'ge_{t-2} + h'Fge_{t-3} + \cdots\,]\,,$$

and

$$y_{2t} = e_t + h'ge_{t-1} + h'Fge_{t-2} + \cdots\,.$$

Note that $y_{1t} = y_{1t|t-1}$, i.e., y_{1t} is a predictable process, i.e., known at time point

$t-1$, and $y_{2t} - y_{2t|t-1} = e_t$. Alternatively, in the decomposition $y_{1t} = c\tau_t + h'z_t$, and $y_{2t} = e_t$, y_{1t} is a predictable process and y_{2t} is a martingale difference. This procedure easily generalizes to a vector-valued data process. They are given in Section 5.

11.3 Contents of Random Walk Components

This section proposes a measure of random walk components in time series, based in part on Aoki [1987]. Consider, as an example, a univariate series with a random walk component

$$y_t - y_{t-1} = e_t + a_1 e_{t-1},$$

where e_t is i.i.d., with mean zero and variance σ^2. Following the preceding section, decompose this into $y_t = y_{1t} + y_{2t}$ with

$$y_{1t} - y_{1t-1} = (1 + a_1)e_t$$

and

$$y_{2t} = - a_1 e_t.$$

Clearly y_{2t} is weakly stationary and y_{1t} is a random walk, since $y_{1t} = y_{10} + (1+a_1)\Sigma_{s=1}^{t} e_s$. From $y_{1t} - y_{1,t-k} = (1+a_1)\Sigma_{s=t-k+1}^{t} e_s$, we see that $var\,(y_{1t} - y_{1,t-k})^2 / k = (1+a_1)^2\sigma^2$, which is independent of k. The second component has $var\, y_{2t} = a_1^2\sigma^2$ and $var\,(y_{2t} - y_{2t-k})/k = 2a_1^2\sigma^2/k \to 0$ as $k \to \infty$. The ratio $var\,(y_t - y_{t-k}) / k\, var\,(y_t - y_{t-1})$, proposed by Cochrane [1987] as a measure of random walk component, has the limiting value $(1+a_1)^2 / (1 + a_1^2)$ as $k \to \infty$. This ratio is greater than one if $a_1 > 0$, i.e., if $y_{1t} - y_{1t-1}$ and y_{2t} are negatively correlated. It seems undesirable to have a measure which indicates that the series contains a random walk content greater than 100%. A measure of the random walk component should be such that it reaches a max of one when y_t is all random walk, i.e., when no weakly-stationary component is present in y_t and it is less than one otherwise. For example, the ratio of $var\,\Delta y_{1t} / (var\,\Delta y_{1t} + var\, y_{2t})$, or the ratio of the conditional variances similar to the one in Section 1, satisfies this criterion.

Next consider a model

$$y_t = y_{t-1} + u_t,$$

where u_t is mean zero and weakly stationary with a rational spectral density. This last assumption means that there is a finite-dimensional state space model for u_t of the form

$$z_{t+1} = Fz_t + gv_t$$

and

$$u_t = h'z_t + v_t ,$$

where v_t is serially uncorrelated with variance σ_v^2. We have discussed in Section 3.2 that

$$W(q^{-1}) = 1 + h'(q^{-1}I - F)^{-1}g \tag{7}$$

is a factor in the spectral density for u_t. In Section 1 of this Chapter, we discussed that u_t is often modeled in the literature as

$$u_t = A(L)v_t ,$$

where some technical condition is imposed on the coefficients of the lag polynomial to ensure its weak stationarity, such as

$$A(L) = 1 + a_1L + a_2L^2 + \cdots, \quad \Sigma a_j^2 < \infty .$$

Instead of specifying the infinite number of coefficients in the lag polynomial individually, our state space model represents $A(L)$ as $W(L^{-1})$. The coefficients are represented in terms of a small number of model parameters by

$$a_j = h'F^{j-1}g, \quad j = 1, 2, \cdots .$$

Because the matrix F is asymptotically stable, the sum of squares of the coefficients is bounded as assumed, and $\Sigma_1^\infty a_j^2 = h'\Theta h$ where Θ is the unique solution of a Lyapunov equation, $\Theta = F\Theta F' + gg'$.

We can decompose y_t into a random walk and a weakly-stationary component by the partial fraction method (as mentioned earlier) by separating out the residue term at $L = 1$,

$$y_t = \frac{A(L)}{1-L} u_t$$

$$= \left\{ \frac{A(1)}{1-L} + A^*(L) \right\} u_t ,$$

where $A^*(L) = [A(L) - A(1)] / (1-L)$ contains only integer power in L. Since $A(L) = W(L^{-1})$, we can use (7) to evaluate $A(1)$ and $A^*(L)$ in closed form as

$$A(1) = 1 + h'(I - F)^{-1}g$$

and

$$A^*(L) = -h'(I - LF)^{-1}F(I - F)^{-1}g.$$

We have now decomposed y_t into

$$y_{1t} - y_{1t-1} = A(1)u_t$$

and

$$y_{2t} = -h'(I - LF)^{-1}F(I - F)^{-1}gu_t.$$

The first component has $var\ \Delta y_t = A(1)^2\sigma_e^2$. The variance of Δy_{2t} can be calculated in the transform domain by the Parseval relation, but here we show a time domain method. Use a representation for y_{2t} given by

$$s_t = Fs_{t-1} - \psi e_t$$

and

$$y_{2t} = h's_t ,$$

where

$$\psi = F(I - F)^{-1}g.$$

Then

$$y_{2t} - y_{2t-1} = h'(s_t - s_{t-1})$$

$$= h'[(F - I)s_{t-1} - \psi e_t].$$

Because s_{t-1} and e_t are uncorrelated,

$$var\ \Delta y_{2t} = h'[(F - I)S(F - I)' + \psi\psi'\sigma_e^2],$$

where

$$S = cov\ s_{t-1} ,$$

which is a constant since s_t is weakly stationary. From the dynamic equation for this state vector S must satisfy a Lyapunov equation given by

$$S = FSF' + \psi\psi'\sigma_e^2,$$

or

$$vec\ S = (I - F \otimes F)^{-1}(vec\ \psi\psi'),\ \sigma_e^2$$

and we can evaluate the first term in the variance as

$$h'(F-I)S(F-I)'h = [\, h'(F-I) \otimes h'(F-I)\,]\ vec\, S.$$

The proposed measure of the random walk component is

$$measure\ of\ the\ random\ walk\ component = 1 / (1+\theta)\,, \tag{8}$$

where $\theta = var\ \Delta y_{1t} \,/\, var\ \Delta y_{2t}$.

In the next section we propose a model

$$\tau_{t+1} = \tau_t + BHz_t + Be_t$$

$$z_{t+1} = Fz_t + Ge_t$$

and

$$y_t = C\tau_t + Hz_t + e_t$$

for a p-dimensional data series with random walk component where B is $1 \times p$, C is $p \times 1$, F is $n \times n$, H is $p \times n$, and G is $n \times p$ for some n. The p-dimensional series e_t is mean-zero serially-weakly stationary, and uncorrelated. The transfer function representation of this model is $y_t = \Theta(q)e_t$, where

$$\Theta(q) = I + (C,\ H)\begin{bmatrix} q-1 & -BH \\ 0 & qI-F \end{bmatrix}^{-1}\begin{bmatrix} B \\ G \end{bmatrix}$$

$$= \left[I + \frac{CB}{q-1} \right] [\, I + H(qI-F)^{-1}G\,]\,.$$

Thus

$$y_t - y_{t-1} = W(q)e_t\,,$$

where

$$W(q) = [\,-(q-1)I + CB\,]\,[\,I + H(qI-F)^{-1}G\,]\,.$$

We can then proceed, as shown previously, to isolate the random walk component, and calculate the measure of its content.

11.4 Cointegration, Error Correction, and Dynamic Aggregation

We recognize that the decompositions discussed in Sections 2 and 3 are nothing but special cases of decomposition by dynamic modes described in Chapter 3.

We thus generalize slightly the decompositions we have discussed so far, and consider separating out or isolating slowly-decaying components and the random walk component, if present, from the data series by using the notion of dynamic aggregation proposed by Aoki [1968], in order to point out that the two aggregation schemes disucssed in Chapter 3 are the natural framework to discuss the notion of cointegration or a common factor model representation. The common factor representation then suggests the two-step algorithm described in the next section. See Aoki [1989] also.

Such a separation of a subset of dynamic modes is often a useful modeling strategy; fast decaying, i.e., transitory dynamic modes and slower ones may be modeled differently, for example, by different algorithms, or the time horizon of interest may be different for the two subsets of modes. Such a decomposition, if possible, may also help us to understand complex phenomena and simplify their approximate representations. When one suspects that qualitatively different mechanisms are responsible for the two types of dynamic responses, then such a decomposition is definitely useful or may provide a useful approximate description of dynamic behavior. As discussed in Chapter 3, such decomposition is routinely observed in the time series literature. For example, macroeconomic time series such as real GNP and CPI are often decomposed into (random) trend components and possibly cyclical or irregular variations about the trends. Here one is implicitly using the frequency components of data series to separate lower-frequency and higher-frequency components, i.e., slower-changing components are called trends and faster-moving components are called cycles or irregular components (relative to the time span of the data set, of course). To be concrete, suppose that the dynamics are linear and not explosive with the dynamic matrix having n eigenvalues, $1 > |\lambda_1| \geq |\lambda_2| \geq \ldots \geq |\lambda_n|$. The eigenvalues are then grouped into two classes, the first containing all eigenvalues with magnitude ρ_c or larger, and the second the rest.

This dichotomy is arbitrary if the critical number ρ_c used to separate eigenvalues has no theoretical basis. When $\rho_c = 1$ is used, then the dichotomy corresponds to the decomposition of series into a random walk trend component plus a weakly-stationary (transitory) component. This decomposition has attracted much attention in the macroeconomic literature, because a dynamic mode with eigenvalue of magnitude one has indefinite memory, i.e., its impulse responses are of the same magnitude and do not decay to zero. In theory, the critical value $\rho_c = 1$ sharply separates unstable and random walk components from the rest, which are transient. In practice, eigenvalues of magnitude close to one may not be distinguishable statistically from those with unit magnitude. These series, called "nearly integrated", are discussed later in this chapter. Especially for economic data with small sample sizes, models constructed using the dichotomy with ρ_c less than 1 and those with ρ_c equal to one may not be statistically distinguishable. For this practical reason, ρ_c may be chosen close to but not equal to one in classifying dynamic modes into trends and the rest.

Suppose a model for the time series is given in the innovation state space form

$$\begin{aligned} z_{t+1} &= Az_t + Be_t \\ y_t &= Cz_t + e_t \ . \end{aligned} \tag{9}$$

Define the matrix P and Q as in Chapter 3. Introduce the new coordinate system by

$$z_t = P\tau_t + Qz_t, \tag{10}$$

where $\tau_t = P'z_t$ is a state vector for the slow components or the trend dynamics which is an aggregation of the whole dynamic system and which retains only the aspect of slow dynamic movements of the original system. Similarly, the vector z_t is an aggregated state vector for the transitory dynamics. We have shown in Chapter 3 that these aggregated vectors are governed by dynamics with a block triangular transition matrix (3.12). In the notation we just introduced they are

$$\tau_{t+1} = \Lambda\tau_t + P'AQz_t + P'Be_t, \tag{11}$$

and

$$\begin{aligned} \eta_{t+1} &= Q'AQz_t + Q'Be_t \\ &= Fz_t + Q'Be_t. \end{aligned} \tag{12}$$

The data vector is decomposed into

$$y_t = CP\tau_t + CQz_t + e_t.$$

In the extreme case of $\rho_c = 1$, the matrix Λ contains only random walk components, e.g., I_k. When the dimension of τ_t, k, is less than that of y_t, then the vector τ_t serves as common factors for the trend components of y_t. Any vector α such that $\alpha'CP = 0$ annihilates the random walk components in y_t, because $\alpha'y_t$ only contains shorter-run modes, i.e., $\alpha'y_t$ is weakly stationary, even though y_t contains nonstationary components $CP\tau_t$. Granger calls $\alpha'y_t$ cointegrated and α a cointegrating factor, Granger [1983]. If some components of the data vector are not cointegrated, then those series do not share the common trend dynamics. Consequently, the corresponding elements in the matrix C are all zero.

The short-run dynamics (12) can be thought of as the dynamic version of the aggregation in the sense used in Aoki [1968]. In the economic literature, the notion of aggregation is introduced to sum (or aggregate) microeconomic relations to produce a macroeconomic relation. Here the n-dimensional dynamic relations (9) are aggregated to produce a lower-dimensional model (12) which approximates the dynamic behavior of (9) in the short-run. The vector $z_t = Q'x_t$ is the aggregation relation because $z_{t+1} = Q'x_{t+1} = Q'(Ax_t + bu_t) = FQ'x_t + Q'Bu_t = Fz_t + Q'Bu_t$ where (10) is used. This is exactly (12b). Substituting for τ the expression $P'z_t$ obtained from (10),

$$y_t = CQz_t + CPP'x_t + v_t$$

$$\approx CQz_t + v_t \ .$$

Namely, $CPP'x_t$ is the error in disaggregating z_t into y_t. The combination $\alpha' y_t$, where α is the null space of $P'C'$, i.e., a cointegrating factor, can be disaggregated exactly except for the noise since

$$\alpha' y_t = \alpha' CQz_t + \alpha' v_t$$

exactly. See Johansen [1988] and Aoki [1989, p.178] also.

Although (12) of Chapter 3, i.e., the common trend representation described above, is an intuitively-pleasing representation in economic applications because it shows how short-run disturbances 'feed' into longer-run phenomena, sometimes one may wish to focus on the effects of equilibrium (or longer-run) relations on short-run dynamics. This is the other aggregation scheme discussed in Chapter 3, that of the error-correction model, given in (13) there. Since τ_t "aggregates" out the shorter-run variables, the relation $\tau_t = 0$ represents a long-run equilibrium. This representation is so called because non-zero values of longer-run, i.e., slower-moving variables τ_t (i.e., error signal indicating deviation from the long-run equilibrium) "feed" into this shorter-run, transitory, or quickly-decaying dynamic relation. The second term in the shorter-run dynamics represents how this deviation from the longer-run (equilibrium) relation $\tau_t = 0$ affects ζ_t. In the literature this is usually put as

$$\Delta\zeta_{t+1} = \zeta_{t+1} - \zeta_t$$

$$= (F - I)\zeta_t + R'AS\,\zeta_t + R'Bu_t . \tag{14}$$

11.5 Two-Step Modeling Procedure

The developments of the previous section suggest the modeling procedure of this section for apparently "nonstationary" or "trending" data series. It consists of the two parts described below.

First Step

Choose K to be 1 or 2 and choose n to be small, and apply the algorithm of Chapter 9. (The dimension of the state vector is constrained by the algorithm to be $n \leq Kp$, but typically one or two should be tried first.) The algorithm will build a model

$$\tau_{t+1} = \rho\tau_t + bw_t \text{ and } y_t = C\tau_t + w_t , \tag{15}$$

where τ_t is a common random trend term and w_t is a usually highly-correlated but weakly-stationary residual. Assume for the moment that w_t is mean zero. The number $n = dim\ \tau_t$ should be equal to the number of common random walks, i.e., the dimension of the random walk vector. If your choice of n, denoted by n^*, is less than n, then w_t in (15) will not be weakly stationary. Visual inspection of the residuals w_t would immediately alert us to this fact. If $n^* > n$, no harm is done. We later show that the extra components in τ_t implied by this choice automatically incorporate parts of the dynamic in the residuals to be modeled in the second step. Suppose that y_t is univariate and we try to model it with $K = 1$and$n^* = 1$. Then (3a) of Chapter 9 yields $\hat{\rho} = \hat{\lambda}_2 / \hat{\lambda}_1$, while the least-squares estimate of ρ in $y_t = \rho y_{t-1} + u_t$ for some u_t is given by $\hat{\rho}_{LS} = \hat{\lambda}_1 / \hat{\lambda}_0$. We discuss their large sample properties later in this chapter. When a balanced model is constructed for a non-stationary series, the derivations in Sec. 11.7 tell us that the estimate $\hat{\rho}$ is $O_p(1)$, but the estimate $\hat{C}$ is $O_p(T^{1/2})$ even though the product $\hat{C}\hat{b}$ is $O_p(1)$. If one wishes to avoid this, then we may either use the estimate $\hat{\rho}$ in the first step and apply the second step to the series $\{y_t - \hat{\rho}y_{t-1}\}$, or use a non-balanced representation by definining the trend variable τ_t by $\Sigma V'R_t^{-1}y_{t-1}^-$, and estiamte matrix C using it as instrument by $(\sum y_t\tau_t')\,(\sum\tau_t\tau_t')^{-1}$. See Phillips and Hansen [1990] for a related approach and disucssions of asypmptotic distributions.

Second Step

Having produced a weakly-stationary series as the residuals of a modeling exercise in the first step, the algorithm of Chapter 9 applied to w_t estimates a model for w_t as

$$z_{t+1} = Fz_t + Ge_t \text{ and } w_t = Hz_t + e_t , \tag{16}$$

where e_t should now be serially uncorrelated, i.e, martingale differences.

Jointly written, y_t is modeled by

$$\begin{bmatrix} \tau_{t+1} \\ z_{t+1} \end{bmatrix} = \begin{bmatrix} \rho & bH \\ 0 & F \end{bmatrix} \begin{bmatrix} \tau_t \\ z_t \end{bmatrix} + \begin{bmatrix} b \\ G \end{bmatrix} e_t , \tag{17}$$

and the data series is decomposed into

$$y_t = C\tau_t + Hz_t + e_t . \tag{18}$$

As explained earlier, τ_t is the common trend term. In some series w_t is not mean zero. Then the second step is applied to $w_t - \bar{w}$ where $\bar{w}$ is the sample mean of w_t to produce

$$y_t = C\tau_t + Hz_t + e_t + \bar{w} , \tag{19}$$

with

$$\begin{bmatrix} \tau_{t+1} \\ z_{t+1} \end{bmatrix} = \begin{bmatrix} \rho & bH \\ 0 & F \end{bmatrix} \begin{bmatrix} \tau_t \\ z_t \end{bmatrix} + \begin{bmatrix} b \\ G \end{bmatrix} e_t + \begin{bmatrix} b \\ 0 \end{bmatrix} \bar{w} .$$

See Section 8 for further discussion of the drift terms. The eigenvalues of the matrix F are less than those of ρ. If ρ is equal to one (or the identity matrix), then taking the first difference of (19), we see that Δy_t is governed by the same equation as w_t in (16) if Cb is equal to one (or the identity matrix mentioned above).

Eq. (17) and (18) imply that we decompose the transfer function from e_t to y_t into the product of "long-run" and short-run transfer functions

$$\begin{aligned} y_t &= [I + C(qI - \rho)^{-1}b]\, w_t \\ &= [I + C(qI - \rho)^{-1}b]\,[I + H(qI - F)^{-1}G]\, e_t . \end{aligned}$$

Thus, the multipliers are also composed of the product of long-run and short-run ones.

Suppose there is only one unit root and let $A(L) = 1 + H(L^{-1}I - F)^{-1}G$. Then the transfer function is rewritten as

$$A(L) + \frac{CbA(L)L}{1-L} = \frac{CbA(1)L}{1-L} + A(L) + CbA^*(L) ,$$

where the last two terms represent weakly stationary components. In this form a close resemblance with the Stock-Watson representation is clear.

This is also clear from the expression of the multiplier for the joint model as the sum of the longer-run multiplier, the shorter-run one, and their cross effects

$$(C \;\; H)\begin{bmatrix} \rho & bH \\ 0 & F \end{bmatrix}^k \begin{bmatrix} b \\ G \end{bmatrix} = C\rho^k b + C\chi_k G + HF^k G \tag{20}$$

where

$$\begin{bmatrix} \rho & bH \\ 0 & F \end{bmatrix}^k = \begin{bmatrix} \rho^k & \chi_k \\ 0 & F^k \end{bmatrix}$$

with

$$\chi_1 = bH$$

$$\chi_{k+1} = \rho^k bH + \chi_k F \quad , \quad k = 1, 2, \cdots$$

The first term in (20) is the multiplier of the model (17) and the last term is that of (18), and the middle term represents the interaction, i.e., the spill-over of short-run disturbances into the trend.

11.6 Dynamic Structure of Seasonal Components

Time series containing seasonal components are often modeled in the literature by

$$(1 - \phi L^s)y_t = e_t \tag{21}$$

with $0 < \phi \le 1$, and s denoting the periods in a year. Here we illustrate the point on a quartered model $s = 4$.

This type of model for seasonal components has been discussed in books by Nerlove, Grether and Carvalho [1979], Harvey [1981], Gourieroux and Monfort [1983], and others, and in journal articles by Cleveland and Tiao [1976], Pierce [1978], Hausman and Watson [1985] and others. Maravall [1987], criticizing this type of model, proposes an alternative approach. An approach similar to Maravall is also found in Aoki [1988]. We follow Maravall and Aoki in this section.

First, consider the model with $\phi = 1$. The transfer function $(1 - L^4)^{-1}$ has peaks at $\omega = 0$, $T/2$, and π. Factor $1 - L^4$ into $(1 - L)\,(1 + L)\,(1 + L^2)$. The factor $(1 - L)$ induces the peak at $\omega = 0$; similarly the factor $(1 + L)$ induces the peak as $\omega = \pi$, and the factor $1 + L^2$ produces the peak at $\pi/2$.

To account for these factors, we decompose y_t into y_{it}, $i = 1, 2$, and 3 corresponding to these individual spectral peaks by

$$y_t = y_{1t} + y_{2t} + y_{3t} + u_t \tag{22}$$

with

$$(1-L)y_{1t} = \theta_1(L)v_{1t},$$

$$(1+L)y_{2t} = \theta_2(L)v_{2t},$$

and

$$(1+L^2)y_{3t} = \theta_3(L)v_{3t},$$

where u_t and v_{it}, $i = 1 \sim 3$ are uncorrelated. Substituting these into (21), with the ϕ parameter set at 1, we obtain an identity

$$\varepsilon_t = (1+L)(1+L^2)\,\theta_1(L)v_{1t} + (1-L)(1+L^2)\,\theta_2(L)v_{2t}$$
$$+ (1-L^2)\,\theta_3(L)v_{3t} + (1-L^4)u_t\ . \tag{23}$$

Since ε_t is a white noise, the lag structures in the right hand side should cancel each other out, i.e.,

$$\theta_1(L) = 1 - \alpha L\ ,$$

$$\theta_2(L) = 1 - \beta L\ ,$$

and

$$\theta_3(L) = 1 - \gamma L - \delta L^2\ .$$

There are four unknown parameters α, β, γ, δ and four unknown variances, σ_u^2, σ_i^2 for the variances of v_{it}, $i = 1 \sim 3$. We have only a system of five equations in (23) to constrain these eight parameters. To achieve the just identified condition, Maravall imposes $\alpha = -1$, $\beta = 1$, $\gamma = 0$, and $\delta = 1$, in (22), i.e.,

$$(1-L)y_{1t} = (1+L)v_{1t}\ ,$$

$$(1+L)y_{2t} = (1-L)v_{2t}\ ,$$

and

$$(1+L^2)y_{3t} = (1+L^2)v_{3t}\ .$$

The motivation for this type of identification constraint is the following: In the decomposition (22), the white noise component in y_t is already captured by u_t and hence y_{it} should not contain further white noise components, i.e., it is not possible to decompose y_{it} further into $y_{it}^* + \eta_{it}$ for some η_{it} and a white noise term η_{it}, $i = 1 \sim 3$. A consequence of this is that the magnitude of the transfer function $\theta_i(L)/(1-L)$ should be zero at some frequency, such as at $\omega = \pi$ for $\theta_1(L)$. The remaining parameter β, γ, and z are similarly specified. Then σ_i^2, $i = 1 \sim 3$ are solved as a function of σ_ε^2.

The total seasonal components are represented by $y_{2t} + y_{3t} = s_t$ for which the transfer function vanishes at $\omega = 1$, i.e.,

$$(1+L)(1+L^2)s_t = (1+L^2)(1-L)v_{2t} + (1+L)(1-L^2)v_{3t}$$

$$= (1-L)(1-\theta_1 L - \theta_2 L^2)n_t ,$$

where n_t is a white noise sequence. The parameters are determined by equating the expressions for auto-covariances on these two alternative expressions. In the Maravall approach the random walk component is driven by a white noise process which is uncorrelated with that which drives the seasonal components.

We now turn to a state space approach to modeling seasonal components which utilizes an approach similar to that of Maravall. Kitagawa and Gersch [1984] adopt slightly different state space formulations. They posit a relation $s_t + s_{t-1} + s_{t-2} = \eta_t$ in their quarterly adjustment model which is governed by

$$\begin{bmatrix} s_t \\ s_{t-1} \\ s_{t-2} \end{bmatrix} = \begin{bmatrix} -1 & -1 & -1 \\ 1 & 0 & 0 \\ 0 & 1 & 0 \end{bmatrix} \begin{bmatrix} s_{t-1} \\ s_{t-2} \\ s_{t-3} \end{bmatrix} + \begin{bmatrix} \eta_t \\ 0 \\ 0 \end{bmatrix} .$$

Note that this dynamic matrix has eigenvalues -1 and $\pm i$.

When transfer functions have poles (roots) on the unit circle, we use the partial fraction expansion (i.e., basically the method of Gilbert described at the end of Chapter 4) to write

$$y_t = [1 + G(L)]E_t$$

with

$$G(L) = \frac{N(L)}{(1-L^4)\,d(L)} ,$$

where $d(1) \neq 0$, and $d(L)$ has poles strictly inside the unit circle as

$$G(L) = \frac{1}{1-L} R_1 + \frac{1}{1+L} R_2 + \frac{R_3 + R_4 L}{1+L^2} + U(L) , \tag{24}$$

where $U(L)$ is due to poles of $d(L)$. The first term in $G(L)$ contains the unit root. The second and the third together represent seasonal components with periods π and $\pi/2$, respectively.

We introduce state variables by

$$\eta_{1t} = (1+L)(1+L^2)y_t = \frac{N(L)}{(1-L)\,d(L)}\ \varepsilon_{1t},$$

$$\eta_{2t} = (1-L)(1+L^2)y_t = \frac{N(L)}{(1+L)\,d(L)}\ \varepsilon_{2t},$$

and

$$\eta_{3t} = (1-L^2)\,y_t = \frac{N(L)}{(1+L^2)d(L)}\ \varepsilon_{3t}.$$

These are modeled by

$$\tau_{t+1} = \tau_t + B_1 w_t\ ,$$

$$\eta_{1t} = C_1 \tau_t + w_t\ ,$$

$$\rho_{t+1} = -\rho_t + B_2 w_t\ ,$$

$$\eta_{2t} = C_2 P_t + w_t\ ,$$

and

$$s_{t+1} = S\ s_t + B_e\ w_t\ ,$$

$$\eta_{3t} = C_3\, s_t + B\ w_t\ ,$$

where

$$z_{t+1} = F\ z_t + G\ e_t\ ,$$

$$w_t = H\ Z_t + e_{t.}$$

These are related to the data series by

$$\begin{bmatrix} \eta_{1t} \\ \eta_{2t} \\ \eta_{3t} \\ \eta_{3t-1} \end{bmatrix} = \begin{bmatrix} 1 & 1 & 1 & 1 \\ 1 & -1 & 1 & -1 \\ 1 & 0 & -1 & 0 \\ 0 & 1 & 0 & -1 \end{bmatrix} \begin{bmatrix} y_t \\ y_{t-1} \\ y_{t-2} \\ y_{t-3} \end{bmatrix}$$

or inverting it

$$y_t = \tfrac{1}{4}\,(\hat{\eta}_{1t} + \hat{\eta}_{2t} + 2\,\hat{\eta}_{3t})$$

$$= e_t$$

$$+\left[\frac{C_1 B_1}{q-1} H\,(I-F)^{-1} G - \frac{C_2 B_2 H_2 (I+F)^{-1} G}{q+1}\right.$$

$$\left. + \frac{R_3 + R_4 q}{q^2+1}\right] e_t + \hat{w}(L) e_t$$

where $\hat{w}\,(L)$ represents a transfer function of a weakly-stationary process. This expression has the general form of the partial fraction expansion (24).

Note that in this approach $e_t/(q-1)$, $e_t/(q+1)$ and $e_t/(q^2+1)$ are only asymptotically uncorrelated with each other.

11.7 Large Sample Properties

We derive large sample properties of the estimates of system parameters obtained in the first step of the two-step modeling algorithm for integrated processes. See Phillips [1987], Phillips and Durlauf [1986], Ahn and Reinsel [1987], and Chan and Wei [1988] for related results on least-squares estimates and Johansen [1988] on statistical analysis of cointegration vectors. Related material is also found in Aoki [1990a].

We begin with a univariate series with a single unit root and estimate its state space model

$$\tau_{t+1} = \rho\tau_t + bw_t\,, \qquad (25)$$

$$y_t = c\,\tau_t + w_t\,,$$

where w_t is some mean-zero weakly-stationary process. The hypothesis is that $\rho = 1$. By eliminating τ_t, this representation is the same as that of

$$y_t = \rho y_{t-1} + u_t\,, \qquad (26)$$

where

$$u_t = (cb - \rho) w_{t-1} + w_t\,,$$

which is more commonly found in the literature. Eq. (25) leads to an expression for y_t

$$y_t = w_t + cbw_{t-1} + cb\,\rho w_{t-2} + \cdots$$

and (26) to that of

$$y_t = u_t + \rho u_{t-1} + \rho^2 u_{t-2} + \cdots$$

Because we can express the two noise processes as $u_t = [\,1 + (cb - \rho)L\,]\,w_t$ and

$w_t = [\,1 + (cb - \rho)L\,]^{-1}u_t$, which is well defined if $|cb - \rho| < 1$, the series $\{u_t\}$ and $\{w_t\}$ are equivalent under this assumption. For example with the US real GNP series, $cb - \rho$ is very close to zero. By assumption, w_t is weakly-stationary so is u_t.

Define

$$S_T = \Sigma_1^T u_t, \quad S_0 = 0,$$

and assume that $Eu_t^2 = \sigma_u^2 < \infty$. We assume also that

$$\sigma^2 = lim_{T \to \infty} E(T^{-1} S_T^2)$$

$$= \sigma_u^2 + 2\Sigma_{k=1}^{\infty} \gamma_k > 0,$$

where

$$\gamma_k = E(u_k\, u_0).$$

Define

$$X_T(r) = S_{j-1} / \sigma\sqrt{T}, \quad (j-1)/T \le r < j/T,$$

and

$$X_T(1) = S_T / \sigma\sqrt{T}.$$

A weakly-stationary series satisfies the "mixing" condition that $P(A_1 \cap A_2) \le \phi P(A_1)P(A_2)$ where A_1 is any set in the σ-field generated by $\{u_t, t \le T_1\}$ and A_2 is any set in the σ-field generated by $\{u_t, t \ge T_1 + 1\}$ and ϕ is a constant independent of T_1. Billingsley [1971 Sec. 8] has shown that under this condition

$$X_T(r) \Rightarrow B(r) \quad \text{as} \quad T \to \infty$$

where $\Rightarrow$ indicates weak convergence and $B(r)$, $0 \le r \le 1$ is a standard Brownian motion, i.e., has mean zero and variance one at $r = 1$. Also see Billingsley [1968, Th.20.1] for an alternative set of conditions under which this convergence holds. Although the map from r to $X_T(r)$ is discontinuous, it is true that

$$h(X_T(r)) \Rightarrow h(B(r)) \quad \text{as} \quad T \to \infty$$

for any continuous function $h(\cdot)$ on $D\,[0, 1]$, the class of functions which are right continuous and with finite left limits. See Billingsley [1968, Chapter 3] or Pollard [1984, p. 89] for further details such as alternative conditions for this type of invariance theorem.

Denote the sample autocovariance by $\hat{\Lambda}_k$,

$$\hat{\Lambda}_k = T^{-1}\Sigma y_{t+k-1}y_{t-1}, \quad k = 0, 1, \ldots,$$

its first difference by ν_k,

$$\nu_k = \hat{\Lambda}_{k+1} - \hat{\Lambda}_k,$$

and the second difference by ω_k,

$$\omega_k = \nu_{k+1} - \nu_k, \quad k = 0, 1, \cdots$$

By direct calculation

$$\begin{aligned} T^{-2}\Sigma_1^T S_t^2 &= \sigma^2 T^{-1}\Sigma_1^T (S_t / \sigma\sqrt{T})^2 \\ &= \sigma^2 \Sigma_i \int_{(i-1)/T}^{i/T} (S_{i-1}^2 / \sigma^2 T)dr \\ &= \sigma^2 \int_0^1 X_T(r)^2 dr , \end{aligned}$$

which weakly converges by the continuity theorem to

$$\hat{\Lambda}_0 / T \Rightarrow \sigma^2 \int_0^1 B(r)^2 dr. \tag{27}$$

Following Phillips [1987], we note for later use that

$$\begin{aligned} \nu_0 &= T^{-1}\Sigma y_t (y_t - y_{t-1}) \\ &= T^{-1}\Sigma y_t u_t . \end{aligned}$$

Express $S_{t-1}u_t$ as $(S_t^2 - S_{t-1}^2 - u_t^2) / 2$ in the above. Then we see that

$$\begin{aligned} \nu_0 &= (2T)^{-1} S_T^2 - (2T)^{-1}\Sigma_1^T u_t^2 \\ &\Rightarrow \frac{\sigma^2}{2} B(1)^2 - \sigma_u^2 / 2 , \end{aligned} \tag{28}$$

because

$$T^{-1}\Sigma_1^T u_t^2 \to \sigma_u^2 = lim\, E(u_t^2 / T).$$

Proceeding analogously we obtain

$$\nu_1 \Rightarrow \frac{\sigma^2}{2} B(1)^2 - \sigma_u^2 / 2 - \gamma_1 , \tag{29}$$

where

$$\gamma_1 = lim\, E(T^{-1}\Sigma u_{t+1}u_t).$$

From (27) and (28) we note that $\hat{\Lambda}_1 / T = (\hat{\Lambda}_0 + \nu_0) / T$ also weakly converges to the same limit as $\hat{\Lambda}_0 / T$, given by (28).

It is convenient to examine the cases of K equal to one and two separately.

K = 1. When $K = 1$, then $H = \hat{\Lambda}_1$. Since $p = 1$, $n^* = 1$ is the only possible value for the dimension, and the estimate of ρ in (25) is, by (3a) of Chapter 9, equal to

$$\hat{\rho} = \hat{\Lambda}_2 / \hat{\Lambda}_1.$$

From (26), the least-squares estimate of ρ is given by

$$\hat{\rho}_{LS} = \hat{\Lambda}_1 / \hat{\Lambda}_0.$$

Both estimates are consistent, and we note that

$$T(\hat{\rho} - 1) = \nu_1 / \hat{\Lambda}_1 / T \Rightarrow \frac{\frac{1}{2} B(1)^2 - \frac{\sigma_u^2}{2\sigma^2} - \frac{\gamma_1}{\sigma^2}}{\int_0^1 B(r)^2 dr}, \tag{30}$$

and

$$T(\hat{\rho}_{LS} - 1) = \nu_0 / \hat{\Lambda}_0 / T \Rightarrow \frac{\frac{1}{2} B(1)^2 - \frac{\sigma_u^2}{2\sigma^2}}{\int_0^1 B(r)^2 dr}. \tag{31}$$

Using τ_t as instrument, we can show that $Tb\delta c$ weakly converges to the same limit as that given in (27) in view of the fact in Section 8. Since the sign of γ_1 is indefinite, no clear inequality exists between (30) and (31). The difference between the least-squares estimate and $\hat{\rho}$ obtained by the stochastic realization estimator converges weakly to

$$T(\hat{\rho}_{LS} - \hat{\rho}) \Rightarrow (\gamma_1 / \sigma^2 \int_0^1 B(r)^2 dr. \tag{32}$$

The estimates $T(\hat{\rho} - 1) / \nu_1$, $T(\hat{\rho}_{LS} - 1) / \nu_0$, $T(\hat{\rho}_{LS} - \hat{\rho}) / \gamma_1$ and $\hat{\Lambda}_k / T$ $k = 0, 1, \cdots$ all weakly converge to the same random variable $1 / \sigma^2 \int_0^1 B(r)^2 dr$.

The parameter σ_u^2 may be estimated by $T^{-1}\Sigma(y_t - \hat{\rho}y_{t-1})^2$, and γ_1 by $T^{-1}\Sigma(y_{t+1} - \hat{\rho}y_t)(y_t - \hat{\rho}y_{t-1})$. Phillips [1987] suggests other estimates for them. We estimate them by building a state space model for $\{w_t\}$ (or equivalently for $\{u_t\}$) and calculating them based on the system parameter estimates obtained in this second step of the algorithm. Take the residual series $\{w_t\}$ of the first step and estimate h, F and G in $z_{t+1} = Fz_t + Ge_t$, and $w_t = hz_t + e_t$, and the covariance $cov\ e_t = \Delta$, as explained in Chapter 9. Let $\mu_k = E(w_{t+k} w_t)$. Its estimate is related by

$$\hat{\mu}_k = \hat{h}\hat{F}^{k-1}(\hat{F}\hat{Z}\hat{h}' + \hat{G}\hat{\Delta})$$

to the estimates of $Z = cov\ z_t$ and Δ. The estimates $\hat{Z}$, $\hat{G}$, and Δ are obtained by solving the Riccati equations as discussed earlier. Then from $u_t = w_t + (cb - \rho)w_{t-1}$, we estimate σ_u^2 and γ_1 by

$$\hat{\sigma}_u^2 = [\,1 + (\hat{c}\hat{b} - \hat{\rho})^2\,]\,\hat{\sigma}_w^2 + 2(\hat{c}\hat{b} - \hat{\rho})\hat{\mu}_1,$$

$$\hat{\gamma}_1 = \hat{\mu}_1 + (\hat{c}\hat{b} - \hat{\rho})\hat{\sigma}_w^2 + (\hat{c}\hat{b} - \hat{\rho})\hat{\mu}_2 + (\hat{c}\hat{b} - \hat{\rho})^2\hat{\mu}_1,$$

$$\sigma_w^2 = hZh' + \Delta$$

$$Z = cov\ z_t, \quad \Delta = cov\ e_t$$

and

$$\sigma^2 = \sigma_u^2 + 2\sum_{i=1}^{\infty} \gamma_i\ ,$$

where

$$\gamma_i = \lambda_i\ [\,1 + (\hat{c}\hat{b} - \hat{\rho})^2\,]\,\mu_i + (\hat{c}\hat{b} - \hat{\rho})(\mu_{i-1} + \mu_{i+1}),\quad i \geq 1.$$

In summing γ_i, we use the relation $\Sigma_i^{\infty}\mu_i = \hat{h}(I - \hat{F})(\hat{F}\hat{Z}\hat{h}' + \hat{G}\hat{\Delta})$.

K = 2. With K chosen to be two, the estimate of the state vector dimension could be 1 or 2. The estimate $\hat{n} = 1$ results if we regard the ratio of the singular value ξ_2 / ξ_1 to be near zero, while $\hat{n}$ is set to 2 if this ratio if not regarded to be negligible.

If n^* is set to two, then the eigenvalues of the matrix are given by the roots of

$$\begin{aligned} O &= |\lambda H - H^A| \\ &= [\,\Lambda_2(\nu_2 - \nu_1) - \nu_1\nu_2\,] + \lambda\,[\,\hat{\Lambda}_2(\nu_1 - \nu_3) + \nu_1(\nu_2 + \nu_3)\,] + \hat{\Lambda}_3(\nu_3 - \nu_2) \\ &= \hat{\Lambda}_2\,\{\,\lambda^2(\nu_2 - \nu_1) + \lambda(\nu_1 - \nu_3) + \nu_3 - \nu_2\,\} \\ &\quad - \lambda^2\nu_1\nu_2 + \lambda\nu_1(\nu_2 + \nu_3) + \nu_2(\nu_3 - \nu_2). \end{aligned}$$

Recalling that $\hat{\Lambda}_2 = O_p(T)$ while $\nu_i = O_p(1)$, $i = 1, 2, \cdots$, the zeros are given by those of the quadratic polynomial in the bracket up to $O_p(T^{-1})$:

$$\lambda = \frac{1}{2(\nu_2 - \nu_1)}\,\{\,(\nu_3 - \nu_1) \pm |\nu_3 - \nu_2 - \nu_2 + \nu_1|\,\},$$

i.e.,

$$\lambda = 1 + O_p(T^{-1}),$$

and

$$\lambda = \omega_2 / \omega_1 = \hat{\gamma}_3 / \hat{\gamma}_2,$$

where we use the relation

$$\omega_1 = \nu_2 - \nu_1 = -\hat{\gamma}_2,$$

and

$$\omega_2 = \nu_3 - \nu_2 = -\hat{\gamma}_3.$$

The eigenvalue close to 1 clearly reflects the dynamics of the random trend. To evaluate the magnitude of the second eigenvalue, note that

$$\hat{\gamma}_2 = [\, 1 + (\hat{c}\hat{b} - \hat{\rho})^2 \,]\, \hat{\mu}_2 + O\,(\hat{d}\hat{b} - \hat{\rho})(\hat{\mu}_1 + \hat{\mu}_3),$$

and

$$\hat{\gamma}_3 = [\, 1 + (\hat{c}\hat{b} - \hat{\rho})^2 \,]\, \hat{\mu}_3 + (\hat{c}\hat{b} - \hat{\rho})(\hat{\mu}_2 + \hat{\mu}_4).$$

The residuals are modeled by a state space model, as shown in (16), in the second step. If the residuals are modeled by a scalar dynamics, i.e., if z_t in (16) is scalar-valued, then $\hat{\mu}_{i+1} = f\,\hat{\mu}_i$ where $|f\,| < 1$, $i = 1, 2, \cdots$. In this case it follows that $|\hat{\gamma}_3 / \hat{\gamma}_2| < 1$. Even when the dimension of z_t is not one, the inequality $|\hat{\gamma}_3 / \hat{\gamma}_2| < 1$ follows, using a suitable norm, and this eigenvalue reflects parts of dynamics attributable to correlated residuals.

One could also directly calculate singular values in this case. Singular values are the positive square roots of the eigenvalues of the matrix HH', which is the same as H^2 when data are scalar,

$$HH' = \begin{bmatrix} \hat{\Lambda}_1^2 + \hat{\Lambda}_2^2 & \hat{\Lambda}_1\hat{\Lambda}_2 + \hat{\Lambda}_2\hat{\Lambda}_3 \\ \hat{\Lambda}_1\hat{\Lambda}_2 + \hat{\Lambda}_1\hat{\Lambda}_3 & \hat{\Lambda}_2^2 + \hat{\Lambda}_3^2 \end{bmatrix}.$$

The eigenvalues are

$$\xi^2 = \tfrac{1}{2}\, [\, \hat{\Lambda}_1^2 + \hat{\Lambda}_2^2 + \hat{\Lambda}_3^2 \pm 2\hat{\Lambda}_2(\hat{\Lambda}_1 + \hat{\Lambda}_3)\sqrt{1 + (\hat{\Lambda}_1 - \hat{\Lambda}_3)^2 / 4\hat{\Lambda}_2^2}\,].$$

Since the expression under the square root is $1 + O_p(T^{-2})$, the larger of the two eigenvalues is equal to

$$\xi_1^2 = 4\hat{\Lambda}_2^2 + 2\Lambda_2(\nu_2 - \nu_1) + \nu_1^2 + \nu_2^2 + O_p(T^{-1}),$$

and the smaller one is given by

$$\xi_2^2 = \frac{(\nu_2 - \nu_1)^2}{2} + O_p(T^{-2}).$$

The eigenvectors are easily seen to be $U = [\, u_1, u_2 \,]$, where

$$u_1' = \frac{1}{\sqrt{2}}(1, \; 1) + O_p(T^{-1})$$

and

$$u_2' = \frac{1}{\sqrt{2}}(1, \; -1) + O_p(T^{-1}).$$

As in (24) and (25), the difference $\nu_2 - \nu_1$ converges weakly, where

$$(\nu_2 - \nu_1)^2 = \hat{\gamma}_2^2$$

and hence

$$\xi_2 = |\hat{\gamma}_2| \,/\, \sqrt{2}\,.$$

The ratio of the squared singular values approaches zero

$$\xi_2^2 \,/\, \xi_1^2 = \hat{\gamma}_2 \,/\, 8\Lambda_1^2 \to 0 \quad \text{in probability}$$

as $T \to \infty$, and

$$T\xi_2 \,/\, \xi_1 \Rightarrow \frac{|\gamma_2| / \sqrt{8}}{\sigma^2 \int_0^1 B(r)^2 dr}\,.$$

When the underlying dynamics for the random walk have a single unit root, the numerical rank of the Hankel matrix will be one for sufficiently large T. For a small-sized sample, $n^* = 2$ is possible. With this choice, the matrix A is 2×2. Its leading (1, 1) element is estimated to be

$$\hat{a}_{11} = \frac{1}{\xi_1^2} u_1' H^A H' u_1$$

$$= 1 + O_p(T^{-1})\,,$$

since $\nu' = u_1' H \,/\, \xi_1$. Similar calculations show that

$$\hat{c}_1 = (\hat{\Lambda}_1 \quad \hat{\Lambda}_2) H u_1 \,/\, c\,\xi_1^{3/2} = \sqrt{\Lambda_1} + O_p(T^{-1/2})\,.$$

When $n^* = 2$ is tried, the rest of the elements of the matrix A are estimated to be equal to

$$\hat{a}_{12} = \xi_1^{-1/2}\, \xi_2^{-3/2}\, u_1' H^A H' u_2\,,$$

$$\hat{a}_{21} = \xi_2^{-1/2}\, \xi_1^{-3/2} u_2' H^A H' u_1\,,$$

and

$$\hat{a}_{22} = \xi_2^{-2} u_2' H^A H' u_2\,.$$

We have already noted that the second eigenvalue of the matrix $\hat{A}$ generated by these additional elements is $\hat{\gamma}_3 / \hat{\gamma}_2$, and is less than one in magnitude.

Drift Term

When a drift term is hypothesized to be present, the least squares estimates of μ and p in a hypothesized model, $y_t = \mu + py_{t-1} + u_t$, amount to subtracting sample means in the autocovariance matrix calculations. Define

$$\Gamma_i = T^{-1}\Sigma(y_{t-1+i} - \bar{y}_{i-1})(y_{t-1} - \bar{y}_{-1}),$$

where

$$\bar{y}_i = T^{-1} \sum_{0}^{T-i} y_{t+i} \quad , \quad i = 0, 1, 2, \cdots$$

Under the null hypothesis that the process is generated by

$$y_t = y_{t-1} + u_t \ , \quad y_0 = 0 \, ,$$

with the same set of assumptions on $\{u_t\}$ as before, the least-squares regression equation

$$y_t = \hat{\mu} + \hat{\rho} y_{t-1} + \hat{u}_t \, ,$$

is estimated by

$$\hat{\rho}_{LS} = \Gamma_1 / \Gamma_0 = \sum_{t=1}^{T} (y_t - \bar{y}_0) y_{t-1} \Big/ \sum_{t=1}^{T} (y_{t-1}^2 - \bar{y}_{-1}^2) \, , \tag{33}$$

and

$$\hat{\mu} = \bar{y}_0 - \hat{\rho}_{LS} \bar{y}_{-1} \ . \tag{34}$$

We assume that $\{u_t\}$ is weakly stationary, even though this assumption can be weakened considerably, as in Phillips and Perron [1986], who have shown that

$$T^{-1/2} \bar{y}_{-1} \Rightarrow \sigma \int_0^1 W(r) dr \ . \tag{35}$$

To see this, note that

$$X_T(t) = \frac{1}{\sigma\sqrt{T}} S_{[Tt]} = \frac{1}{\sigma\sqrt{T}} S_{j-1} ,$$

where $(j-1)/T \le t < j/T$, and hence

$$T^{-1/2}\bar{y}_{-1} = \sigma\Sigma_j \int_{(j-1)/T}^{j/T} X_t(t)dt = \sigma\int_0^1 X_T(t)dt$$

$$\Rightarrow \sigma\int_0^1 W(r)dr \qquad (36)$$

as $T \to \infty$.

From (34), the bias of the estimate is given by

$$\hat{\rho}_{LS} - 1 = \frac{\Sigma_t (y_t - y_{t-1})y_{t-1} - \bar{y}_{-1}(\bar{y}_0 - \bar{y}_{-1})}{\Sigma_t y_{t-1}^2 - \bar{y}_{-1}^2}$$

$$= \frac{\Sigma_t u_t y_{t-1} - \bar{y}_{-1} T^{-1} S_T}{\Sigma_t y_{t-1}^2 - \bar{y}_{-1}^2} ,$$

where

$$T^{-2}\Sigma y_{t-1}^2 \Rightarrow \sigma^2 \int_0^1 W(r)^2 dr , \qquad (37)$$

$$T^{-1}\Sigma_t u_t y_{t-1} \Rightarrow \tfrac{1}{2}[\sigma^2 W_1^2 - \sigma_u^2] , \qquad (38)$$

and

$$T^{-1/2} S_T \Rightarrow \sigma W_1 . \qquad (39)$$

The bias term is thus seen to converge weakly to

$$T(\hat{\rho}_{LS} - 1) \Rightarrow \frac{\tfrac{1}{2}(W_1^2 - \sigma_u^2/\sigma^2) - W_1\int_0^1 W(r)dr}{\int_0^1 W(r)^2 dr - (\int_0^1 W(r)dr)^2} , \qquad (40)$$

because $T^{-1}\bar{y}_{-1}S_T = (T^{-1/2}\bar{y}_{-1})(T^{-1/2}S_T)$, and from (37) and (39), the continuity theorem states that this converges weakly to $\sigma^2 W_1 \int_0^1 W(r)dr$.

Let

$$\upsilon_i = \Gamma_{i+1} - \Gamma_i , \quad i = 0, 1, \cdots .$$

For example

$$\upsilon_0 = T^{-1}\Sigma u_t y_{t-1} + T^{-1}\bar{y}_{-1} y_T .$$

Thus, from (37), (38) and (39)

$$\upsilon_0 \Rightarrow \tfrac{1}{2}[\sigma^2 W_1^2 - \sigma_u^2] + \sigma^2 W_1 \int_0^1 W(r)dr \quad .$$

Similarly

$$\upsilon_1 = T^{-1}\Sigma u_{t+1} y_{t-1} - T^{-1}\bar{y}_{-1} y_T \Rightarrow \tfrac{1}{2}\sigma^2 W_1^2 - \tfrac{1}{2}\sigma_u^2 - \gamma_1 \, ,$$

where we use the identity

$$u_{t+1} y_{t-1} = \tfrac{1}{2}(y_{t+1}^2 - y_t^2) - u_t u_{t+1} - u_{t+1}^2 \, ,$$

i.e.,

$$\upsilon_1 - \upsilon_0 = -\gamma_1 \quad .$$

Here we use the fact that $T^{-1}\Sigma u_t^2 \rightarrow \sigma_u^2$, a.s., and $T^{-1}\Sigma u_{t+1} u_t \rightarrow \gamma_1$, a.s. The matrices Λ's are related to Γ_i by

$$\Gamma_i = \Lambda_i - \bar{y}_{-1}\bar{y}_{i-1} \quad .$$

As in (37)

$$T^{-1/2}\bar{y}_i \Rightarrow \sigma \int_0^1 W(r)dr \quad , \quad i = 0, 1, \cdots$$

Thus

$$\Gamma_i / T = \Lambda_i / T - (T^{-1/2}\bar{y}_{-1})(T^{-1/2}\bar{y}_i)$$

$$\Rightarrow \sigma^2 \int_0^1 W^2(r)dr - (\sigma \int_0^1 W(r)dr)^2 \quad , \quad i = 0, 1, \cdots \quad ,$$

since $\Lambda_i = \Lambda_0 + \upsilon_0 + \cdots + \upsilon_{i-1}$ and υ's are $O_p(1)$ while Λ_0 is $O_p(T)$.

From (34), (39) and (40) the intercept is such that

$$\sqrt{T}\,\hat{\mu} = T(1 - \hat{\rho}_{LS})T^{-1/2}\bar{y}_{-1} + T^{-1/2} y_T$$

$$\Rightarrow \sigma W_1 + (\sigma \int_0^1 W(r)dr)\,(N_2 / D_2) \, ,$$

by the continuity theorem, where N_2 and D_2 are, respectively, the numerator and the denominator in (40).

The variance σ_u^2 may be estimated by

$$\hat{s}^2 = T^{-1}\Sigma(y_t - \hat{\mu} - \hat{\rho} y_{t-1})^2$$

$$= (1-\hat{\rho})^2\Gamma_0 + 2(1-\hat{\rho})T^{-1}\Sigma u_t y_{t-1} - 2(1-\hat{\rho})T^{-1}\bar{y}_{-1}y_T$$

$$+ T^{-1}\Sigma u_t^2 - T^{-2}y_T^2 + O_p(T^{-1}) \quad ,$$

where (34) is used and the subscript LS is dropped from $\hat{\rho}_{LS}$. All terms except $T^{-1}\Sigma u_t^2$ are $o_p(T^{-1})$.

Actually, it is not necessary to evaluate σ_u^2 explicitly. The bias term $(\sigma^2 - \sigma_u^2 - \gamma_1) / \sigma^2\int_0^1 W(r)^2 dr$ can be directly estimated as shown below, and the table of $(W_1^2 - 1) / \int_0^1 W^2(r)dr$ tabulated by Fuller [1976] or Evans and Savin [1981] can be accordingly adjusted to obtain the distribution for $T(\hat{\rho}-1)$. Since $\sigma^2 = \sigma_u^2 + 2\sum_{i=1}^{\infty}\gamma_i$, the numerator is calculated as

$$\sigma^2 - \sigma_u^2 - \gamma_1 = \gamma_1 + 2\sum_2^{\infty}\gamma_i$$

$$= 2h(I-F)^{-1}m - \gamma_1 \quad ,$$

because of $\gamma_i = hF^{i-1}m$. The integral $\sigma^2\int_0^1 W(r)^2 dr$ can be estimated by Λ_0 / T, $\upsilon_1 / T(\hat{\rho}-1)$ with $K = 1$ or by $\sigma_1 / 2T$ with $K = 2$.

11.8 Drifts or Linear Deterministic Trends?

This section develops the effects on the parameter estimates of misspecifying data-generating processes. For later use, we reproduce here asymptotic behavior of some sample functions of the random walk process $\{S_t\}$, $S_t = S_{t-1} + u_t$, $S_0 = 0$ and $\{u_t\}$. They are easy to calculate. See Phillips and Perron [1986, Lemma 2.3], for example.

Fact.

$$T^{-2}\Sigma_1^T S_t^2 \Rightarrow \sigma^2\int_0^1 W(t)^2 dt,$$

$$T^{-3/2}\Sigma_1^T S_t \Rightarrow \sigma\int_0^1 W(t)dt,$$

$$T^{-5/2}\Sigma_1^T tS_t \Rightarrow \sigma\int_0^1 tW(t)dt,$$

$$T^{-1/2}\Sigma u_t \Rightarrow \sigma W(1),$$

$$T^{-1}\Sigma u_t S_{t-1} \Rightarrow \tfrac{1}{2}[\sigma^2 W(1)^2 - \sigma_u^2],$$

and

$$T^{-3/2}\Sigma t u_t \Rightarrow \sigma\int_0^1 t dW(t) = \sigma[W(1) - \int_0^1 W(t)dt],$$

where

$$\sigma_u^2 = lim_{T\to\infty} T^{-1}\Sigma_1^T E(u_t^2),$$

and

$$\sigma^2 = \sigma_u^2 + 2\Sigma_i E(u_{t+i}u_t).$$

We have already encountered some of them when we discussed large-sample properties in Section 7. Others can be derived analogously. To derive the third result, write jS_j as $T(j/T)\sigma\sqrt{T}X_T(t)$ and recognize $(j/T)X_T(t)$ as $tX_T(t) + O_p(T^{-1})$. In the sequel we use these expressions without explicitly referring to them.

No Drift Term. When the data-generating process is a true random walk $y_t = y_{t-1} + u_t$, the estimate of ρ in the hypothesized model $y_t = \rho y_{t-1} + u_t$ has already been discussed. If a model with a drift term is hypothesized, $y_t = \mu + \rho y_{t-1} + u_t$, and μ and ρ are estimated by

$$\hat{\mu} = \bar{y} - \hat{\rho}\bar{y}_{-1},$$

where

$$\hat{\rho} - 1 = [T^{-1}\Sigma u_t y_{t-1} - (\bar{y} - \bar{y}_{-1})\bar{y}_{-1}] / (T^{-1}\Sigma y_{t-1}^2 - \bar{y}_{-1}^2),$$

$\bar{y} = T^{-1}\Sigma_1^T y_t$, and $\bar{y}_{-1} = T^{-1}\Sigma_t^T y_{t-1}$. In this case $y_t = S_t$ and we obtain

$$T(\hat{\rho} - 1) \Rightarrow N_2 / D_2,$$

where

$$N_2 = \tfrac{1}{2}(W_1^2 - \sigma_u^2/\sigma^2) - W_1\int_0^1 W(t)dt,$$

and

$$D_2 = \int_0^1 W(t)^2 dt - [\int_0^1 W(t)dt]^2,$$

and

$$\sqrt{T}\hat{\mu} \Rightarrow \sigma W_1.$$

The situation with a hypothesized deterministic trend is discussed later.

Drift Term. When the true data generating process has a drift term

$$y_t = \mu + y_{t-1} + u_t, \quad y_0 = 0, \quad \mu \neq 0,$$

and a model without drift, $y_t = \rho y_{t-1} + u_t$, is hypothesized, the term depending on μ totally dominates the least-squares estimate of ρ (or, for that matter, the estimate of the algorithm in Chapter 9). Note that the least squares estimate is such that

$$T(\hat{\rho} - 1) = T\left[\frac{\Sigma y_t y_{t-1}}{\Sigma y_{t-1}^2} - 1\right]$$

$$= \frac{\mu \bar{y}_{-1} + T^{-1}\Sigma u_t y_{t-1}}{T^{-2}\Sigma y_{t-1}^2} \rightarrow \frac{3}{2} \quad \text{a.s} ,$$

since

$$\bar{y}_{-1} = T^{-1}\Sigma y_{t-1} = \frac{\mu(T-1)(T-2)}{2T} + T^{-1}\Sigma_1^{T-1} S_t$$

$$= \frac{\mu}{2} T + O_p(T^{1/2}) ,$$

and

$$T^{-2}\Sigma y_{t-1}^2 = \mu^2 T/3 + 2\mu \Sigma t S_t / T^2 + \Sigma S_t^2 / T^2$$

$$= \mu^2 T / 3 + O_p(T^{1/2}) .$$

The expression $T(\hat{\rho} - 1) - 3/2$ is $O_p(T^{-1/2})$, and

$$\sqrt{T}\,(T(\hat{\rho} - 1) - 3/2) \Rightarrow -3(\sigma/\mu)\,[\,3\textstyle\int tW dt - W(1)\,] .$$

When the model is hypothesized to be $y_t = \mu + \rho y_{t-1} + u_t$, and ρ estimated by the least-squares method, then $T(\hat{\rho} - 1) = T^{-1}\Sigma u_t y_{t-1} / T^{-2}\Sigma y_{t-1}^2$, and we see that $T(\hat{\rho} - 1) \rightarrow 0$, a.s., because the term in it which depends on μ is $O_p(T^{-1/2})$. By straightforward calculation, we derive that

$$T^{3/2}(\hat{\rho} - 1) \Rightarrow 3(\sigma/\mu)[W(1)/2 - \int_0^1 W(t)dt\,] .$$

The least-squares estimate $\hat{\mu}$ is given by

$$\hat{\mu} = \bar{y}_0 - \hat{\rho}\bar{y}_1 = \bar{y}_0 - \bar{y}_{-1} + \frac{(\Sigma u_t y_{t-1})\bar{y}_{-1}}{\Sigma y_{t-1}^2} ,$$

and hence

$$\sqrt{T}\,(\hat{\mu} - \mu) \Rightarrow \frac{\sigma}{4}\,[\,W(1) + 6\int_0^1 W(t)dt\,] .$$

The effects on the least-squares estimates of the relations between the hypothesized and true data generating processes are tabulated in Table 1.

Deterministic Trend Term. One is usually interested in testing the hypothesis of a random walk with a drift against that the series is stationary around a linear deterministic trend. The hypothesis of a unit root with a trend in logarithmic series implies that the rate of change of the series is forever monotonic. Since this is patently unrealistic, at least for economic time series, we do not examine this possibility.

Now we use a random-walk data-generating process with and without drift and examine the consequence of fitting regression models which assume the presence of deterministic trend terms. First we take $y_t = y_{t-1} + u_t$, $y_0 = 0$, as the data-generating process but fit a regression model $y_t = \beta[t - (T+1)/2] + \rho y_{t-1} + u_t$. The least-squares method yields

$$\hat{\beta}\Sigma y_{t-1}[t-(T+1)/2] + \hat{\rho}\Sigma y_{t-1}^2 = \Sigma y_t y_{t-1},$$

and

$$\hat{\beta}\Sigma[t-(T+1)/2]^2 + \rho\Sigma y_{t-1}[t-(T+1)/2] = \Sigma y_t[t-(T+1)/2].$$

Rewrite the right-hand side as

$$\Sigma y_t y_{t-1} = \Sigma y_{t-1}^2 + \Sigma u_t y_{t-1},$$

and

$$\Sigma y_t[t-(T+1)/2] = \Sigma y_{t-1}[t-(T+1)/2] + \Sigma u_t[t-(T+1)/2],$$

and rewrite the equations for $\hat{\beta}$ and $\hat{\rho}-1$. Upon solving them we obtain

$$\hat{\beta} = \{(\Sigma y_{t-1}^2)[\Sigma u_t[t-(T+1)/2]]$$
$$- [\Sigma y_{t-1}[t-(T+1)/2)][\Sigma u_t y_{t-1}]\}/D, \tag{41}$$

and

$$\hat{\rho} - 1 = \{[\Sigma[t-(T+1)/2]^2]][\Sigma u_t y_{t-1}]$$
$$- [\Sigma y_{t-1}[t-(T+1)/2]][\Sigma u_t[t-(T+1)/2]]\}/D,$$

with

$$D = \Sigma[t-(T+1)/2]^2\Sigma y_{t-1}^2 - [\Sigma y_{t-1}[t-(T+1)/2)]]^2. \tag{42}$$

Thus $T^{-5}D \Rightarrow \sigma^2\{1/12\int_0^1 W(t)^2 dt - [\int tWdt - ½\int Wdt]^2\}$. The numerator for $\hat{\beta}$ is such that

$$T^{1.5}\,\hat{\beta} \Rightarrow N_6 / D\ ,$$

where

$$N_6 = \sigma\left[\frac{W_1}{2} - \int W dt\right]\int W^2 dt - ½[W_1^2 - \sigma_u^2/\sigma^2]\,[\int_0^1 tW dt - \int_0 W dt\ ],$$

with

$$D = 1/12\int_0^1 W^2 dt - [\int_0^1 tW dt - ½\int_0^1 W dt]^2,$$

and

$$T(\hat{\rho} - 1) \Rightarrow N_7 / D\ ,$$

with

$$N_7 = 1/24(W_1^2 - \sigma_u^2/\sigma^2) - \left[\frac{W_1}{2} - \int W dt\right](\int tW dt - ½\int W dt)\,.$$

Next, assume that the true data-generating process is $y_t = \mu + y_{t-1} + u_t$, $\mu \neq 0$, $y_0 = 0$, to which we fit the same model with the deterministic trend. Because $y_t - y_{t-1} = \mu + u_t$, a new term is added to the numerator of (41)

$$-\mu(\Sigma y_{t-1})\Sigma y_{t-1}\,[t - (T+1)/2)\,]/D,$$

and to that of (42)

$$\mu(\Sigma y_{t-1})\Sigma[t - (T+1)/2\,]^2/D.$$

The denominator does not change. Noting that $y_t = \mu t + S_t$ we calculate

$$\Sigma y_{t-1}^2 = \mu^2 T^3/3 + 2\sigma\mu T^{5/2}\int_0^1 tX_T(t)dt + O_p(T^2),$$

$$\Sigma y_{t-1}[t - (T+1)/2\,] = \frac{\mu T^3}{12} + \sigma T^{5/2}\,[\int_0^1 tW dt - ½\int_0^1 W dt\ ],$$

$$\Sigma y_{t-1} = \mu T^2/2 + \sigma T^{3/2}\int_0^1 X_T(t)dt + O_p(T)\,,$$

and

$$\Sigma u_t y_{t-1} = \sigma\mu T^{3/2}\int_0^1 t dW + (T/2)\,[\,\sigma^2 X_T(1)^2 - \sigma_u^2\,] + O_p(T^{½})\,.$$

Now $D = \dfrac{\mu^2 T^6}{48} + O_p(T^5)$, and hence

$$\sqrt{T}\,(T\hat{\beta} + 2\mu) \Rightarrow 4\sigma[W_1 - 4\int_0^1 W dt + 6\int_0^1 tW dt\ ],$$

and

$$\sqrt{T}\ [T(\hat{\rho} - 1) - 2] \Rightarrow 12(\sigma/\mu)^2\int_0^1 W dt\ .$$

The results are summarized in Table 1. Table 1 shows that the estimate $\hat{\rho}$ goes wrong when a constant drift term is present in the data-generating mechanism which is not allowed for in a hypothesized model. Perron [1986] provides details on the distributions of test statistics.

DGP (True Model)	Hypothesized Model: $y_t = \rho y_{t-1} + u_t$	$y_t = \mu + \rho y_{t-1} + u_t$	$y_t = \beta\left(t - \frac{T+1}{2}\right) + \rho y_{t-1} + u_t$
$y_t = y_{t-1} + u_t$	$T(\hat{\rho}-1) \Rightarrow \frac{N_1}{D_1}$	$T(\hat{\rho}-1) \Rightarrow \frac{N_2}{D_2}$ $\sqrt{T}\,\mu \Rightarrow \sigma\left[W_1 + \int_0^1 W(r)dr \frac{N_2}{D_2}\right]$	$T^{1.5}\hat{\beta} \Rightarrow \frac{N_6}{D}$ $T(\hat{\rho}-1) \Rightarrow \frac{N_7}{D}$
$y_t = \mu + y_{t-1} + u_t$	$T(\hat{\rho}-1) \to 3/2$ $\sqrt{T}\,[T(\hat{\rho}-1) - 3/2] \Rightarrow (3\sigma/\mu)N_5$	$T^{1.5}(\hat{\rho}-1) \Rightarrow N_3$ $T^{1/2}(\hat{\mu}-\mu) \Rightarrow N_4$	$T^{1.5}(T\hat{\beta}+2\mu) \Rightarrow N_8$ $T^{1.5}[T(\hat{\rho}-1) - 2] \Rightarrow N_9$
	$N_1 = 1/2(W_1^2 - \sigma_u^2/\sigma^2)$ $D_1 = \int_0^1 W(t)^2 dt$ $N_5 = W_1 - 3\int_0^1 tW dt$	$N_2 = N_1 - W_1\int_0^1 W(t)dt$ $D_2 = D_1 - \left(\int_0^1 W(t)dt\right)^2$ $N_3 = (3\sigma/\mu)\left[\frac{W_1}{2} - \int_0^1 W(t)dt\right]$ $N_4 = \frac{\sigma}{4}W_1 + \frac{3\sigma}{2}\int_0 W(t)dt$	See text for N's and D's

Two-Step Estimator. The two-step procedure of Section 5 first constructs a model

$$\tau_{t+1} = \rho\tau_t + bv_t$$

$$y_t = C\tau_t + v_t ,$$

i.e., $y_{t+1} = \rho y_t + u_{t+1}$ with $u_t = v_t + (cb - \rho)u_{t-1}$. In the case of non-zero drift, u_t is replaced by $\mu + u_t$, and v_t will have a non-zero mean θ, i.e., $u_t = v_t + (cb - \rho)u_{t-1}$ and $\mu = \theta + (cb - \rho)\theta$. When $cb - \rho = 0$, $u_t = v_t$ and $\mu = \theta$. When cb equals ρ, u_t is equal to v_t, and μ to θ.

The second step subtracts out the non-zero (sample) mean from v_t and fits

$$z_{t+1} = Fz_t + Ge_t ,$$

$$v_t - \theta = hz_t + e_t .$$

Jointly then, $y_t = \theta + c\tau_t + hz_t + e_t$. Thus, the constant θ is the drift term in state space representation.

With $K = 1$, the parameter ρ is estimated by Λ_2 / Λ_1, where

$$\Lambda_i = T^{-1}\Sigma y_{t+i-1} y_{t-1}', \quad i = 1, 2.$$

When the data generating process has a drift term in it, the true model is $y_t = \mu + y_{t-1} + u_t$, and

$$\Lambda_1 = \mu\bar{y}_{-1} + T^{-1}\Sigma y_{t-1}^2 + T^{-1}\Sigma u_t y_{t-1}.$$

As shown in Section 8, the second term in Λ_1 is $O_p(T^2)$. The autocovariance matrix Λ_2 is related to Λ_1 by

$$\Lambda_2 = \Lambda_1 + \mu\bar{y}_{-1} + T^{-1}\Sigma u_{t+1} y_{t-1}.$$

Thus the error in the estimate is given by

$$\hat{\rho} - 1 = (\mu\bar{y}_{-1} + T^{-1}\Sigma u_{t+1} y_{t-1}) / \Lambda_1.$$

The largest term in the numerator is $O_p(T)$ and that in the denominator is $O_p(T^2)$. Hence we conclude that

$$\sqrt{T}\ [\,T(\hat{\rho} - 1) - 3/2\,] \Rightarrow 3(\sigma / \mu)\ [\,2\int_0^1 W dt - W_1 - 3\int_0^1 tW dt\,]. \quad (43)$$

Referring to Table 1, we observe that this result is similar to the one for the regression model when $y_t = \rho y_{t-1} + u_t$ is hypothesized.

When the sample mean is subtracted in calculating Λ's, we derive, instead,

$$\sqrt{T}\ [\,T(\hat{\rho} - 1) - 6\,] \Rightarrow 12(\sigma / \mu)\ [\,8\int_0^1 W dt - W_1 - 12\int_0^1 tW dt\,].$$

Alternatively, when drift is suspected to be present, its least-squares estimate may be subtracted, thus converting the model into

$$\eta_t = \rho\eta_{t-1} + w_t \ ,$$

where

$$\eta_t = y_t - \hat{\mu} \ ,$$

and

$$w_t = u_t + \mu - \hat{\mu} \ .$$

Then w_t is weakly-stationary with mean-zero when u_t is.

11.9 Regime Shifts

Discontinuous changes of parameter values in data-generating processes are called regime shifts. This is a particular kind of switching regression, as discussed in Goldfeld and Quandt [1976], for example, in which parameter values are known to have shifted once in the data set, e.g., $\theta = \theta_1$ for $t < t_c$ and $\theta = \theta_2$ for $t > t_c$, with unknown t_c (or a finite number of times with unknown switching time instants).

Suppose that the variance of the innovations shifts from one value $\Delta = \Delta_1$, to another $\Delta = \Delta_2$ at an unspecified time t_c. From Section 7.3 the (maximized) likelihood function conditional on the switching time t_c is

$$L(y^T \mid t_c) = const \mid \Delta_1 \mid^{-t_c/2} \mid \Delta_2 \mid^{-T_2/2} exp - ½\, tr\ (\Delta_1 S_1 + \Delta_2 S_2) \ ,$$

where $S_1 = \Sigma_1^{t_c} e_t e_t{}'$ and $S_2 = \Sigma_{t_c+1}^{T} e_t e_t{}'$ with $T_2 = T - t_c$. We follow Goldfeld and Quandt [1976, p. 7] and maximize $L(y^T \mid t_c)$ with respect to t_c. Let t_c^* be its value which maximizes this likelihood function. We then say the regime has shifted at time $t_c^* + 1$. If one has some prior information on the likely value of t_c, this information should be incorporated as a prior probability distribution in the Bayesian framework. See Zellner [1984] for example.

Example The US real GNP from 1948.1 to 1986.1 has been used to locate a regime shift. We know that the Bretton Woods Agreement broke down in August 1971 when then-President Nixon decided to discontinue the fixed exchange rate for the US dollar. In 1973 oil prices increased dramatically. Our prior belief in the timing of a regime shift is therefore that if there is one regime shift, then it is likely to have occurred in the early 1970s.

The likelihood function achieves its maximum at $t_c = 92$, i.e., at 1971.1:

$$\frac{L(t_c=93)}{L(t_c=92)}=\frac{(.01995)^{59}(.0953)^{91}}{(.0017)^{58}(.1008)^{92}}<1,$$

and

$$\frac{L(t_c=91)}{L(t_c=92)}=\frac{(.10995)^{59}(.0953)^{91}}{(.0931)^{90}(.11406)^{60}}<1.$$

Dividing the total sample into [1, 91] and [92, 152], the first segment is modeled by

$$\tau_{t+1}=.9841\tau_t+2.063v_t\ ,$$

$$y_t=.482\tau_t+v_t\ ,$$

$$v_t=(.7785\quad .1436)z_t+e_t\ ,$$

$$z_{t+1}=\begin{bmatrix} .429 & -.569 \\ .569 & .484 \end{bmatrix} z_t+\begin{bmatrix} -.585 \\ .402 \end{bmatrix} e_t$$

$$cov\ e_t=.931\times 10^{-4}\ ,$$

and the second segment is modeled by

$$\tau_{t+1}=.974\tau_t+4.20v_t\ ,$$

$$y_t=.238\tau_t+v_t\ ,$$

$$v_t=(.598,\quad .0208)z_t+e_t\ ,$$

$$z_{t+1}=\begin{bmatrix} .612 & -.337 \\ .337 & .956 \end{bmatrix} z_t+\begin{bmatrix} -.402 \\ .064 \end{bmatrix} e_t\ ,$$

$$cov\ e_t=.114\times 10^{-3}\ .$$

In both segments, $T(\hat{\rho}-1)$ is of the order -1.5. It is not clear whether or not the unit root is present in these two subperiods. It is probably better to allow for gradual changes rather than a sudden discontinous change. See also Section 12.3 for other examples.

11.10 Nearly Integrated Processes

Consider a univariate AR(1) process

$$y_t = \rho y_{t-1} + u_t, \quad y_0 = 0, \tag{44}$$

or its state space version $x_t = \rho x_{t-1} + b w_{t-1}$, $y_t = h x_t + w_t$, where ρ is close to, but not equal to, one. Such a process is called nearly integrated and has been analyzed by Phillips [1987], Chan and Wei [1988], and Chan [1988]. Here we assume that w_t is a weakly-stationary process. Then u_t is also a weakly-stationary process when $|ch - \rho| < 1$.

Let $\rho = e^{-c/T}$ and generate a double array $y_{t,T}$ by

$$y_{t,T} = e^{-c/T} y_{t-1,T} + u_t, \quad t = 1, \cdots T \quad , \quad y_{0T} = 0, \tag{44}$$

where c is a fixed non-negative constant. Define

$$J_c(t) = \int_0^t e^{-c(t-s)} dW(s),$$

where $W(\cdot)$ is a standard Brownian motion. This process is called an Ornstein-Uhlenbeck process and is generated by the stochastic differential equation

$$dJ_c(t) = -cJ_c(t)dt + dW(t), \quad J_c(0) = 0,$$

which corresponds to the Langevin equation. See Arnold [1974, p. 134]. For a fixed $t > 0$, $EJ_c(t) = 0$ and its variance is given by

$$EJ_c(t)^2 = (1 - e^{-2ct}) / 2c.$$

Since it is a Gaussian process, its distribution is $N[0, (1 - e^{-2ct}) / 2c]$. Chan [1988] has shown that $J_c(t)$ is equal to $e^{c(1-t)} W(B_c(t))$ in distribution, where

$$B_c(t) = e^{-2c}(e^{2tc} - 1) / 2c \ ,$$

since it is normal, and mean zero and its variance is equal to that of $J_c(t)$,

$$E(e^{2c(1-t)} W^2(B_c(t))^2) = e^{2c(1-t)} \cdot B_c(t) = (1 - e^{-2ct}) / 2c.$$

This implies that J_c is governed by

$$dJ_c(t) = -cJ_c(t)dt + e^{c(1-t)} dW(B_c(t)).$$

Then

$$\int_0^1 J_c(t) dJ_c(t) + c \int_0^1 J_c(t)^2 dt = \int_0^1 e^{c(1-t)} J_c(t) dW(B_c(t))$$

$$= \int_0^1 e^{2c(1-t)} W(B_c(t)) dW(B_c(t)). \tag{45}$$

Integration by parts of (45) yields

$$J_c(t) = W(t) - c \int_0^t e^{-c(t-s)} W(s) ds. \tag{46}$$

Since $dJ_c(t)^2 = 2J_c(t)dJ_c(t) + (dJ_c(t))^2$ and $(dJ_c(t))^2 = (dW_t)^2 = dt$, by Ito's lemma, Arnold [1974, p. 91], we have from (44)

$$dJ_c(t)^2 = 2J_c(t)[-cJ_c(t)dt + dW(t)] + dt;$$

or, integrating it from 0 to 1, we obtain

$$J_c(1)^2 = 1 - 2c\int_0^1 J_c(t)^2 dt + 2\int_0^t J_c(t) dW(t). \tag{47}$$

Solve (44)

$$y_{t,T} = \Sigma_1^t e^{-(t-j)c/T} u_j .$$

Define

$$X_T(t) = \frac{1}{\sigma\sqrt{T}} S_{j-1}, \quad (j-1)/T \le t < j/T, \quad 0 \le t \le 1,$$

where

$$S_j = u_1 + \cdots + u_j .$$

Noting that

$$u_j = \sigma\sqrt{T} \int_{(j-1)/T}^{j/T} dX_T(t),$$

we derive as in Phillip [1987]

$$\begin{aligned} T^{-1/2} y_{[Tt]} &= \sigma\Sigma_{j=1}^{[Tt]} e^{-c([Tt]-j)/T} \int_{(j-1)}^{j/T} dX_T(s) + O_p(T^{-1/2}) \\ &= \sigma\Sigma_{j=1}^{[Tt]} \int_{(j-1)/T}^{j/T} e^{-c(t-s)} dX_T(s) + O_P(T^{-1/2}) \\ &= \sigma\int_0^t e^{-c(t-s)} dX_T(s) + O_p(T^{-1/2}) . \end{aligned}$$

Carrying out this integral by parts, we note that

$$\begin{aligned} \int_0^t e^{-c(t-s)} dX_T(s) &= e^{-c(t-s)} X_T(s) \Big|_0^t - ce^{-ct} \int_0^t e^{cs} X_T(s) ds \\ &= X_T(t) - \int_0^t e^{-c(t-s)} X_T(s) ds \\ &\Rightarrow W(t) - c\int_0^t e^{-c(t-s)} W(s) ds \\ &= J_c(t), \end{aligned}$$

where (46) is used. This establishes that

$$T^{-1/2} y_{[T\gamma]} \Rightarrow \sigma J_c(t). \tag{48}$$

This equation generalizes $T^{-1/2} S_{[Tt]} \Rightarrow \sigma W(t)$.

Proceeding analogously, we derive

$$T^{-2} \Sigma y_t^2 = \sigma \int_0^1 \left[\int_0^t \int_0^t e^{-c(t-u)-c(t-s)} dX_T(s) dX_T(u) \right] dt + O_p(T^{-1/2})$$

$$\Rightarrow \sigma^2 \int_0^1 \left[\int_0^t \int_0^t e^{-c(t-u)-c(t-s)} dW_s dW_u \right] dt = \sigma^2 \int_0^1 J_c(t)^2 dt ,$$

as $T \to \infty$ where the continuous mapping theorem is employed, since

$$\int_0^t e^{-c(t-s)} dX_T(s) = X_T(t) + c \int_0^t e^{-c(t-s)} X_T(s) ds$$

is a continuous functional on $D(0, 1)$. This equation generalizes $T^{-2} \Sigma S_t^2 \Rightarrow \sigma^2 \int_0^1 W(t)^2 dt$. From $T^{-1/2} y_T = \sigma J_c(1)$, we derive $T^{-1} y_T^2 = \sigma^2 J_c(1)^2$. Squaring (43) and summing over t and noting that $\rho^2 = 1 - 2c/T + o(T^{-1})$ we obtain

$$T^{-1} \Sigma (y_t^2 - y_{t-1}^2) = -2cT^{-2} \Sigma y_{t-1}^2$$

$$+ 2T^{-1} \Sigma y_{t-1} u_t + T^{-1} \Sigma u_t^2 + o_p(T^{-1})$$

and hence

$$2T^{-1} \Sigma y_{t-1} u_t \Rightarrow \sigma^2 J_c(1)^2 + 2c\,\sigma^2 \int_0^1 J_c(t)^2 dt - \sigma_u^2$$

$$= 2\sigma^2 \int_0^1 J_c(t) dW(t) + \sigma^2 - \sigma_u^2 , \tag{49}$$

where (47) is used. This is a generalization of

$$T^{-1} \Sigma u_t S_{t-1} \Rightarrow ½ \left[\sigma^2 W_1^2 - \sigma_u^2 \right].$$

In the expression for the bias of the least-squares estimate

$$\hat{\rho}_{LS} - \rho = \Sigma y_{t-1} u_t \,/\, \Sigma y_{t-1}^2 ,$$

the denominator Σy_{t-1}^2 is such that

$$T^{-2} \Sigma y_t^2 \Rightarrow \sigma^2 \int_0^1 J_c(t)^2 dt ,$$

and from (48)

$$T^{-1}\Sigma y_{t-1}u_t \Rightarrow \sigma^2\int_0^1 J_c(t)dW_t + \tfrac{1}{2}(\sigma^2 - \sigma_u^2)$$

where

$$\sigma^2 = E(u_1^2) + 2\Sigma E(u_1 u_k),$$

and

$$\sigma_u^2 = E(u_1^2).$$

Thus

$$T(\hat{\rho}_{LS} - \rho) \Rightarrow \frac{\sigma^2\int_0^1 J_c(t)dW_t + \tfrac{1}{2}(\sigma^2 - \sigma_u^2)}{\sigma^2\int_0^1 J_c(t)^2 dt},$$

which corresponds to the expression for $T(\hat{\rho}_{LS} - 1)$ derived previously. With $K = n = 1$, the algorithm of Chapter 9 produces $\hat{\rho} = \Lambda_2 / \Lambda_1$. Then

$$\hat{\rho} - \rho = (\Lambda_2 - \rho\Lambda_1) / \Lambda_1 ,$$

where

$$\Lambda_2 - \rho\Lambda_1 = T^{-1}\Sigma y_t u_{t+2} ,$$

and

$$\Lambda_1 = \Lambda_\rho - \rho\Lambda_0 + \rho\Lambda_0$$
$$= \rho\Lambda_0 + T^{-1}\Sigma y_t u_{t+1} .$$

Proceeding analogously, we derive

$$T^{-1}\Sigma y_t u_{t+2} \Rightarrow \sigma^2\int_0 J_c(t)dW_t + \tfrac{1}{2}(\sigma^2 - \sigma_u^2) - \gamma_1 .$$

Since $T^{-1}\Sigma y_t u_{t+1}$ is $O_p(1)$ and Λ_0 is $O_p(T)$, $T(\hat{\rho} - \rho)$ has the same limiting distribution as $T(\hat{\rho}_{LS} - \rho)$.

Chan [1988] has numerically tabulated percentiles for a statistic $(\Sigma y_t^2)^{1/2}(\hat{\rho}_{LS} - \rho)$.

12. Numerical Examples

This chapter presents examples of models estimated from post World War II macroeconomic time series data for the United States of America, the United Kingdom, West Germany, and Japan. Actual macroeconomic time series used in model construction are described in the data appendix. Models of dimension n constructed using K sub-blocks of Hankel matrices are denoted by $mK.n$, $m2.3$, for example. In these examples J is taken to be equal to K. Recall the constraint that $n \leq Kp$ where p is the number of a series in the data set. These models thus utilize information contained in the first $2K - 1$ covariance matrices of the data vectors. Estimated models are then used to obtain dynamic multipliers in some cases, utilizing the procedure described in Section 7.4.

12.1 West Germany

A three-dimensional series consisting of the industrial production index (1980 = 100), M1, and consumer price index has been used.

We apply the two-step procedure described in Section 11.5 to the data set consisting of the industrial production index (IPI), M_1 and the consumer price index after dividing each component of the series by its initial value so that its logarithm is zero. The first step estimates $m1.1$ with

$$\hat{\rho} = .9908, \quad \hat{b} = (-.244 \quad .476 \quad 1.689),$$

and

$$\hat{c} = (.0246 \quad .274 \quad .506)'.$$

Note that the value of the first component of c is about 1/10 of those of the remaining components. This indicates that the IPI shares very little of the common trend which is present for the other two series. This is clearly visible from the plot of the scaled series in which each component is divided by its own first value. In the second step, the model $m2.2$ is chosen, since its residuals are nearly uncorrelated, and its multiplier profiles are smoother than those of $m2.3$, even though on the residual characteristics alone, the three-dimensional model is slightly better. Eigenvalues of the short-run transition matrix are $.966 \pm j.0425$, and -.280. The innovation is practically white noise with

$$cov\ e = \Delta = \begin{bmatrix} 3.161 & .042 & -.077 \\ * & .182 & -.013 \\ * & * & .802 \end{bmatrix} \times 10^{-4}.$$

The matrix $\Delta\Delta_d^{-1}$ shows that IPI shocks do not affect the other two series much, while M_1 shocks affect IPI and CPI shocks affect IPI. The Durbin-Watson statistic (D.W.) of the residuals are, respectively, 1.92, 1.73, and 1.83. The R^2s are, respectively, .871, .999, and .999. A three-equation structural model can be identified in the manner described in Section 7.4. The only surprise in the shape of the impulse responses seems to be the responses to money supply shocks. The CPI gradually increases, but the IPI series becomes slightly negative and gradually returns to zero over 40 quarters.

12.2 United Kingdom

The series consists of IPI, M1, and RPI (retail price index). Each of the three series, arranged in the order IPI, RPI, and M1, is divided by its initial data values, so that the initial value of the logarithms is zero. The first step estimates m1.1 with

$$\hat{\rho} = .9828$$

$$\hat{b} = (.906 \quad 2.313 \quad .0658)$$

$$\hat{c} = (-.00783 \quad .4206 \quad .3631)'.$$

The trace of Λ_0 is .32, and the standard error of c is about .08. Note that the IPI series has almost no common trend component with the other two series.

In the second step, we choose $m3.5$ because its residuals are the closest to being martingale differences, among several models we estimated. R^2s are .84, .999, and .999, respectively. Here, the matrix $\Delta\Delta_d^{-1}$ indicates that exogenous disturbances to any one series do not induce any significant disturbances, except that a weak feedback exists between IPI and M shocks. The D.W. statistics for the three residual components are 1.97, 1.95, and 1.95. The estimated parameters in the second step are such that the eigenvalues are $.8638 \pm j.0366$, $.6397 \pm j.0847$, and $-.6936$ and

$$cov\ e = \Delta = \begin{bmatrix} 2.341 & .00301 & -.254 \\ .0030 & .4032 & .0322 \\ -.254 & .0332 & 3.077 \end{bmatrix} \times 10^{-4}.$$

Here, the matrix $\Delta\Delta_d^{-1}$ indicates that disturbances to any one series do not induce any significant disturbances in the other series, except that a weak feedback exists between IPI and M shocks.

12.3 The United States of America

A Money Stock

The quarterly money stock data from first quarter 1947 to second quarter 1982, a total of 141 data point, is split into two subsets, and separate models are estimated for the two subperiods. As in the US real GNP example we earlier presented, we maximize the likelihood function with respect to the date of the regime shift. Given that one regime shift occurred in the specified quarter, the date of the shift is estimated to be the 4th quarter 1970 by maximizing the likelihood function. The likelihood function is proportional to

$$|\Delta_1|^{-N_1/2}\,|\Delta_2|^{-N_2/2}\,exp{-\tfrac{1}{2}}(\Delta_1^{-1}S_1+\Delta_2^{-1}S_2)\,,$$

where Δ_i is the innovation variance and S_i is the sum of the squared residuals of the data set i, $i = 1, 2$. For example, with a break at the 4th quarter 1970, $\Delta_1 = .151 \times 10^{-1}$, $\Delta_2 = .454 \times 10^{-2}$, $S_1 = .2197$ and $S_2 = .0913$.

The ratio of the likelihood functions of the neighboring period is such that

$$\frac{L(71.1)}{L(70.4)} = \frac{(.151\times10^{-1})^{47.5}\,(.454\times10^{-2})^{23}}{(.154\times10^{-1})^{48}\,(.496\times10^{-2})^{22.5}}\,exp{-\tfrac{1}{2}}(R_1-R_0)<1\,,$$

where

$$R_1 = \frac{.2197}{.154\times10^{-1}} + \frac{.0913}{.496\times10^{-2}}\,,$$

$$R_2 = \frac{.223}{.151\times10^{-1}} + \frac{.0889}{.454\times10^{-2}}\,,$$

and

$$\frac{L(70.3)}{L(70.4)} = \frac{(.151\times10^{-1})^{47.5}\,(.454\times10^{-2})^{23}}{(.157\times10^{-1})^{47}\,(.447\times10^{-2})^{23.5}}\,exp{-\tfrac{1}{2}}(R_1-R_0)<1\,,$$

with

$$R_{-1} = \frac{.2242}{.157\times10^{-1}} + \frac{.0900}{.447\times10^{-2}}\,.$$

Therefore, the likelihood function with the break at the fourth quarter 1970 is the local maximum. Note that this shift in monetary regime took place one quarter earlier than that of the US real GNP series.

Money Stock and CPI

The monthly bivariate series composed of the M_2 and *CPI* from January 1972 to September 1984 is used, after the data are scaled so that the initial values are one for both components. Even though the *CPI* seems to grow faster after 1979 and the operating procedures of the Federal Reserve Board were changed in October 1979, we have treated the series assuming no regime shift. Actually, the plot of the money stock residuals by the model m1.1 estimated by the first step exhibit irregular movements after October 1979. For this bivariate series, model 1.2 at the first stage and m2.2 at the second stage have been chosen.

Model m1.2

$$\tau_{t+1} = A\,\tau_t + Bv_t$$

$$y_t = c\,\tau_t + v_t$$

is estimated with

$$\hat{A} = \begin{bmatrix} .9903 & .0225 \\ .0007 & .9583 \end{bmatrix}, \quad \lambda(\hat{A}) = .9908,\ .9578,$$

$$\hat{B} = \begin{bmatrix} .757 & 1.085 \\ -4.625 & 4.048 \end{bmatrix},$$

$$\hat{C} = \begin{bmatrix} .478 & -.207 \\ .567 & .174 \end{bmatrix}.$$

Here the second eigenvalue is not likely to be one since $\tau(\lambda - 1)$ is -6.46. Therefore the two components are cointegrated. There is likely to be only one unit root.

Model 2.2

$$\zeta_{t+1} = F\,\zeta_t + Ge_t$$

$$v_t = H\,\zeta_t + e_t$$

is estimated to have the matrices

$$\hat{F} = \begin{bmatrix} .9447 & -.1143 \\ .040 & -.1746 \end{bmatrix},$$

$$\hat{G} = \begin{bmatrix} -.507 & -.1101 \\ 1.19 & -1.59 \end{bmatrix},$$

$$\hat{H} = \begin{bmatrix} -1.044 & -.2170 \\ -1.175 & -.055 \end{bmatrix},$$

$$\Delta = cov\,e = \begin{bmatrix} .297 & .032 \\ .032 & .088 \end{bmatrix} \times 10^{-4}.$$

US Consumer Price Index

The monthly US Consumer Price Index (CPI) from January 1972 to July 1987, a total of 187 data points, has been modeled by the two-step algorithm. Although visual inspection of the plot seems to suggest a break in the lags at about January 1982, the whole data set is treated here assuming no regime shift. Data are scaled by dividing the CPI by its initial value, so that the logarithm starts with $y_1 = 0$. The first step builds

$$\tau_{t+1} = \rho \tau_t + b v_t$$

$$y_\tau = c\,\tau_t + v_t$$

with

$$\hat{\rho} = .9936,$$

$$\hat{b} = 1.591,$$

$$\hat{c} = .6596.$$

The value $T(\hat{\rho} - 1)$ is -1.20. If the residuals are uncorrelated, we can't reject the hypothesis that ρ is .8 or closer to 1 at about 10% confidence level according to Chan's table [1988]. Here the residuals are correlated and we cannot really use his table. We use the estimated value of ρ as is. Note that $\hat{b}\hat{c} = 1.0019$, and hence v_t and u_t in a hypothesized model $y_{t=1} = \rho y_t + u_{t+1}$ are likely to be statistically indistinguishable. The second step produces a two-dimensional model for v_t as

$$z_{t+1} = F z_t + G e_t$$

$$v_t = H z_t + e_T.$$

With $K = 2$ and $n = 1$, the correlation at lag 11 is 0.105 and all others are less than 0.1 in magnitude. The D.W. is 2.113. The estimated parameters are $\hat{F} = .858$, $\hat{G} = 1.688$, $\hat{H} = .286$, and $cov\ e = .66 \times 10^{-5}$. With $K = 6$ and $n = 2$

$$\hat{F} = \begin{bmatrix} .9467 & .0896 \\ .0876 & .4920 \end{bmatrix}$$

has the eigenvalues .961 and .476,

$$\hat{G} = \begin{bmatrix} -1.201 \\ 1.518 \end{bmatrix}, \quad \hat{H} = (-.2685 \quad .1010).$$

The innovation is such that $cov\ e = .64 \times 10^{-5}$ and its serial correlation coefficients are all less than 0.1 --except that at lag 20, it is .13. The D.W. is 2.146. Varying K or increasing n does not make the residual substantially less correlated.

Real GNP, CPI, and M2

Possible advantages of using a two-step procedure can be illustrated by jointly modeling quarterly data consisting of real GNP, CPI, and M2 from first quarter 1973 to third quarter 1989. The first step estimates the transition (dynamic) parameter ρ to be about .980. If this is taken to be an indication of the presence of a unit root and the first differences of the logarithms are taken, then they are as shown in Figure 1, while the residuals of the first step shown in Figure 2 are less like white noises. Solid lines are real GNP, broken ones are for CPI, and the dashed lines represent M2. The label P stands for the CPI time path. A three-variable structural model similar to the one discussed in Section 7.4 is identified. The matrix S introduced there is given by

$$S = \begin{bmatrix} 1 & -.854 & -.559 \\ -.184 & 1 & 0 \\ 0 & 0 & 1 \end{bmatrix}.$$

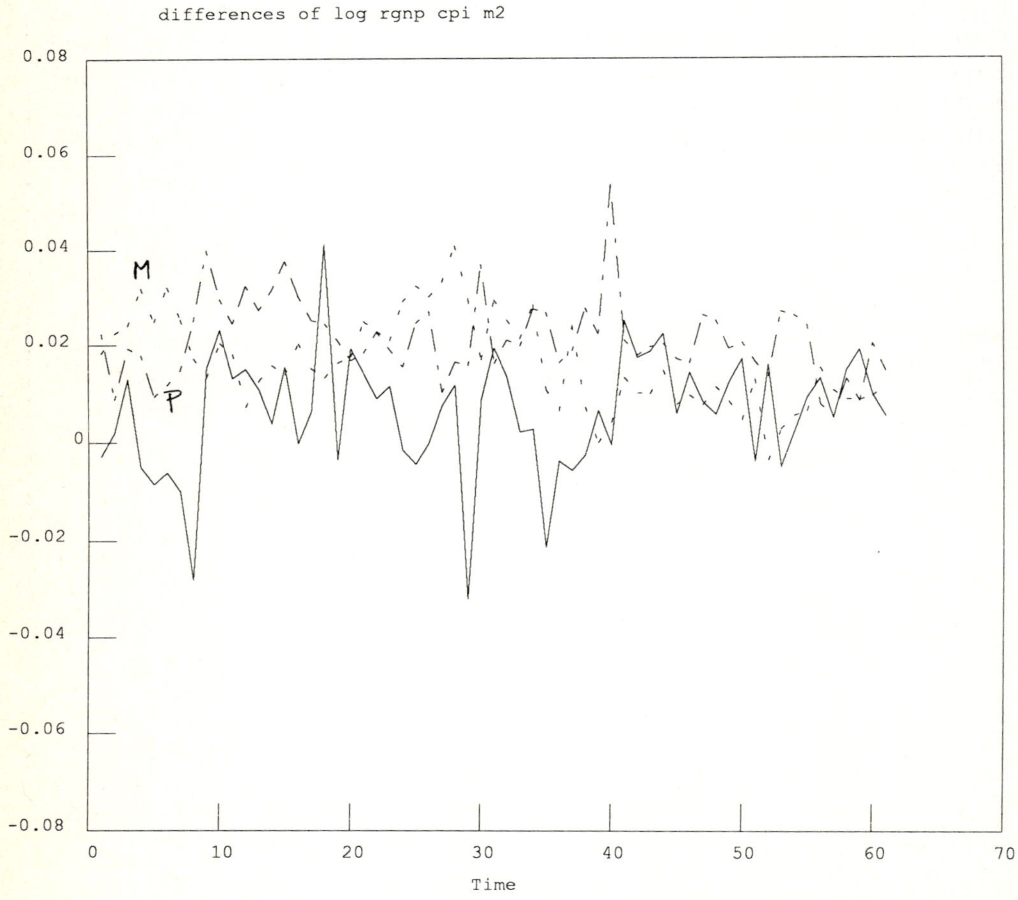

Fig. 1 The First Difference of the loganithms of real GNP, CPI and M2

The impulse responses to structural shocks are plotted in Figures 3 through 5. The variances are 1.34×10^{-4}, $.50 \times 10^{-4}$, and 2.75×10^{-4}, respectively.

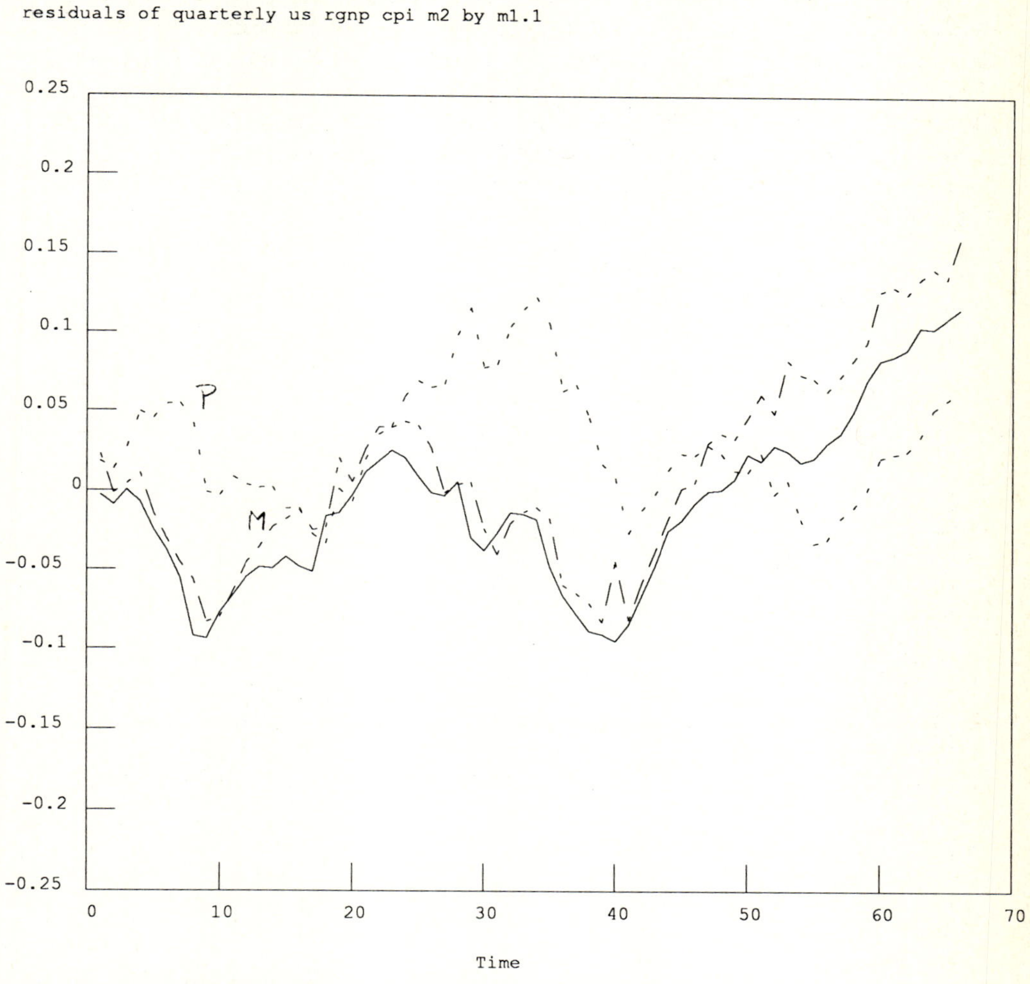

Fig. 2 Residuals of the first step by m1.1 of the three series

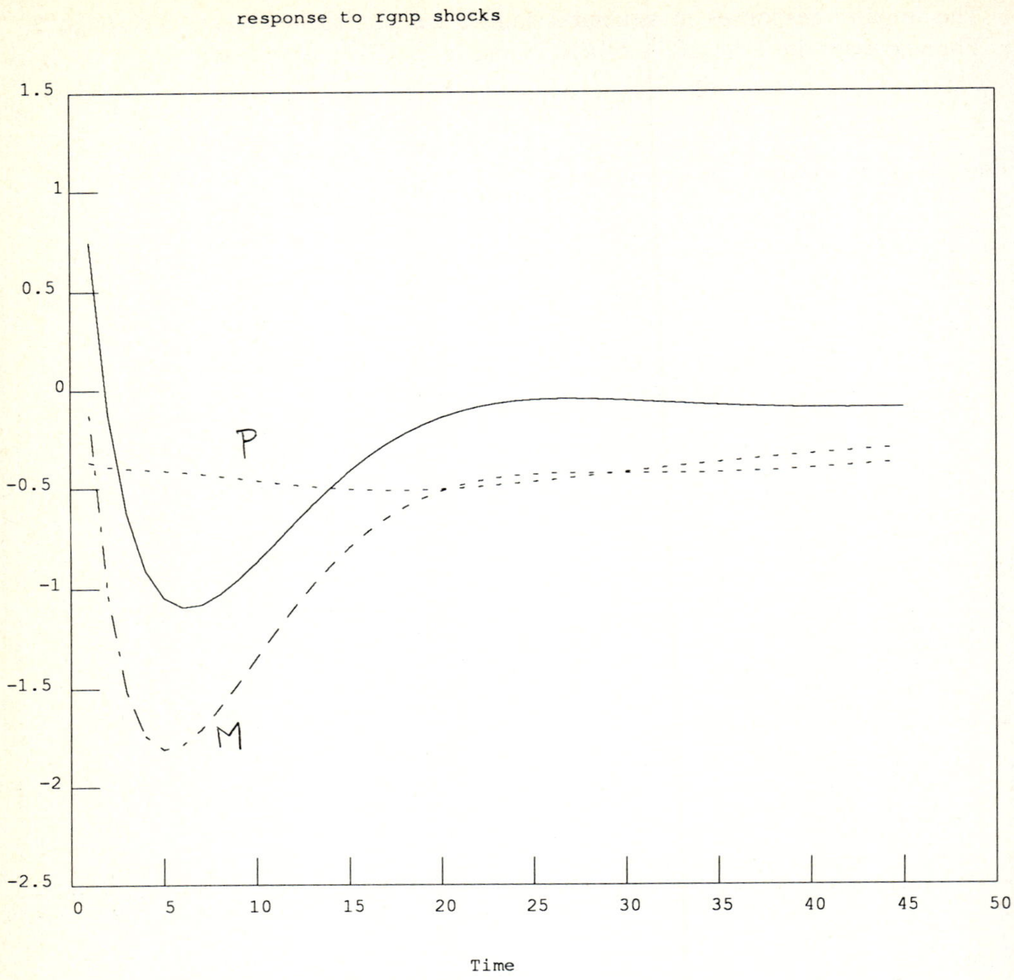

Fig. 3 Responses to real GNP shocks

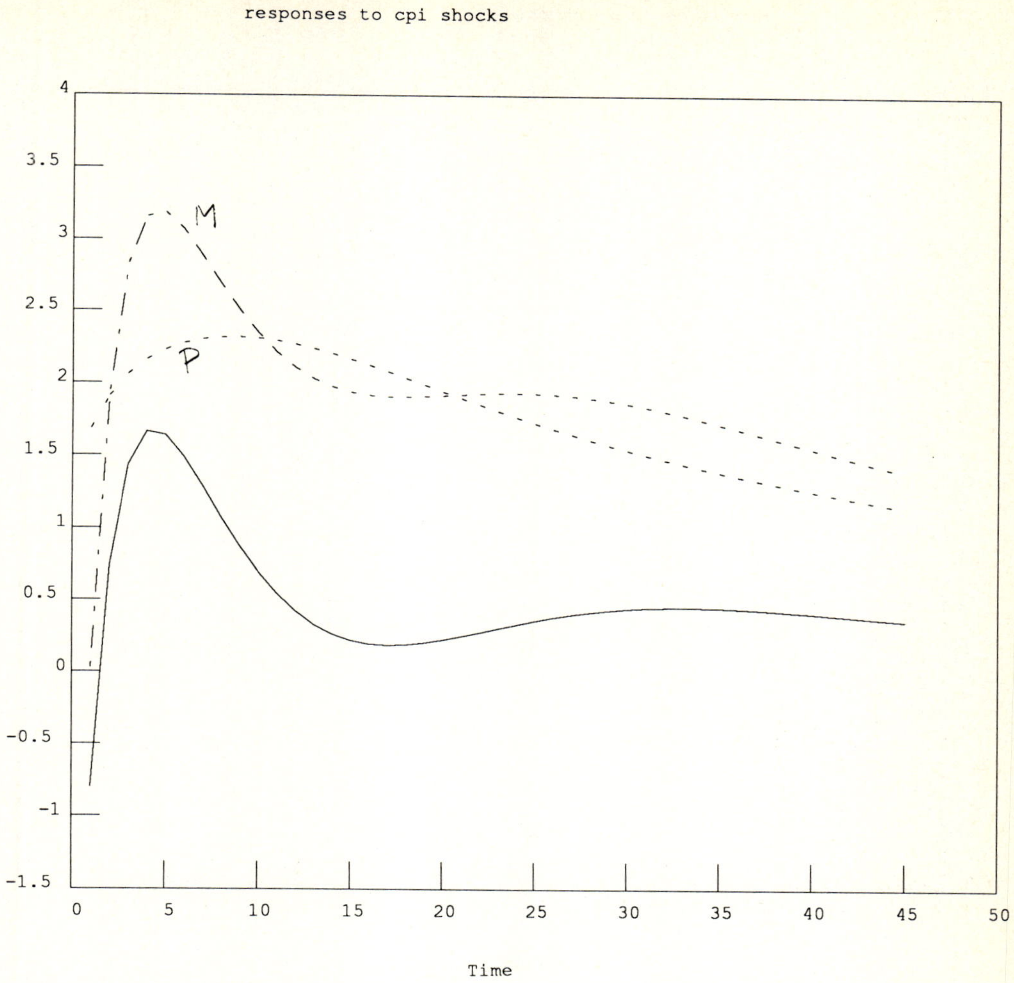

Fig. 4 Responses to CPI shocks

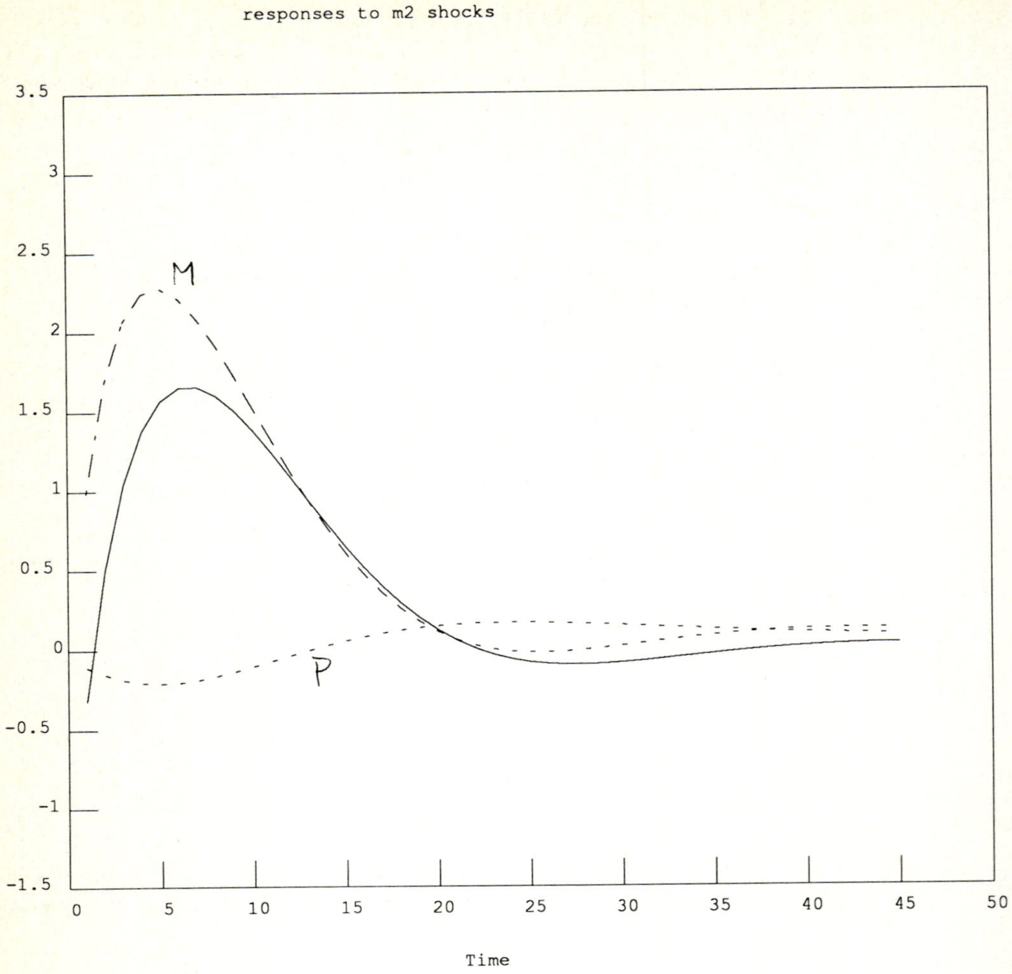

Fig. 5 Responses to M2 shocks

12.4 The US and West German Real GNP Interaction

Examination of the likelihood ratios seems to show that both US and West German real GNP underwent a regime shift in early 1971, if one such shift is assumed to have taken place. For this reason, we use the data for the period 1973.1 to 1989.3, a total of 66 data points, to avoid the regime shift and examine dynamic interactions of these two real GNP series. The real GNP series of the USA, West Germany, and Japan are plotted in Figure 6. Each component is scaled by its first value. The lower two lines are the US (solid line) and West German (broken line) real GNP's. The top line shows the Japanese real GNP. The statistic $T(\hat{\rho}-1)$ is about -2.24, and it is very definite if the unit root exists. Here we do not impose the unit root condition. Comparing the residuals from the first step without imposing the unit root and the results of taking the difference of the logarithms, the former seems to reveal more structure than the latter. We model the residuals of the first step by m3.2.

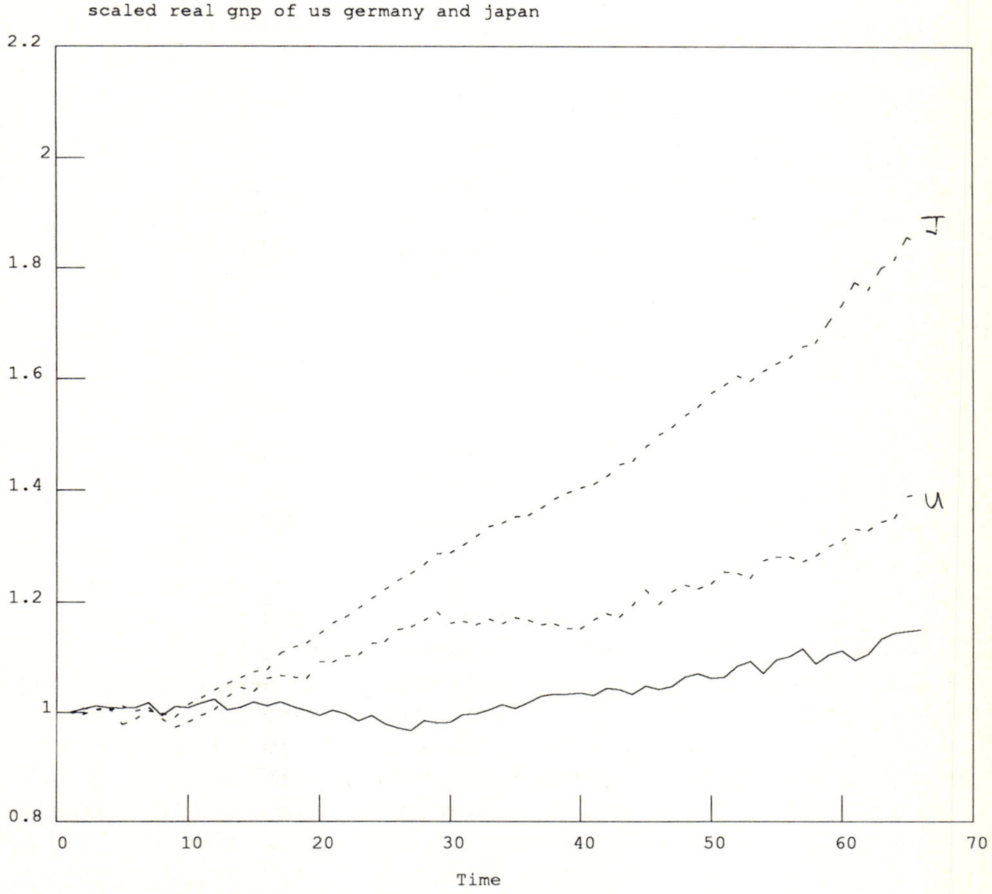

Fig. 6 Real GNP of the USA, West Germany and Japan

The first step produces a scalar model $m\,1.1$

$$\tau_{t+1} = \rho\tau_t + bv_t \;,$$

$$y_t = ct_t + v_t$$

where

$$\hat{\rho} = .966$$

$$\hat{b} = (\,-.053 \;\; 5.846\,)\,,$$

$$\hat{c} = \begin{bmatrix} ..0482 \\ .1674 \end{bmatrix}.$$

Note that

$$\hat{b}\hat{c} = .976.$$

The second step estimates $m\,3.2$

$$z_{t+1} = Fz_t + Ge_t$$

$$v_t = Hz_t + e_t$$

with

$$\hat{F} = \begin{bmatrix} .918 & -.043 \\ 0.036 & -.044 \end{bmatrix}, \; \lambda(\hat{F}) = .916, \text{ and } -.0043\,,$$

$$\hat{G} = \begin{bmatrix} -2.941 & .835 \\ 1.172 & -3.605 \end{bmatrix},$$

$$\hat{H} = \begin{bmatrix} -.282 & -.026 \\ -.026 & .166 \end{bmatrix},$$

$$cov\,\Delta = \begin{bmatrix} .145 & .0118 \\ .0118 & .137 \end{bmatrix} \times 10^{-3}.$$

The residuals of the second step are nearly uncorrelated, with D.W. statistics of 2.04 and 1.89 respectively. The covariance matrix of the innovation shows that the mutual interaction between these two countries is small and nearly symmetrical. When a two-dimensional structural model is fitted, then the matrix S introduced in Section 7.4 turns out to have an off-diagonal element of about .077. Dynamic multiplier profiles indicate that shocks to German real GNP produce monotonically-decaying responses in the two GNP series, while shocks to US GNP produces a peak in the German GNP series after about six quarters, as shown in Figure 7, although their own responses are monotonically decaying.

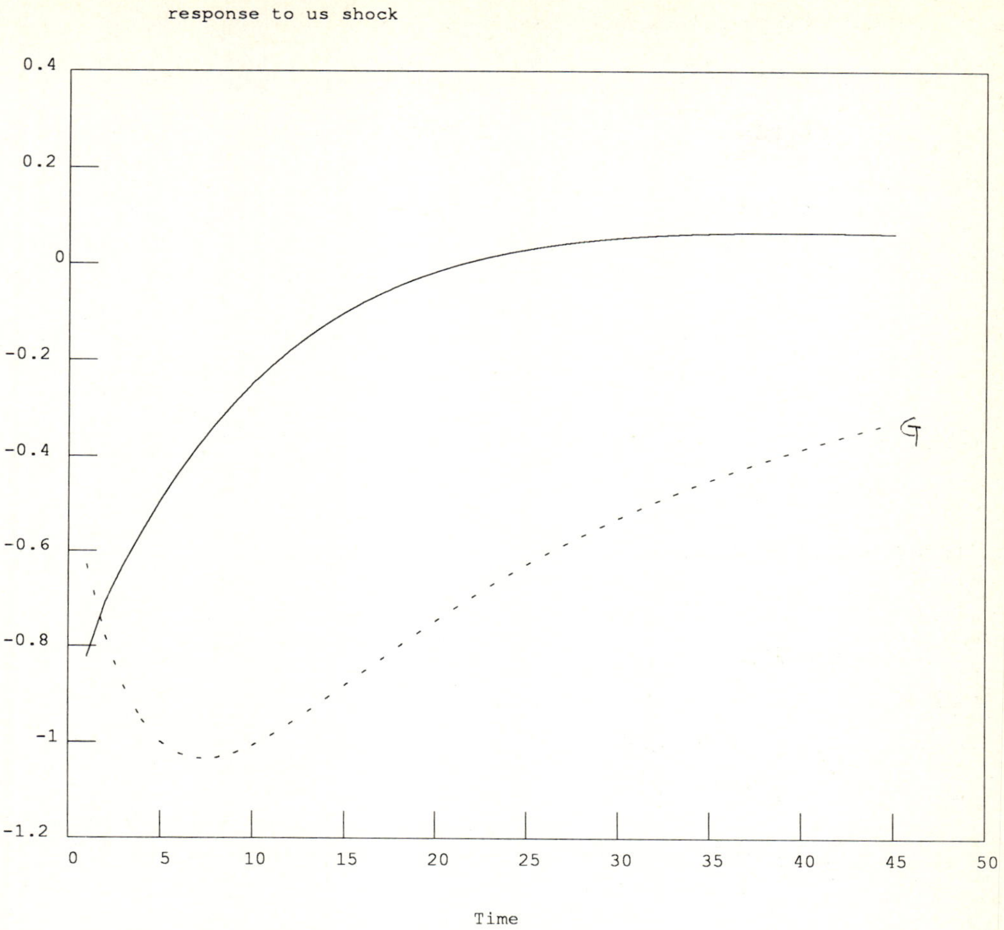

Fig. 7 Responses of the US (solid line) and German real GNP to shocks to the US real GNP

These response patterns are different qualitatively from those of the US-Japan real GNP responses we later describe.

12.5 The US and West German Real GNP and Unemployment Rate

Here is an example in which the series are clearly nonlinear. There are obvious nonlinearities in the data series, as one can see in Figure 8, which plots the unemployment rates in the two countries from 1972.2 to 1987.2. The solid line is for the US data. To avoid mixing data points before and after a possible regime shift in early 1971, which seem to be present in the real GNP series of both countries, we use the data from 1971.1 for the US and 1971.2 for West Germany with the

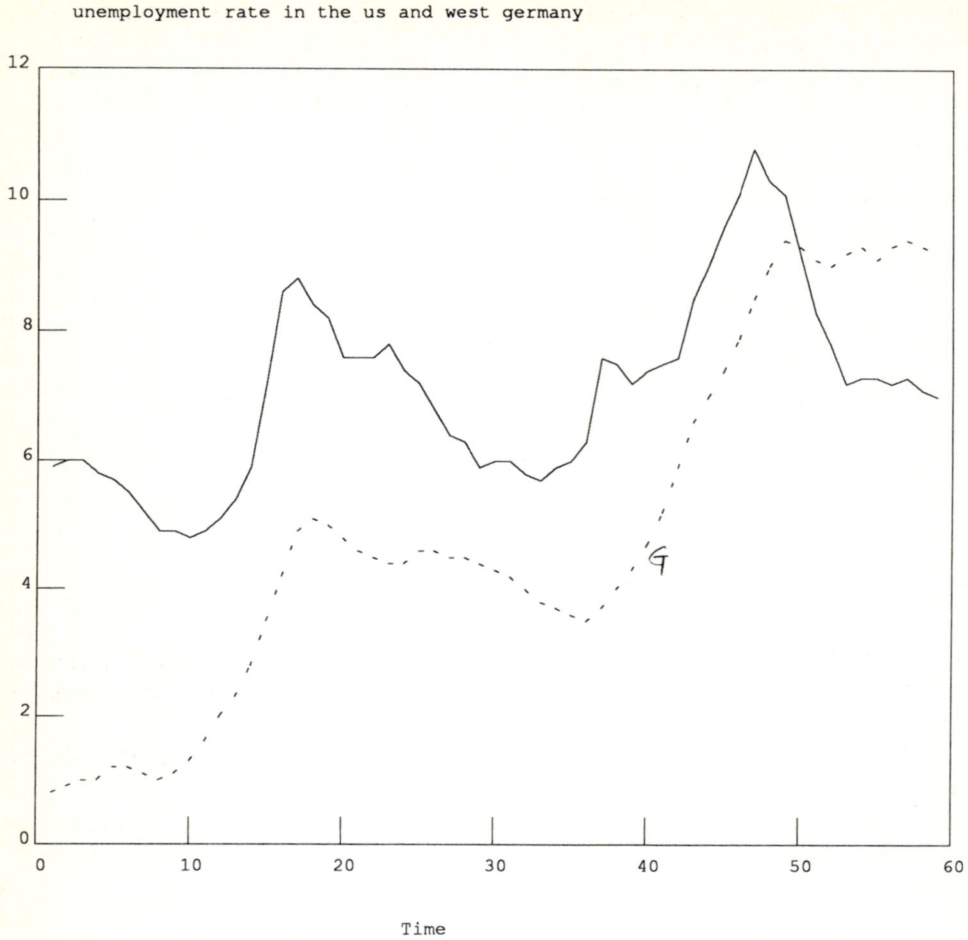

Fig. 8 Unemployment rates in the US (solid line) and West Germany

total data points of 60 and 61, respectively. The raw data are such that $var\ (ln\ y) = 1.1 \times 10^{-2}$ and $var\ (ln\ u) = 4.04 \times 10^{-2}$ for the USA and $var\ (ln\ y) = 2.4 \times 10^{-4}$ and $var\ (ln\ u) = .53$ for the West Germany. Note that the variance of $\ln y$ is much smaller, in West Germany than that in the USA, while that of $\ln u$ is much larger.

After experimenting with different scale factors and noting the effects on the residual serial uncorrelatedness, we scale the logarithm of the unemployment rate by $10^{.35}$ and $10^{1.5}$, respectively. This nonlinear transformation allows the first step of the modeling procedure to define reference time paths about which linear approximations are made in the second step to describe deviations from the reference paths. The first-step estimates of ρ are such that $T(\hat{\rho} - 1)$ is -2.935 and -3.81, respectively, for the US and West German data sets. Since we cannot definitively state that ρ is one, we use the estimated value as is in the second step and the multiplier calculation. For the US, $\ln y$ and $\ln u \,/\, 10^{.35}$ is modeled by $m\,1.1$ at the first step, given by

$$\hat{\rho} = .9511,\quad \hat{b} = (.7199,\ \ .5353),\quad \hat{c}' = (.951\ \ \ .647)\,,$$

and the second step is modeled by $m\,3.3$, with

$$F = \begin{bmatrix} .6725 & -.3344 & -.0063 \\ .3440 & 1.0409 & .00055 \\ -.0046 & .1596 & -.1412 \end{bmatrix},\quad \lambda(\hat{F}) = .857 \pm j.285,\ \ -.142,$$

$$\hat{G} = \begin{bmatrix} -.428 & 2.072 \\ .403 & -1.782 \\ -1.814 & .455 \end{bmatrix},\quad \hat{H} = \begin{bmatrix} -.449 & -.148 & -.0386 \\ .5764 & .0973 & -.0234 \end{bmatrix},$$

and

$$\Delta = cov\ e = \begin{bmatrix} .165 & -.207 \\ -.207 & .574 \end{bmatrix} \times 10^{-3}.$$

The magnitude of the complex eigenvalue is .903. The elements in the matrix $\Delta\Delta_d^{-1}$ indicate that exogenous shocks on real GNP induce more disturbances on the unemployment rate than the reverse.

The West German data set is transformed into $\ln y$ and $\ln u \,/\, 10^{1.5}$ and is modeled by

$$\hat{\rho} = .9375,\quad \hat{b} = (1.611\ \ \ 3.346),\quad \hat{c}' = (.1384\ \ \ .2186)$$

in the first step and by $m\,3.3$ at the second step with

$$\hat{F} = \begin{bmatrix} .7948 & -.2206 & .0504 \\ .2598 & .9245 & .1316 \\ -.0410 & .0743 & .4433 \end{bmatrix},\quad \lambda(\hat{F}) = .860 \pm j.214,\ \ .442$$

$$\hat{G} = \begin{bmatrix} -1.154 & 1.866 \\ 1.233 & -2.534 \\ -3.689 & -2.169 \end{bmatrix},\quad \hat{H} = \begin{bmatrix} -.633 & -.185 & .038 \\ .294 & -.092 & -.063 \end{bmatrix}$$

and

$$\Delta = cov\ e = \begin{bmatrix} .371 & -.122 \\ -.112 & .242 \end{bmatrix} \times 10^{-5}.$$

In West Germany, exogenous real GNP shocks induced fewer disturbances on the unemployment rate than in the USA. The magnitude of the complex eigenvalues is .887. The German response is a little slower than the US reduction. The German GNP response monotonically decreases, while that of the US peaks after about 2 quarters. The main difference in the responses to the exogenous increase in the unemployment rates is the sluggishness of the response of the German real GNP. It takes about 10 quarters for the German GNP to come back to the original level after a sudden drop. The US real GNP recovers only after about 7 quarters.

12.6 The US and Japan Real GNP Interaction

The same data series as that in Section 4 is used to estimate a structural model of dynamic interactions of the real GNP of the USA and Japan. The parameter ρ estimated in the first step is about the same, .972. The main difference between this model and that in Section 4 is that here the sign of Δ_{12} is positive, whereas in

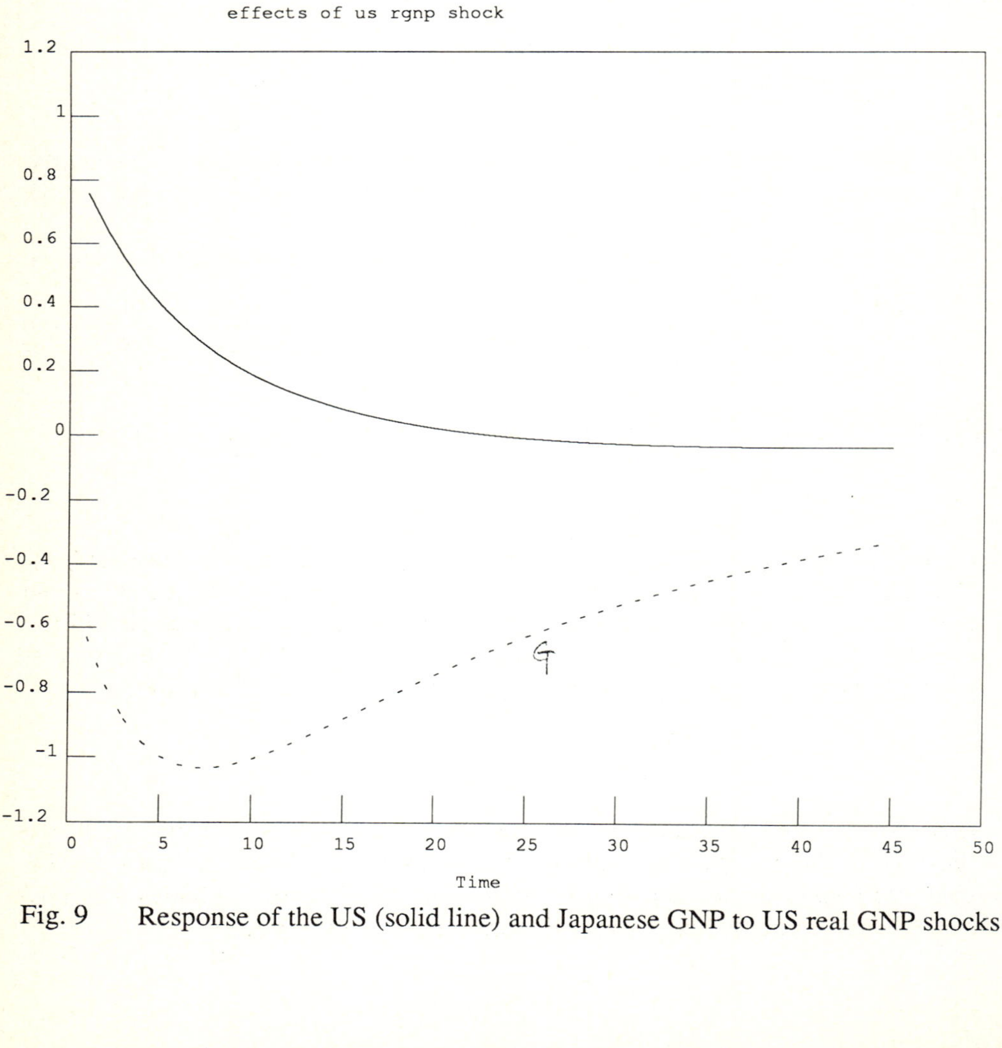

Fig. 9 Response of the US (solid line) and Japanese GNP to US real GNP shocks

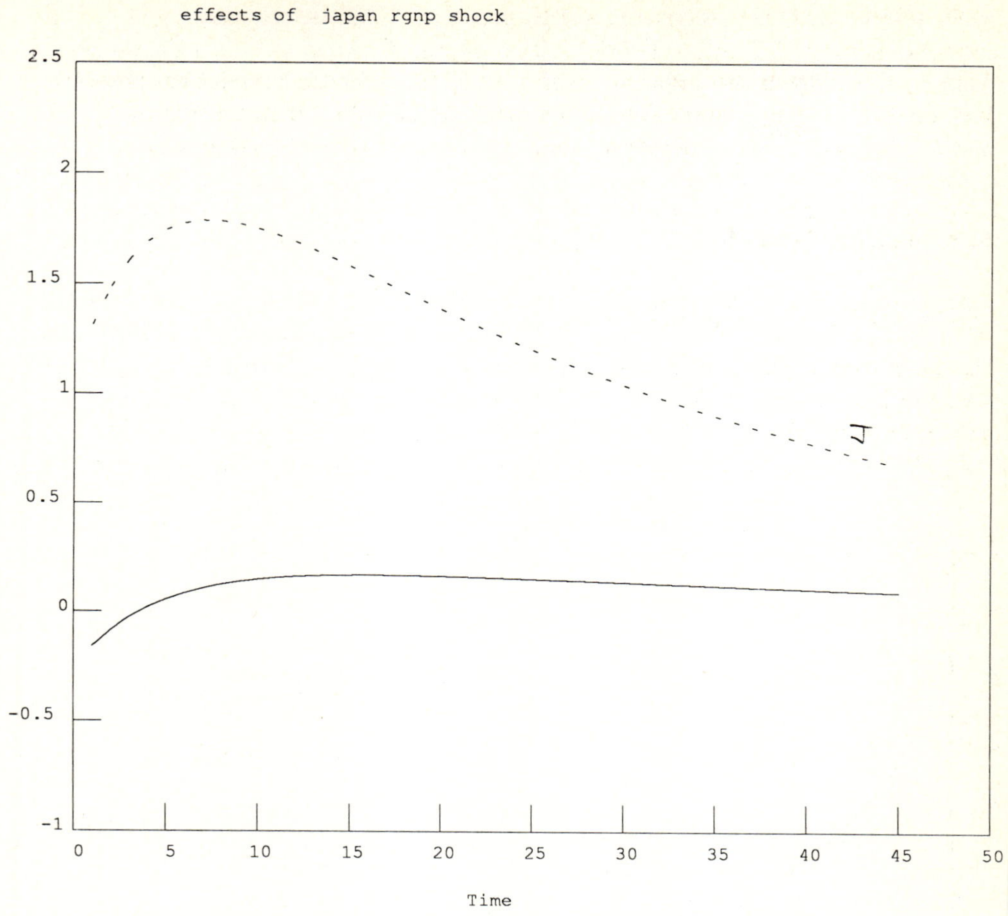

Fig. 10 Response of the US (solid line) and Japanese real GNP to Japanese real GNP shocks

the US-German case it is negative. When a symmetric matrix S is assumed, the off-diagonal element is estimated to be about -.342. The sign is opposite to that of Section 4, and the magnitude is much larger. When we constrain the ratio of the variances $\sigma_{us}^2 / \sigma_J^2$ to be 1.1 (which is about the correct ratio of the variances when the matrix S is constrained to be symmetric), the matrix S becomes

$$S = \begin{bmatrix} 1 & -.274 \\ -.159 & 1 \end{bmatrix} .$$

However, the general shapes of the impulse responses remain substantially the same, even though the magnitude of the effects of US shocks on Japanese real GNP are somewhat smaller with the nonsymmetric S matrix. Figures 9 and 10 show the impulse responses when the symmetric S matrix is used.

12.7 The USA, West Germany, and Japan Real GNP Interaction

This section shows the impulse responses of the structural model discussed in Section 7.4. The estimated system matrices are as shown there. Figures 11, 12, and 13 show the impulse responses with the matrix S given in Section 7.4.

12.8 Further Examples

Further examples of the procedure are found in Aoki [1987a, b, c, and 1989], Aoki and Havenner [1986], Bossaerts [1986], Cerchi and Havenner [1988], and Havenner and Criddle [1989].

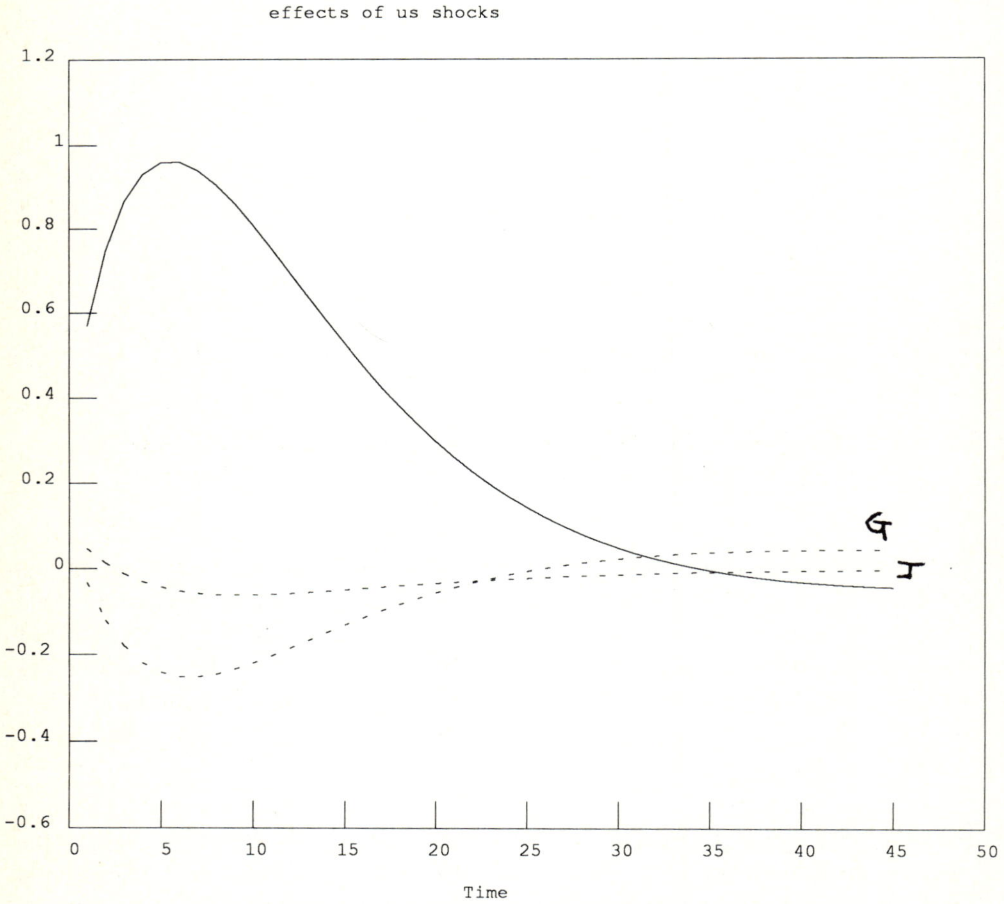

Fig. 11. Response of the three real GNPs to US real GNP shocks

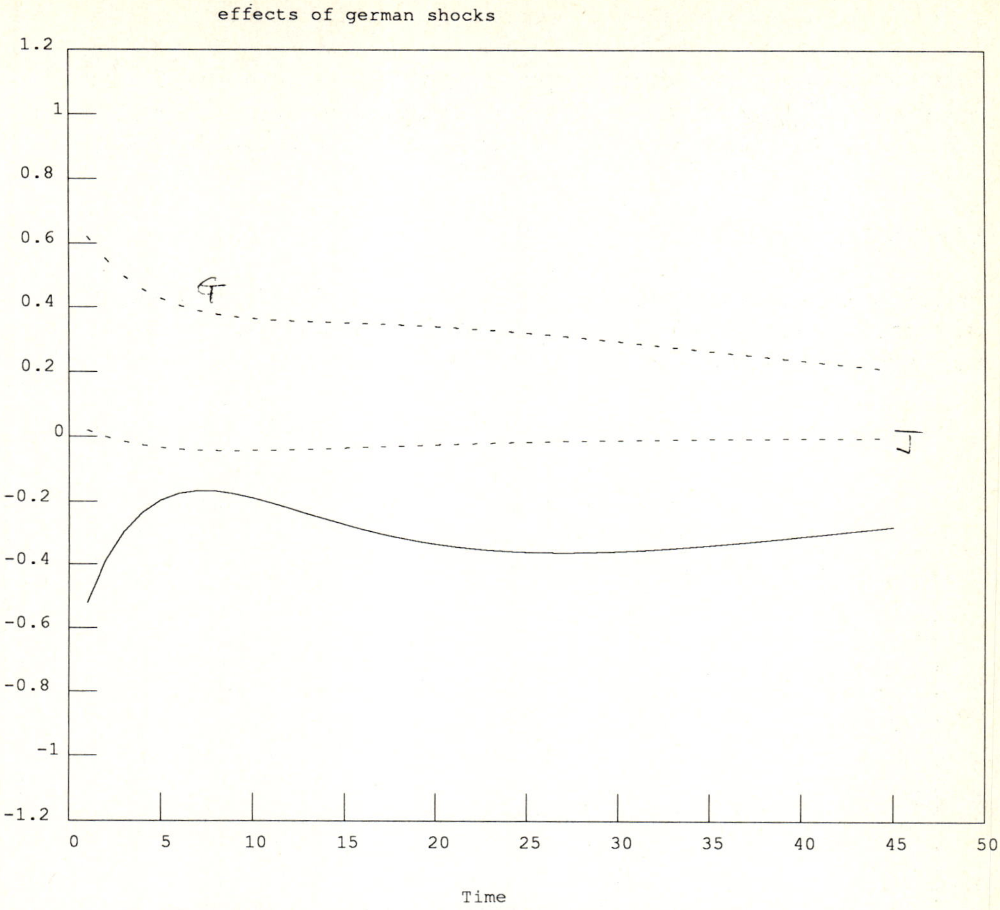

Fig. 12. Response of the three real GNPs to German real GNP shocks

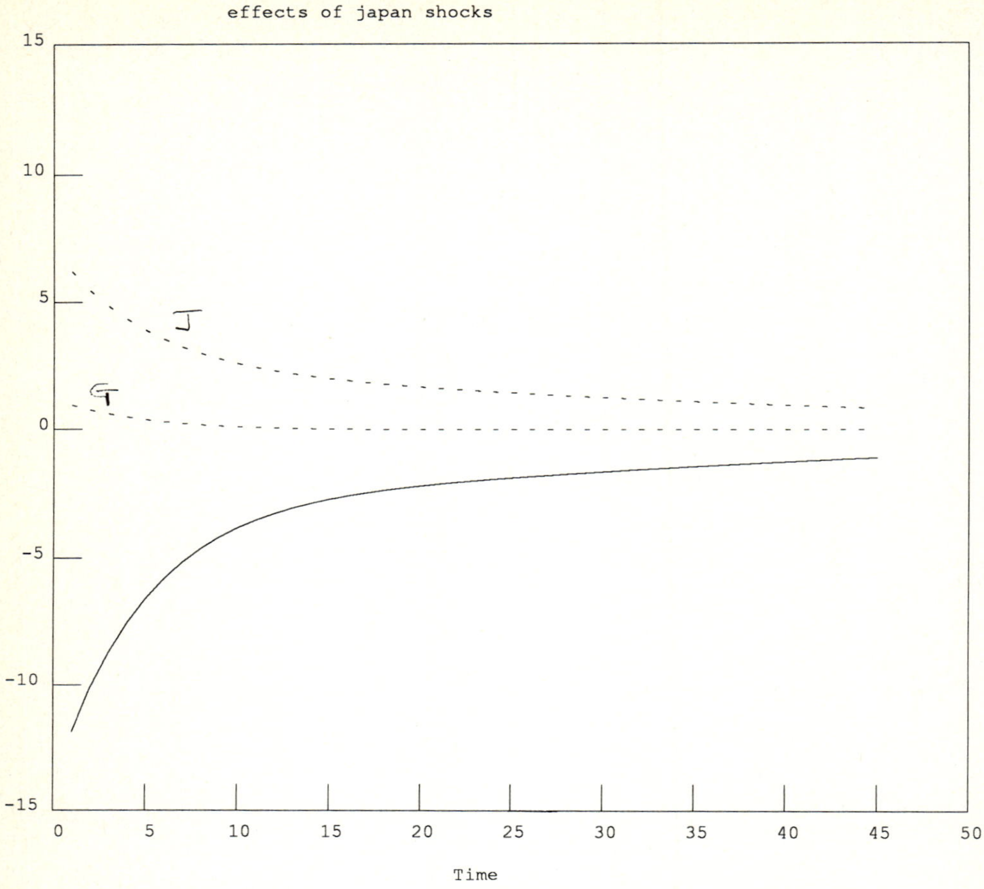

Fig. 13. Response of the three real GNPs to Japanese real GNP shocks

Appendices

A.1 Geometry of Weakly Stationary Stochastic Sequences

The set of mean-zero random vectors with finite variances can be converted into a Hilbert space by defining an inner product of two members x and y by $(x, y) = Exy'$. Two random vectors are orthogonal if their inner product is zero, i.e., if they are uncorrelated.

Suppose that y_t is related to a sequence of basic or elementary random variables, $\{\varepsilon_t\}$, $E\varepsilon_t = 0$, $E\varepsilon_t\varepsilon_s{}' = \sigma^2\delta_{t,s}$, by a moving average process

$$y_t = \theta(L)\varepsilon_t \tag{1}$$

where

$$\theta(L) = \sum_{j=0}^{\infty} \theta_j L^j, \ \theta_0 = 1.$$

For y_t to be a covariance stationary process, we assume that $\sum_{j=0}^{\infty} \theta_j^2$ is finite.

Consider predicting y_{t+m}, $m > 0$, using the information set $I_t = \{y_t, y_{t-1}, y_{t-2}, \ldots\}$ by a linear combination of current and past y's. Assume that $\theta(L)$ is invertible in the sense that $\theta(L)^{-1} = \psi(L)$ exists such that $\psi(L) = \sum_{j=0}^{\infty} \psi_j L^j$, with $\sum_{j=0}^{\infty} \psi_j^2 < \infty$. In other words, the $\{y_t\}$ process of (1) can equivalently be expressed as an autoregresive process

$$\psi(L)y_t = \varepsilon_t. \tag{2}$$

The simple example $y_t = \varepsilon_t - \varepsilon_{t-1}$ shows that not all moving average processes are invertible. However, invertible processes do constitute an important class of stochastic processes. For such processes, the information set I_t defined above is equivalent to the one containing $I_t = \{\varepsilon_t, \varepsilon_{t-1}, \cdots\}$. Then a linear prediction of y_{t+m} based on I_t is of the form

$$y_{t+m\,|t} = A(L)\varepsilon_t$$

where

$$A(L) = a_0 + a_1 L + a_2 L^2 + \cdots.$$

The best linear prediction is, by definition, the one that minimizes the prediction error variance.

$$\sigma_y^2 = E[(y_{t+m} - y_{t+m|t})^2 \mid I_t]$$
$$= E[\{\theta(L)\varepsilon_{t+m} - A(L)\varepsilon_t\}^2 \mid I_t].$$

Separate out the future ε's from those in the information set I_t by writing

$$\theta(L)\varepsilon_{t+m} = \sum_{j=0}^{m-1} \theta_j \varepsilon_{t+m-j} + \sum_{j=m}^{\infty} \theta_j \varepsilon_{t+m-j}$$

$$= \sum_{j=0}^{m-1} \theta_j \varepsilon_{t+m-j} + \theta_m(L)\varepsilon_t{}'$$

where

$$\theta_m(L) = \sum_{i=0}^{\infty} \theta_{m+i} L^i \,.$$

Then

$$\sigma_y^2 = \sum_{j=0}^{m-1} \theta_j^2 \sigma^2 + [\{\theta_m(L) - A(L)\}\varepsilon_t]^2.$$

The choice $A(L)=\theta_m(L)$ clearly minimizes the σ_y^2, i.e.,

$$y_{t+m|y} = \theta_m(L)\varepsilon_t \tag{3}$$

is the best least-squares predictor of y_{t+m} based on information contained in I_t.

Sometimes a notation $[\cdot]_+$ is used to express θ_m(L) as $[\theta(L)/L^m]_+$ where $[\cdot]_+$ collects only non-negative powers of L, dropping all expressions with negative powers of L. Then the best least squares predictor is given by $[\theta(L)/L^m]_+\varepsilon_t$. Non-negative powers of L refer to the current and past values. Recalling that z^{-1} corresponds to L, $[\ ]_+$ can be equivalently defined as non-negative powers of z^{-1} for any expression as a (formal) power series in z^{-1}. For example, write

$$F(z) = \sum_{-\infty}^{\infty} f_n z^{-n} = [F(x)]_+ + [F(x)]_-$$

where

$$[F(z)]_+ = \sum_{n=0}^{\infty} f_n z^{-n},$$

and

$$[F(z)]_- = \sum_{n=1}^{\infty} f_{-n} z^n.$$

The operation $[\ ]_+$ picks out the causal (i.e., realizable) portion of the transfer function $[\ ]$. When basic or fundamental noise sequences are involved, $[\ ]_+$ realizes the orthogonal projection onto the subspace spanned by them.

Examine further the expression in (3). The variable to be predicted, y_{t+m}, can be written as θ(L) $L^{-m}\varepsilon_t$ because $L^{-m}\varepsilon_t$ is ε_{t+m}. The best predictor, $[\theta(L)/L^m]_+\varepsilon_t$, drops the random variables ε_{t+m}, ε_{t+m-1}, ..., ε_{t+1} from $\theta(L)L^{-m}\varepsilon_t$. Because ε_{t+1} ,...,ε_{t+m} are uncorrelated with the ε's that are in the information set I_t, eliminating these uncorrelated random variables is equivalent to taking the orthogonal projection of them onto the subspace spanned by the ε's in I_t. The operator $[\theta(L)/L^m]_+$ is thus the orthogonal projection operator. Put alternately, the best predictor $y_{t+m|t}$ is such that the predictor error $y_{t+m} - y_{t+m|t}$ is orthogonal to (uncorrelated with) all the ε's in I_t, or to $y_t, y_{t-1,...,}$ by the equivalence of the subspace spanned by $\varepsilon_t, \varepsilon_{t-1}, ...$ with that spanned by $y_t, y_{t-1}, \tau = 0, 1, ...$.

This fact is sometimes referred to as the orthogonality principle. Consider a collection of random variables, $x_1, ..., x_n$, which is used to estimate another random variable y in the minimum mean square sense, i.e.,

$$\underset{\{a_i\}}{Min}\, (y-\sum a_i x_i, y-\sum a_i x_i) = \underset{\{a_i\}}{Min} E\, (y - \sum_{i=1}^{n} a_i x_i)^2.$$

Examining a slight change in a_j from its optimal value a_j^0, note that the coefficients are optimal if and only if

$$0 = (y - \sum_{i=1}^{n} a_i^0 x_i, x_j),\ j = 1,...,n.$$

Expressed in words, the error of the optimal estimator, $y - \sum_{i=1}^{n} a_i^0 x_i$, is orthogonal to every vector x_j, $j = 1, \cdots, n$. See Achieser [1956, Chapter 1], for example, for further discussion.

A.2 THE z-TRANSFORM

The z-transform of a sequence $\{x_n\}$ is formally defined by

$$X(z) = \sum_{-\infty}^{\infty} x_n z^{-n} . \tag{1}$$

It is a linear operation. The z-transform of a weighted sum of two sequences is the same weighted sum of the respective z-transforms of the two sequences.

The one-sided z-transform is defined by $X(z) = \sum_{n=0}^{\infty} x_n z^{-n}$, on the implicit assumption that only that part of the sequence $\{x_n\}$ beyond the initial time n = 0 is non-zero. An example of two-sided z-transforms is the covariance generating function of a weakly stationary stochastic time series which is discussed in the Appendix 14 in this book.

Individual members of the sequence can be recovered from X(z) by calculating the coefficient of z^{-k}, for example by

$$x_k = \frac{1}{2\pi j} \int_{|z|=1} X(z) z^{k-1} dz , \tag{2}$$

where the integral is carried out around the unit disc, $|z| = 1$. That the z-transform in the system literature is equivalent to the lag-transform used in econometrics if z^{-1} is substitued for L will soon become evident. This fact is also discussed in Aoki (1976, p.48), for example.

A typical one-sided z-transform arises in characterizing dynamic (impulse) responses of linear systems. Although dynamic systems are described or characterized in many ways, one common way is to specify a dynamic system's impulse response sequence i.e., a sequence of dynamic multipliers, because the (system) response to any other input (exogenous) sequences from the zero initial state, called the zero-state response, is describable by

$$Y(z) = H(z) U(z)$$

where

$$H(z) = \sum_{0}^{\infty} h_i z^{-i}$$

is the z-transform of the impulse responses.

Taking the z-transform is a linear operation, e.g., the z-transform of the sequence, $\{ax_n + by_n\}$, made up of a sum of scalar multiples of two other sequences $\{x_n\}$ and $\{y_n\}$, equals $aX(z) + bY(z)$, where $X(z)$ and $Y(z)$ are the z-transform of these two sequences, respectively. Properties of z-transforms can be discussed on at least two levels. At a purely formal level, the variable z^{-1} merely serves as a place marker in the representation of a sequence as a power series. For example, z^{-7} is associated with x_7, and serves to extract x_7 from $X(z)$. This

is certainly covenient. For example, the z-transform of an infinite sequence of 1s can be formally represented as $1/(1-z)$. The role of z or z^{-1} as place markers is evident in the definitions of generating functions in statistics, probability and other disciplines. On this level, one does not worry about the convergence of the formal series associated with the z-transforms. Infinite sequences are merely conveniently and compactly represented as formal power series. For example, this view is useful in relating two series that are defined by convolution $c_i = \sum_j a_{i-j} b_j$, because the z-transform of $\{c_j\}$, which is $A(z)B(z)$, where $A(z)$ and $B(z)$ are the z-transforms of $\{a_i\}$ and $\{b_j\}$ respectively, can be used to recover c_j.
Equation (1) shows that $z^{-1}x(z)$ corresponds to a sequence $\{ y_n \}$ where $y_n = x_{n-1}$ because of the relation

$$z^{-1}x(z)z^{n-1} = x(z)z^{n-2}$$

in the integrand of (2). In other words, the muliplication by z^{-1} is a backward shift operation $z^{-1}x_n = x_{n-1}$. The lag operator L in econometrics is the same as muliplication by z^{-1}. To see the effects of initial conditions in one-sided z-transforms, let $h_n = Lx_n = x_{n-1}$, and note that

$$H(z) = \sum_{n=0}^{\infty} h_n z^{-n}$$

$$= \sum_{n=0}^{\infty} x_{n-1} z^{-n} = x_{-1} + z^{-1} \sum_{m=0}^{\infty} x_m z^{-m} = x_{-1} + z^{-1} X(z).$$

If x_{-1} is zero, then $H(z) = z^{-1}X(z)$. Let $f_n = x_{n+1}$. Then $F(z) = \sum_{n=0}^{\infty} f_N z^{-n} = \sum_{n=0}^{\infty} x_{n+1} z^{-n} = z \sum_{n=0}^{\infty} x_{n+1} z^{-(n+1)} = zX(z) - zx_0$. These two expressions show that multiplication by z corresponds to forward shift, and multiplication by z^{-1} means backward shift in the time domain.

Using one-sided z-transforms we can solve difference equations by converting them into algebraic ones, just as Laplace transforms allow us to solve differential equations algebraically. First and second order difference equations have been solved this way in Appendix 1. Some more examples are listed below:

Example 1. Given $Y(z) = \sum_0^{\infty} y_n z^{-n}$, $\{y_{n+3}\}$ produces $z^3 Y(z) - z^3 y_0$ - $z^2 y_1 - z^3 y_0$ as its z-transform.

Example 2. The zero-initial state solution of $y_{k+n} + a_{n-1}\, y_{k+n-1} + \cdots + a_0 y_k = g_k$ has its z-transform $Y(z) = G(z) / [z^n + a_n z^{n-1} + \cdots + a_0]$, where G(z) and y(k) are the (one-sided) z-transforms of $\{g_k\}$ and $\{y_k\}$ respectively.

Example 3. Consider a state space model $s_{t+1} = As_t + bx_t$, and $y_t = cs_t + dx_t$. The z-transforms of these two equations are $zS(z) = AS(z) + bU(z) + zs_0$, and $Y(z) = cS(z) + dU(z)$. When $S(z)$ is substituted out from these two equations, one obtaines the relation between the z-transforms of the endogenous sequence and of the exogenous sequence as $Y(z) = [c(zI - A)^{-1}b + d]U(z) + zc(zI - A)^{-1}s_0$. The first part of this expression is the z-transform of the zero-state response, and the second is called zero-input response because it is generated by the response of the system due to the initial state s_0 with no input.

On the second and more sophisticated level, z-transforms are treated as defining analytic functions in some region of the complex plane. In some cases, the formal power series of z do converge in some region of the complex plane thus defining analytic functions. If the infinite series converge for $z = e^{j\omega T}$ then the z-transform evaluated at this value, $X(e^{j\omega T})$, may be interpreted as the Fourier transform of a sampled sequence of a continuous function of time with sampling interval T. By identifying z with $e^{j\omega T}$, we recognized (2) as the formula for calculating the coefficient of the Fourier series expansion

$$x_k = \frac{T}{2\pi}\int_{-\pi/T}^{\pi/T} X(e^{j\omega T})e^{j\omega kT}\,d\omega.$$

The z-transforms are thus related to the Fourier transforms when some specific values are assigned to z. To see this we proceed as follows. Suppose that a function of time, $x(t)$, has the Fourier transform $\hat{X}(\omega)$ where

$$\hat{X}(\omega) = \int_{-\infty}^{\infty} x(t)e^{-j\omega t}\,dt.$$

Its inverse transform is

$$x(t) = \frac{1}{2\pi}\int_{-\infty}^{\infty} \hat{X}(\omega)e^{j\omega t}\,d\omega.$$

The value of $x(\cdot)$ sampled periodically with a time interval T is given by

$$x_n = x(nT) = \frac{1}{2\pi}\int_{-\infty}^{\infty} \hat{X}(\omega)e^{jn\omega T}\,d\omega.$$

Dividing the interval of integration into segments of length T each, let us rewrite the above as

$$x_n = \frac{T}{2\pi}\sum_{-\infty}^{\infty}\frac{1}{T}\int_{(2m-1)\pi/T}^{(2m+1)\pi/T} \hat{X}(\omega)e^{j\omega nT}\,d\omega.$$

If we change the variable of integration from ω to $\omega' = \omega - 2\pi m/T$, we can rewrite the above as

$$x_n = \frac{T}{2\pi}\sum_{-\infty}^{\infty}\frac{1}{T}\int_{-\pi/T}^{\pi/T}\hat{X}(\omega' + 2\pi m/T)e^{j\omega' nT}\,d\omega',$$

where we use $e^{j2\pi mn} = 1$.

Suppose the expression

$$X^*(\omega) = \frac{1}{T}\sum_{-\infty}^{\infty} X(\omega + 2\pi m\ /\ T)$$

is well defined. Then the value of $x(\cdot)$ at $t=nT$ can be related to $X^*(\omega)$ by

$$x_n = \frac{T}{2\pi}\int_{-\pi/T}^{\pi/T} X^*(\omega)e^{j\omega nT}\,d\omega\,. \tag{3}$$

The function $X^*(\omega)$ is periodic with period $2\pi/T$ because

$$X^*(\omega + 2\pi/T) = \frac{1}{T}\sum_{-\infty}^{\infty}\hat{X}(\omega + 2\pi(m+1)/T) = X^*(\omega).$$

A periodic function of t with period Ω can be represented by a Fourier series

$$x(t) = \sum_{-\infty}^{\infty} c_n e^{-j2\pi nt/\Omega}$$

where integration of both sides from - $\Omega/2$ to $\Omega/2$ after multiplying both sides by $e^{j2\pi mt/\Omega}$ yields

$$\int_{-\Omega/2}^{\Omega/2} x(t)e^{j2\pi mt/\Omega}dt = \sum_{-\infty}^{\infty} c_n \int_{-\Omega/2}^{\Omega/2} e^{j2T(m-n)t}\,dt = c_m\Omega$$

where we use

$$\frac{1}{\Omega}\int_{-\Omega/2}^{\Omega/2} e^{j2\pi nt/\Omega}\,dt = \begin{cases}1, & n=0\\ 0, & n\neq 0\end{cases}.$$

Thus

$$c_m = \frac{1}{\Omega}\int_{-\Omega/2}^{\Omega/2} x(t)e^{j2\pi mt/\Omega}\,dt.$$

Now fomrally represent $X^*(\omega)$ by a Fourier series. Since $X^*(\omega)$ is periodic with period $2\pi/T$, its m-th Fourier coefficient, $x(mT)$, is given by

$$\frac{T}{2\pi}\int_{-\pi/T}^{\pi/T} X^*(\omega)e^{j2\pi mT/2\pi}\,d\omega = \frac{T}{2\pi}\int_{-\pi/2}^{\pi/2} X^*(\omega)e^{jm\omega T}\,d\omega\,. \tag{4}$$

Now compare this with (1) to see that x_n is the n-th Fourier coefficient of $X^*(\omega)$,

i.e.,

$$X^*(\omega) = \sum_{-\infty}^{\infty} x_n e^{-jn\omega T}$$

is the Fourier series of $X^*(\omega)$.

Define a function $X(z)$ by

$$X(z) = \sum_{-\infty}^{\infty} x_n z^{-n}.$$

We recognize $X(z)$ as the z-transform of $x(\cdot)$. Thus

$$X^*(\omega) = X(e^{j\omega T}).$$

To recover $x(nT)$ form $x(z)$, set z to $e^{j\omega T}$ and note that $d\omega = dz/jT$ to rewrite (4) as

$$x_n = \frac{1}{2\pi j} \int_{|z|=1} X(z) z^{n-1} dz.$$

This is the inverse z-transform given as (2).

Fourier series are also defined for functions defined on the closed interval $[-\pi, \pi]$. The Fourier coefficients are defined by

$$c_n = \frac{1}{2\pi} \int_{-\pi}^{\pi} f(x) e^{-jnx} dx,\ n = 0,\ \pm 1, \ldots,$$

if the integral exists. We follow Helson [1983] to characterize a class of analytic functions as follows: Let H^2 denote the class of analytic functions f in $|z| \le 1$ for which the function $f_r(\theta) = f(re^{i\theta})$ is bounded in L^2 - norm as $r \to 1$, i.e., $//f_r//_2 = (\int_{-\pi}^{\pi} |f(re^{i\theta})|^2 d\theta)^{1/2}$ remains bounded as $r \to 1$. The space H^2 then is identified with a closed subspace L^2 of the circle:

$$H^2 = (f \in L^2 : \int_{-\pi}^{\pi} f(\theta) e^{in\theta} d\theta = 0,\ n = 1,\ 2,\ \cdots).$$

In other words, the element $f \in H^2$ has the one-sided z-transform $f(z) = \sum_0^{\infty} (a_n z^{-n})$, where $z = e^{-j\theta}$, because the Fourier coefficients vanish for negative integers. The shift operator S can be defined on a space of square summable sequences of complex numbers by

$$S(a_0,\ a_1,\ a_2,\ \cdots) = (0,\ a_0,\ a_1,\ \cdots)$$

where

$$\sum |a_i|^2 < \infty,$$

or on H^2 by

$$(Sf)(\theta) = e^{i\theta} f(\theta) \quad (= z^{-1} f(z)).$$

A.3 Discrete and Continuous Time System Correspondence

Observations or data on some dynamic phenomena are available not continously but only intermittently. For example, economic activities go on all the time but records are kept, or become available, only at some discrete intervals, such as daily, weekly or annually. Even when data are gathered continuously, we may not be capable of, or interested in dealing with them continuously in time.

Given a dynamic phenomenon to model, then, one usually has a choice of representing it either in continuous or discrete time domain. A straightforward way to obtain a discrete time representation is to sample the behavior periodically, i.e., to take a sequence of snapshots of the phenomena and relate them by means of difference equations. See Aoki [1976, p.8] for such an approach.

Alternatively, continuous-time and discrete-time domain descriptions can be constructed to preserve the input-output characterization, i.e., transfer function (matrix) descriptions. We describe this procedure following Glover (1984). A discrete-time system

$$x_{t+1} = Ax_t + Be_t \;, \text{ and } y_t = Cx_t + De_t \;, \tag{1}$$

has the transfer function (matrix)

$$T(z) = D + C(zI - A)^{-1}B \;. \tag{2}$$

Changing the variable from z to s by

$$s = (z-1)/(z+1) \;, \text{ or } z = (s+1)/(s-1) \;,$$

which maps inside the unit circle in the complex plane into the left-half of the complex plane, and conversely, define the transfer function for the continous-time system by

$$S(s) = T\left(\frac{s+1}{s-1}\right) . \tag{3}$$

Some simple algebra shows that this is the transfer function of a continous-time system

$$\dot{x} = Fx + Gu \;, \text{ and } y = Hx + Ju \;, \tag{4}$$

i.e., $S(s) = J + H(sI-F)^{-1}G$, if the system matrices are defined by

$$F = (I+A)^{-1}(A-I) \;, \; G = \sqrt{2}(I+A)^{-1}B \;, \tag{5}$$

$$H = \sqrt{2}C(I+A)^{-1} \;, \text{ and } J = D - C(I+A)^{-1}B \;.$$

If the matrix A is asymptotically stable, these equations are well-defined.

This process can be reversed by starting from a continous-time system description (4) with the transfer function (matrix) $S(s)$, (5) can be solved for the system matrices of the discrete-time system (1), i.e., on the assumption that $(I-F)^{-1}$ exists, they are given by

$$A=(I-F)^{-1}(I+F)\,,\quad B=\sqrt{2}(I-F)^{-1}G\,, \tag{6}$$

$$C=\sqrt{2}H(I-F)^{-1}\ \text{,and}\ D=J+H(I-F)^{-1}G\ .$$

In other words, (5) and (6) define the one-to-one correspondence for stable dynamic systems in the continous- time domain and the discrete-time domain. This reverse assignment also preserves the transfer function, i.e., $S[(z-1)/(z+1)]=T(z)$ obtains.

More remarkably, these two systems of dynamic equations generate two Hankel matrices constructed from the respective Markov parameters which have the same set of singular values. This fact can be established by showing that the two Lyapunov equations for the observability and reachability grammians have the same solutions because the Hankel matrix is the product of these two matrices, as shown in Chapters 5 and 9. For example, start from the discrete-time domain description, and consider the Lyapunov equation for G_c, which is the solution of

$$AG_cA' - G_c = -BB'\ .$$

When the system matrices of (4) are substituted into the above, this equation yields another Lyapunov equation

$$FG_c + G_cF' = -GG'\ ,$$

which is the equation for the reachability grammian of (2). The other Lyapunov equation can be similarly shown to lead to the same equation for the continous-time system.

Finally, letting $z=e^{i\theta}$, the two transfer functions are related by

$$T(e^{i\theta}) = S(i\tan\theta/2)\ ,$$

showing that the McMillan degre of F and S are the same, i.e., if $S(s)$ is approximated by another transfer function with McMillan degree k i, $\hat{S}_k(s)$, and if $\hat{T}_k(z)$ is defined by

$$\hat{T}_k(z) = \hat{S}_k[(z-1)/(z+1)]\ ,$$

then, the i-th singular value of the difference of the transfer function is the same, i.e.,

$$T(e^{i\theta})-\hat{T}_k(e^{i\theta}) = S(i\tan\theta/2)-\hat{S}_k(i\tan\theta/2)\ ,$$

and

$$\sigma_i[T(z)-\hat{T}_k(z)] = \sigma_i[S(s)-\hat{S}_k(s)]\ .$$

A.4 Some Useful Relations for Matrix Quadratic Forms

Here we collect some useful formulas for ease of reference. See Bellman [1960] for more complete coverage, and some of the proofs.

1. Completion of squares can be used to find the minimum and the minimizing variable in

$$\min_x [a + 2b'x + x'Qx] + a - b'Qb,$$

where $Q' = Q$ and the minimizing x is given by $-Q^{-1}b$. This result can be seen by writing the expression in [.] as $a + (x + Q^{-1}b)'Q(x + Q^{-1}b) - b'Qb$.

2. The extrema of $x'Qx$ subject to the normalization condition $x'x$ is related to the minimum and maximum eigenvalues of the matrix Q by

$$\lambda_{\min}(Q) \leq x'Qx / x'x \leq \lambda_{\max}(Q).$$

3. The minimum of the same expression subject to a linear constraint $Ax = z$, where A is an m-by-n matrix of rank n, is achieved by $x = Q^{-1}A'(AQ^{-1}A')^{-1}z$.

4. The Lyapunov equation $X - A\ X\ A' = P$, where the matrix P is symmetric and positive definite, arises in several context. In this book we have encountered this type of equation in connection with grammians for observability and reachability. Here we show that stability analysis using Lyapunov functions gives rise to this type of equation. Using a positive definite symmetric matrix Q, consruct a scalar valued function by $V(x) = x'Qx$. This is an example of a function, called Lyapunov function. It is positive everywhere except at the origin, which is the unique minimum of this function. Given a dynamic system, $x_{t+1} = A\ x_t$, the Lyanpunov function is monotonically decreasing with time if

$$V(x_{t+1}) - V(x_t) = x_t'(A'QA - Q)x_t$$

if negative for all non-zero x_t, i.e., if the matrix algebraic equation $A'QA - Q = -P$ has a positive definite solution for any positive definite matrix P.

By writing this equation using the Kronecker product notation as $(I - A \times A)vecQ = vecP$, we see that the solution exists if the product of any two eigenvalues of A is not equal to 1, because the eigenvalues of $I - A \times A$ are $1-\lambda_i\lambda_j$ for all i and j. The condition that all eigenvalues of A are less than one in magnitude, i.e., that A is asymptotically stable, is sufficient for the existence of positive definite solutions.

The solution is clearly symmetric. Next, we show that the solution is unique if A is stable. Suppose that there are two solutions Q_1 and Q_2. The difference obeys the equation,$A'(Q_1 - Q_2)A = Q_1 - Q_2$. To convey the main idea, assume that the eigenvalues of the matrix A are all distinct. Let $A\ v = \lambda\ v$. Since

$$\lambda A' (Q_1 - Q_2) v = (Q_1 - Q_2) v$$

if $(Q_1 - Q_2) v$ is not zero,(and there is at least one eigenvector for which this holds, otherwise Q_1 and Q_2 will be identical, contrary to our assumption), then it is an eigenvector of A, with an eigenvalue λ^{-1}, which is greater than one in magnitude if λ is less than one in magnitude as assumed. Hence $Q_1 = Q_2$ follows.

By iterating the equation, the solution Q may be expressed as an infinite series

$$Q = P + A'PA + (A')^2PA^2 + \cdots .$$

This series is well defined for a stable A. To see this, note that

$$v' [\Sigma_{h=0}^{\infty} (A')^h\ P\ A^h]\ v = (v'\ P\ v)\Sigma_0^{\infty} |\lambda|^{2h} = (v'\ P\ v)/(1 - |\lambda|^2) < \infty,$$

for any eigenvector of A. This argument establishes the finiteness of the infinite sum when A has distinct eigenvalues and that the solution matrix is positive definite. Even with some repeated eigenvalues, the sum can be shown to be bounded and P to be positive definite. The converse is also true, because

$$v' (A'\ Q\ A - Q)v = (|\lambda|^2 - 1)v'|Q\ v = -v'\ P\ v < 0$$

implies that the magnitude of eigenvalues of A is less than one, i.e., A is asymptotically stable.

A.5 Computation of Sample Covariance Matrices

Sample covariances of $\{y\}$ are commonly calculated from a finite data $y_0, \ldots, y_{T-1}$ by

$$\hat{\Lambda}_k = \frac{1}{T} \sum_{i=0}^{T-k-1} y_{i+k} y_i' , \qquad k = 0, \ldots, T-1 . \tag{1}$$

This estimate is consistent, but biased. However, this approximation leads to an approximate spectrum which satisfies positive semi-definiteness. The next example, due to van Zee [1981], shows that the unbiased estimates obtained by replacing N^{-1} by $(N-k)^{-1}$ may lead to an approximate spectrum which is indefinite. For this reason, the former is preferred.

Example. Let $T = 3$ and $(y_0, y_1, y_2) = (1, 0, -1)$. With $(N-k)$ replacing N, sample covariances are

$$\hat{\Lambda}_0 = 2/3 , \quad \hat{\Lambda}_1 = 0 ,$$

and

$$\hat{\Lambda}_2 = -1 .$$

However, this approximation to the Hankel matrix

$$H_3 = \begin{bmatrix} 2/3 & 0 & -1 \\ 0 & 2/3 & 0 \\ -1 & 0 & 2/3 \end{bmatrix}$$

is indefinite. (It has one negative eigenvalue). Now (1) calculates the approximate covariances as

$$\hat{\Lambda}_0 = \frac{2}{3} , \qquad \hat{\Lambda}_1 = 0 , \qquad \hat{\Lambda}_2 = \frac{-1}{3} .$$

The matrix is semi-definite:

$$\hat{H}_3 = \begin{bmatrix} 2/3 & 0 & -1/3 \\ 0 & 2/3 & 0 \\ -1/3 & 0 & 2/3 \end{bmatrix} \geq 0 .$$

To see that the positive semi-definiteness is preserved with this approximation, let

$$\hat{H}_T = \begin{bmatrix} \hat{R}_0 \hat{R}_1 , \ \ldots , \ \hat{R}_{T-1} \\ \hat{R}_1 & & \cdot \\ \cdot & & \cdot \\ \cdot & & \cdot \\ \cdot & & \cdot \\ \hat{R}_{T-1} , \ \ldots , & & \hat{R}_0 \end{bmatrix} .$$

This matrix can be written as a product, $(1/T)\hat{Y}\hat{Y}'$, which shows that $\hat{H}_T \geq 0$, where

$$\hat{Y} = \begin{bmatrix} 0 \cdots 0 \, y_0 y_1 \cdots y_{T-1} \\ \\ y_0 \cdots y_{T-1} \, 0 \cdots 0 \end{bmatrix} \quad : T \times (2T-1) .$$

If the infinite-dimensional matrix Z is defined with $\hat{Z}$ as its $T \times T$ submatrix in its upper left corner and zero everywhere else, and if Y is similarly extended so that $Z = YY'/T$, then $Z \geq 0$. An alternative proof based on partial realization is given by Kimura [1982].

A.6 Properties of Symplectic Matrices

Section 7.2 describes a non-iterative method for solving the Riccati equation to determine the covariance matrix of the innovation vector. The first step of the solution procedure builds the matrix Φ defined there from the system matrices, A, M, and C. The matrix Φ is called symplectic, since it satisfies the definitional relation

$$\Phi' J = J\Phi^{-1}$$

where J is

$$J = \begin{pmatrix} 0 & I \\ -I & 0 \end{pmatrix}.$$

This appendix collect some properties of symplectic matrices. Kucera [1972] discusses symplectic matrices in the optimal control context.

1. If λ is an eigenvalue of Φ, then so is its inverse. To see this, start from the definitional relation of the eigenvalue-eigenvector pair

$$(u_i' \; v_i')\, \Phi' = \lambda_i (u_i' \; v_i')$$

and from the definitional relation of the symplectic matrix, rewrite this eigenvalue-eigenvector relation as

$$(u_i' \; v_i')\, J\Phi^{-1} = \lambda_i \; (u_i' \; v_i')J \; ,$$

which is the same relation as

$$(-v_i' \; u_i')\Phi^{-1} = \lambda_i(-v_i' \; u_i').$$

Therefore, if λ is an eigenvalue of Φ with eigenvector $(u' \; v')'$, then $(-v' \; u')'$ is an eigenvector with the eigenvalue $1/\lambda$.

If the matrix Φ has n eigenvalues

$$\Phi \begin{pmatrix} u_i \\ v_i \end{pmatrix} = \lambda_i \begin{pmatrix} u_i \\ v_i \end{pmatrix}, \quad i = 1,\dots,n \; ,$$

such that the n by n matrix composed of the n vectors, u_i, $i = 1,\dots,n$ is nonsingular, then the matrix $V\,U^{-1}$ is a solution of the Riccati equation, where the matrix U collects the vectors $u's$, and similarly for V. This has been established in Section 7.2 using the real Schur form.

2. The solution of the Riccati equation derived above is symmetric if 1 or -1 is not one of the eigenvalues of the matrix Φ.

To establish this, define $N = U'V - V'U$, and note that it can be written as $Z'JZ$, where Z is a $2n$ by n matrix built up from U and V. Noting that $\Phi Z = Z\Lambda$, and $\Phi^{-1}Z = Z\Lambda^{-1}$, where Λ is the matrix in the real Schur decomposition referred to above, the matrix N satisfies an algebraic equation

$$\Lambda' N - N\Lambda^{-1} = Z'(\Phi' J - J\Phi^{-1})Z = 0 .$$

In the last equality, the definition of the symplectic matrix has been used. If the eigenvalues λ_i and λ_j^{-1} are not equal, where λ_i and λ_j are any eigenvalues of Λ, then N must be zero. (One way to see this is by expressing the matrix equation by the Kronecker product.) From $N = 0$ follows the symmetry of the ratio VU^{-1}.

3. The matrix Φ has no zero eigenvalue if the determinant of ψ, defined in Section 7.5, is not zero. This follows straightforwardly by writing the definitional relation

$$\Phi \begin{bmatrix} u \\ v \end{bmatrix} = 0$$

in terms of the n by n submatrices, which reduces to $\psi\, u = 0$.

4. An eigenvalue λ is said to be an uncontrollable eigenvalue of the pair (A, M), if

$$w'A = w'\lambda \text{ , and } w'M = 0$$

holds for some non-zero vector w.

Similarly, λ is said to be unobservable of the pair (A, C), if

$$Az = \lambda z \text{ , and } Cz = 0$$

holds for some nonzero vector z. These definitions makes sense because these vectors will lie in the null space of the appropriate gramians, i.e., the gramians have non-trivial null spaces.

5. Apply these definitions to the eigenvector of Φ with u=0 to see that $v'\psi = \lambda^{-1}v'$, and $v'\hat{Q}=0$. That is, the inverse of any uncontrollable (non-zero) eigenvalue of the pair $(\psi, \hat{Q})$ is an eigenvalue of the matrix Φ, with an eigenvector of the form $(0, v')'$r, where v is the eigenvector associated with λ.
By setting v to zero in the eigenvector-eigenvalue relation for Φ, note that Du=0, and $\psi\, u = \lambda u$, i.e., unobservable eigenvalues of the pair (ψ, D) are the eigenvalues of the matrix Φ with the associated eigenvectors of the form $(u', 0)'$, where u corresponds with the eigenvalue λ. From $u'Du$=0 follows $M'u$=0, and hence $\psi u = A'u$, i.e., $u'A = \lambda u'$. Conversely, if u' is a row eigenvector of A, and $M'u$=0, then $(u' \; 0)'$ is an eigenvector of Φ with the same eigenvalue.

6. Eigenvalues with magnitude one have some special features. Let λ be an eigenvalue of Φ with magnitude one and the associated eigenvector $(u' \ v')'$. Writing the definitional relation out

$$(\psi - \hat{Q}\psi^{-T}D)\,u + \hat{Q}\psi^{-T}v = \lambda u\ , \tag{1}$$

and

$$-\psi^{-T}Du + \psi^{-T}v = \lambda v\ . \tag{2}$$

Multiply (2) from the left by $\hat{Q}$ and subtract it from (1) to obtain

$$\psi u + \lambda\hat{Q}v = \lambda u\ . \tag{3}$$

Multiply (2) from the left by ψ' to obtain

$$-Du + v = \lambda\psi' v\ . \tag{4}$$

Mutliply (3) from the left by $\bar{\lambda}v'$ and (4) from the left by u' and combine it with the result of the first to yield

$$u'Du = v'\hat{Q}v\ . \tag{5}$$

Here the fact that $|\lambda|^2 = 1$ has been used.

Suppose next that the vector u satisfies the relation $Du = 0$. Since Λ_0 is positive definite, this assumption is the same as $M'u$=0. Eq.(5) implies that $\hat{Q}v$ is also zero, or equivalently that Cv is zero. From (3), then, it follows that $\psi u = \lambda u$, and hence $A'u = \lambda u$, i.e., u is an eigenvector of ψ with the same eigenvalue of magnitude 1. Applying the definition in 4 above, the eigenvalue with magnitude one is an uncontrollable eigenvalue of the pair (A, M). (2) shows that v' is a left eigenvector of ψ with eigenvalue $1/\lambda$. From the definition in 4, this eigenvlaue is unobservable by the pair (A, C). On the other hand, $\hat{Q}v = 0$ implies that Du is zero. Then $\psi u = \lambda u$, and $v'\psi = \lambda^{-1}v'$ follow. Conversely, if an eigenvalue of magnitude 1 is either an uncontrollable eigenvalue of the pair (ψ, D), or an unobservable eigenvalue of the pair $(\psi, \hat{Q})$, then Φ has an eigenvalue of magnitude 1. From the definition in 4, this eigenvlaue is unobservable by the pair (A, C).

We have established that the matrix Φ has an eigenvalue of magnitude one if and only if it is either an uncontrollable eigenvalue of (A, M), or an unobservable eigenvalue of (A, C).

7. If $u'A = \lambda u'$, then, from the eigenvalue-eigenvector relation of Φ, and setting v to zero,

$$u'\psi' = u'A - u'M\Lambda_0^{-1}C = u'A\ ,$$

if $Du = 0$.

On the other hand, if $Av = \lambda^{-1}v$, then the relation $-Du + v = \lambda\psi' v$ reduces to

$$M\Lambda_0^{-1}(M'u - Cv) = 0\ . \tag{6}$$

With eigenvalues of unit magnitude, the relation

$$u'Du = v'\hat{Q}v$$

also holds. Combine these two equations to deduce a third

$$v'C'\Lambda_0^{-1}(M'u - Cv) = 0 .$$

Multiply (6) from the left by u' and subtract the result from the last equation above to see that

$$M'u = Cv .$$

Conversely, this relation implies that $Av = \lambda^{-1}v$, and $\psi u = \lambda u$. If $Du = 0$ in addition, then u is an eigenvector of the matrix A with an eigenvalue with unit magnitude. Setting $M'u=0$ shows that if an eigenvalue with unit magnitude is uncontrollable, then its inverse is unobservable, and conversely.

Example. A scalar example will help gain some insight. Consider the case where the matrix Φ has a pair of complex eigenvalues, the case with two real eigenvalues being quite straightforward. By using the real part and the imaginary part of the complex eigenvector as the basis vector, the relation

$$\Phi\,(u \;\; v) = (u \;\; v)\begin{bmatrix} \mu & \nu \\ -\nu & \mu \end{bmatrix} ,$$

holds where the vectors u and v are two-dimensional. Denote their components by subscript i, i=1, 2. Because the matrix Φ is symplectic, the cross product v_1u_2 equals v_2u_1. This can easily be seen by using the fact that $\Phi^{-1} = J\ \Phi'J'$ to express that u is the real part of the eigenvector as

$$u = J\ \Phi'J'\,(u \;\; v)\begin{bmatrix} \mu \\ -\nu \end{bmatrix} ,$$

then multiply both sides from the left by J. The left hand side becomes $(u_2 - u_1)$. Now multiply both sides from the left with a row vector v'. The left hand side becomes $v_1u_2 - v_2u_1$. The right hand side becomes the same expression multiplied by $-\mu^2 - \nu^2$ implying the claimed equality of the cross products for non-zero eigenvalues.

Denote the components of the 2×2 matrix Φ by a, b, c, and d for short. The eigenvector-eigenvalue relation then becomes

$$a\ u_1 + b\ u_2 = \mu u_1 - \nu v_1$$

and

$$c\ u_1 + d\ u_2 = \mu\, u_2 - \nu\, v_2 .$$

Multiply the first relation by u_2 and the second by u_1 and subtract the second from the first. The result is a quadratic equation for the ratio u_2 / u_1, which can easily be seen to be the same as the scalar version of the vector Riccati equation.

Eigenvalues of symplectic matrices consist of pairs of mirror images with respect to the unit circle, i.e., λ and $1 / \lambda$. Therefore, a two- by-two sypmlectic matrix with a pair of complex eigenvalues must be located on the unit circle, having the magnitude 1.

A.7 Common Factors in ARMA Models

A common facrtor in the polynomials representing the AR and the MA parts of an ARMA model, or equivalently a common factor in the numerator and the denominator of its transfer function (matrix) implies that the model is either not controllable or not observable. To see this simply, let d, y_t and e_t be scalars, and b and c be vectors in the state space representation of the model

$$x_{t+1} = A\ x_t + b\ e_t\ ,$$

and

$$y_t = c\ x_t + d\ e_t\ .$$

Its transfer function (which is the same as that of the original ARMA representation giving rise to the state space model above) is

$$G(q) = d + c\ (qI - A)^{-1}\ b = \psi(q)\ /\ \phi(q)\ ,$$

where

$$\phi(q) = |qI - A|, \text{ and } \psi(q) = \det \begin{bmatrix} qI - A & b \\ -c & d \end{bmatrix} .$$

In other words, the ARMA representation of the model is $\phi(q)\,y_t = \psi(q)\,e_t$.

Suppose now that the polynomials $\phi(q)$ and $\psi(q)$ have a common factor $(q-q_1)$. Vanishing $\psi(q_1)$ implies the existence of vectors ξ and η, not both zero such that

$$\begin{bmatrix} q_1 I - A & b \\ -c & d \end{bmatrix} \begin{bmatrix} \xi \\ \eta \end{bmatrix} = 0$$

holds. If the vector η is zero, then ξ is not zero, and hence $(q_1 I - A)\,\xi = 0$ and $c\,\xi = 0$. These two relations imply that $\xi'[c'\ c'A'\ \cdots\ c'(A')^{n-1}] = 0$, with a non-zero vector ξ. In other words, the model is not observable. Next, suppose that η is not zero. Then $(q_1 I - A)\xi + b\eta = 0$. Let $\phi(q) = (q-q_1)\hat{\phi}(q)$, and multiply the above relation by $\hat{\phi}(q)$. The result is the equality $\hat{\phi}(q) b \eta = 0$. Because we know $\phi(A) = 0$ by the Cayley-Hamilton theorem, we have shown that $\hat{\phi}(A)\,b = 0$, i.e., the model is not controllable. The converse is straightforward to establish. See Kailath [1980] for example.

We next show that a non-controllable dynamic system driven by white noise produces a probability distribution of the observation data that is confined to a subspace, i.e., a singular probability distribution function. To see this simply, let y_t be a p-dimensional vector governed by the difference equation

$$y_{t+1} = A\ y_t + b\ e_{t+1}\ ,\ \ y_0 = 0\ ,$$

where e_t is the usual mean zero white noise sequence. Assume that the rank of

$[b \; AB \; \cdots \; A^{p-1}b]$ is less than p. By this non-controllability assumption, there is a non-zero p-dimensional vector α such that $\alpha[b \; Ab. \; \cdots \; A^{p-1}b]=0$. From the Cayley-Hamilton theorem, the vector α is orthogonal to all the column vectors of the form $A^k b$ for all non-negative integers k. The dynamic equation implies that $y_t=\sum_{k=0}^{t} A^k be_{t-k}$, and hence $E(\alpha' y_t et-k)=0$ for all non-negative integers k. Therefore, $\alpha' y_t$ is equal to zero in the mean square sense. The probability distribution function for y is confined to a subspace in the space of all mean zero, finite variance random variables. The rank of the (sample) covariance of $\{y_t\}$ is thus less than p.

A non-controllable dynamic system contains a controllable subsystem which can be taken to correspond to a subvector of y by a suitable change of coordinate if necessary. This subvector may also be identified by a suitable partition of the covariance matrix of y. Let $cov \; y=\Sigma$ and denote the set of eigenvectors and the diagonal matrix of eigenvalues by Γ and Λ respectively , $\Sigma\Gamma=\Gamma\Lambda$, where Λ has only q non-zero eigenvalues where q is equal to the dimension of the controllable subsystem.

The characteristic function for y is given by $E(e^{it'y})=exp(-t'\Sigma t\,/\,2)$. Define θ by $t=\Gamma\theta$, and denote $\Gamma' y$ by ν. Then the characteristic function can be rewritten as $E(e^{i\theta'\nu})$ which is equal to $\exp(-\theta'\Lambda\theta/2)=exp(-\sum_{i=1}^{q} \lambda_i \theta_i^2\,/\,2)$, showing that $\nu_1, \nu_2, \cdots \nu_q$ are independently distributed and ν_j , $j=q+1, \; \cdots \; p$ are zero with probability one, because the latters' characteristic functions are all ones. See Aoki [1967, Appendix III] for example.

A.8 Singular Value Decomposition Theorem

Numerical examination of solutions of algebraic equations (overdetermined or otherwise) is incomplete unless some "condition numbers" are calculated to indicate degree of robustness of solutions or ill-posedness of the problem formulation. Here we illustrate use of the singular value decomposition theorem as a tool for static sensitivity analysis. The fact that any m-by-n matrix A is expressible as $A = U\Sigma V'$ where $U'U = I_m$, $V'V = I_n$ and $\Sigma = diag(\Sigma_r \;\; 0)$, $\Sigma_r =$ diag $(\sigma_1, \ldots, \sigma_r)$ and where r = rank A, is known as the singular value decomposition theorem. (See Stewart [1973] or Golub and Reinsch [1970] for example.) Using this theorem we can easily establish the following facts.

(i) $A' = V\Sigma U'$

(ii) $A'A = V\Sigma'\Sigma V,\ AA' = U\Sigma\Sigma' U'$

(iii) Let A, B be $m \times n$ matrices. Then

$$\|A - B\| = \max_{x \neq 0} \|(A - B)x\| / \|x\|$$

$$= \text{The largest singular value of } (A - B).$$

(iv) $A^+ = V\Sigma^+ U'$, where $\Sigma^+ = diag(\Sigma_r^{-1}, 0)$,

is the Moore-Penrose pseudo-inverse. (See Aoki [1967], Appendix II also.)

(v) The condition number of A is defined to be σ_1/σ_n.

(vi) Let A be an $n \times n$ matrix with eigenvalues $\lambda_1, \ldots, \lambda_n$ arranged in decreasing order of magnitude. Then $\sigma_1 \geq |\lambda_i| \geq \sigma_n$ and $cond(A) \geq |\lambda_1 / \lambda_n|$.

As a simple application of the singular value decomposition theorem, consider the sensitivity of solutions of algebraic equations. Let $Ax=b$ where A is an $n \times n$ matrix. A slight error in specifying b produces an error in the solution x. Denoting them as Δb and Δx, they are related by $A\Delta x = \Delta b$. Using the singular value decomposition $A = U\Sigma V'$, define $b = U\hat{b}$ and $x = V\hat{x}$. Do likewise for the perturbation terms with Δ's. Then $\sigma_i \hat{x}_i = \hat{b}_i$ and $\sigma_i \Delta\hat{x}_i = \Delta\hat{b}$. Since the norms of unhatted vectors are the same as the ones for hatted vectors, we note that $\sigma_n \leq \|b\| / \|x\| \leq \sigma_1$ and $\sigma_n \leq \|\Delta b\| / \|\Delta x\| \leq \sigma_1$, we can bound the ratio of relative changes as

$$(cond\ (A))^{-1} \leq \frac{\|\Delta x\| / \|x\|}{\Delta b / \|b\|} \leq cond\ (A).$$

Suppose now that A is an $m \times n$ matrix with rank $(A) = r$. The singular value decomposition with $b = U\hat{b}$ and $y = V\hat{x}$ shows that

$$\sigma_i \hat{x}_i = \hat{b}_i \ , \qquad i = 1, \ldots, r \ ,$$

and

$$0 = \hat{b}_i \ , \qquad i = r + 1, \ldots, n.$$

The solution, then, is given by $\hat{x}_i = \hat{b}_i / \sigma_i$, $1 \leq 1' \leq r$ and $\hat{x}_i$ is undertermined for $i > r + 1$.

From $|b - Ax|^2 = \sum_{i=1}^{r} |\hat{b}_i - \sigma_i \hat{x}_i|^2 + \sum_{r+1}^{m} |\hat{h}_i|^2 \geq \sum_{r+1}^{m} |\hat{h}_i|^2$, it is clear that $\hat{x}_i = \hat{b}_i / \sigma_i$ is the least square solution and $\hat{x}_i = 0$, $i \geq r+1$ produces the minimum norm solution, i.e., $x = A^{+}b$.

A.9 Hankel Matrices

This appendix describes how Hankel matrices arise as part of the description or solution of some well-known problems.

Impluse Responses. A deterministic counterpart to the prediction problem is to calculate future values of observations from past input sequences, i.e., by assuming no more exogenous sequences are applied to the system. For the model (1) of Chapter 4, the observations $y_{t+1}, y_{t+2}, \cdots, y_{t+N}$ are related to the state vector x_{t+1}, because this state vector summarizes all the information contained in the exogenous inputs that the model transmits into future observations,

$$\begin{bmatrix} y_{t+1} \\ y_{t+2} \\ \cdot \\ \cdot \\ y_{t+N} \end{bmatrix} = \boldsymbol{O}_N \, x_{t+1} \,,$$

where

$$\boldsymbol{O}_N{}' = (C' \, (CA)' \, \cdots \, (CA^{N-1})') \,,$$

and the state vector is related to the inputs by

$$x_{t+1} = \boldsymbol{C}_{t+1} \begin{bmatrix} n_t \\ n_{t-1} \\ \cdot \\ \cdot \\ \cdot \\ n_0 \end{bmatrix} ,$$

where

$$\boldsymbol{C}_{t+1} = (B \;\, AB \;, \ldots, \; A^t B) \,,$$

if the initial state is set to zero. Specifically, this expression shows how the information contained in the input sequence is transmitted to x_{t+1}. Putting these two equations together, future observations are related to the current and past inputs via the Hankel matrix, which is defined by the product of O_N and C_{t+1} in

$$\begin{bmatrix} y_{t+1} \\ \cdot \\ \cdot \\ \cdot \\ y_{t+N} \end{bmatrix} = \boldsymbol{H}_{N,t+1} \begin{bmatrix} n_t \\ \cdot \\ \cdot \\ \cdot \\ n_0 \end{bmatrix} .$$

The elements of this matrix are the Markov parameter (matrices). They also appear as the coefficients of the Laurent expansion of the transfer function matrix of (1) of Chapter 4, i.e., $C\,(zI - A)^{-1} B$.

Identification. The next scalar example illustratres the context in which the Hankel matrix with correlation coefficients as elements appears in some identification problems.

The unknown parameters in the model

$$[1+\phi(L)]\, y_t = [1+\theta(L)]\varepsilon_t$$

where

$$\phi(L) = \sum_{i=1}^{p} \phi_i L^i \text{ , and } \theta(L) = \sum_{i=1}^{q} \theta_i L^i \text{ , } q \le p \text{ ,}$$

are related to the correlations, $E(y_t\, y_{t-h}) = R_h$, by

$$(R_s\ R_{s+1}, \ldots, R_{s+p-1}) \begin{bmatrix} \phi_p \\ \cdot \\ \cdot \\ \cdot \\ \phi_1 \end{bmatrix} = -R_{s+p} \text{ , } s = 1, \ldots, p \text{ .}$$

These relations, when collected into a matrix form, produce a Hankel matrix.

Padé Approximation of Transfer Functions. Hankel matrices arise in yet another way in approximating a given impulse response sequence by that of a model with a rational transfer function. Although the procedure described below is not a recommended way of constructing a lower-order approximate model (because of numerical instability), its simplicity conveys the idea of the nature of this type of approximation, which consists in matching the first few impulse responses of the system with those of the model with a rational transfer function. The problem is to choose a rational approximation $b(z)/a(z)$ to a given sequence $\{h_i , i=1, 2, \cdots\}$, where $a(z) = 1+a_1z^{-1} + \cdots + a_n z^{-n}$, and $b(z) = b_0 + b_1z^{-1} + \cdots + b_n z^{-n}$. The first n impulse responses can be exactly matched by choosing the coefficients of $a(z)$ by solving

$$\begin{bmatrix} h_1 & \cdot & h_n \\ \cdot & \cdot & \cdot \\ \cdot & \cdot & \cdot \\ \cdot & \cdot & \cdot \\ h_n & \cdot & h_{2n-1} \end{bmatrix} \begin{bmatrix} a_n \\ \cdot \\ \cdot \\ \cdot \\ a_1 \end{bmatrix} = - \begin{bmatrix} h_{n+1} \\ \cdot \\ \cdot \\ \cdot \\ h_{2n} \end{bmatrix} .$$

The coefficients of $b(z)$ can be obtained by solving another matrix equation, which is not a Hankel matrix.

Phase-Canonical Transformation. A somewhat more technical use of a Hankel matrtix arises in transforming a controllable state space model into phase canonical form. Given a scalar model

$$x_{t+1} = Ax_t + bn_t \ , \ \text{and} \ y_t = cx_t \ ,$$

is called phase-canonical if the system parameters are given as

$$A = J - e_n a' ,$$

where

$$a' = (a_1, a_2, ..., a_n),$$

$$b' = e_n' = (0\,0 \ \cdots \ 0\,1),$$

and the matrix J has all zero elements except for one's above the diagonal line. This matrix is a shift matrix, e.g., $J\ e_n = e_{n-1}$ where e_{n-1} is a vector with elements all zero except for a single one at the (n-1)-th position.

Any reachable state space model

$$z_{t+1} = F\ z_t + g\ n_t \ , \ \text{and} \ y_t = d\ z_t \ ,$$

can be put into the phase-canonical form by a nonsingular transformation T defined by

$$T = (g\ Fg \ \cdots \ F^{n-1}g)H ,$$

where H is defined by

$$H^{-1} = (b\ Ab \ \cdots \ A^{n-1}b) .$$

This matrix turns out to be a Hankel matrix with the first column vector given by $(a_1\, a_2 \ \cdots \ a_{n-1}\ 1)'$, and the suceeding column vectors generated by shifting the immediate preceeding column vector up by one component and replacing the bottom element with zero, i.e., by multiplying by J.

Toeplitz Matrices. When column elements of a Hankel matrix are reversed, a Toeplitz matrix results. Conversely, listing the columns of a Toeplitz matrix in reverse order produces a Hankel matrix. Let S denote a special symmetric and idempotent matrix which has ones along the counter-diagonal line, and all other elements being zero, $S^2 = I$. Then $H = T\ S$, for any Hankel and Toeplitz matrices H and T. Rearrangement of the rows of a Toeplitz matrix in reverse order produces a different Hankel matrix, i.e., ST is a Hankel matrix different from TS.

Clearly, Toeplitz matrices can be obtained by the pre- or post-multiplication of a Hankel matrix by S as well. By calculating the singular value decompositions of T and H, they are found to be the same.

A.10 Spectral Factorization

Spectrum

Let $\{y_t\}$ be a mean zero weakly stationary process with the covariance matrix

$$Ey_n y_m{}' = \Lambda_{n-m}.$$

Note that

$$Ey_m y_n{}' = \Lambda_{m-n} = \Lambda_{n-m}{}'. \tag{1}$$

The z-transform of the covariance sequence (which may be regarded as a covariance generating function) is defined by

$$S(z) = \sum_{-\infty}^{\infty} \Lambda_n z^{-n}.$$

By a slight abuse of notation $S(\omega)$ will be used to denote the above expression when $z{=}e^{j\omega T}$ is substituted in, where T is the sampling interval of the data points. The assumption that $\sum \Lambda_k^2 < \infty$ ensures the $S(e^{j\omega T})$ is well-defined in mean-square sense. (An alternate definition with z replaced by its inverse is also found in the literature. In either definition, the symbol z merely serves as a place marker.)

From the definitional relation the z-transform of the covariances or the covariance generating function satisfies a relation

$$S(z^{-1}) = \sum_{-\infty}^{\infty} \Lambda_n z^n$$

$$= \sum_{-\infty}^{\infty} \Lambda'_n z^{-n}$$

$$= S'(z)..$$

Alternatively, the relation (1) shows that the covariance generating function can be written as a sum

$$S(z) = J(z) + J(z^{-1})',$$

where Re $J(e^{j\omega T}) > 0$. The function $U(z)$ is positive on $|z| = 1$. When $\Lambda's$ are scalar, $J(e^{j\omega T})$ is even in ω:

$$S(e^{j\omega T}) = \Lambda_0 + \sum_1^{\infty} \Lambda_n e^{-j\omega nT} + \sum_{-1}^{-\infty} \Lambda_n e^{-j\omega nT}$$

$$= \Lambda_0 + \sum_1^\infty \Lambda_n e^{-j\omega nT} + \sum_1^\infty \Lambda_n e^{j\omega nT}$$

$$= \Lambda_0 + 2\sum_1^\infty \Lambda_n \cos\omega nT.$$

The function $S(z)$, called spectrum, which is the z-transform of a sequence of covariance matrices, satisfies the next three properties:

(i) $S'(1/z)=S(z)$ (sometimes called parahermitian)

(ii) Analytic on $|z| = 1$,

and

(iii) $S(z) > 0$ on $|z| = 1$.

The sum form of $S(z)$ shows that (i) is true. Functions of the complex variable z satisfying (ii) and (iii) are called positive real functions.

Now regard a mean zero stationary stochastic process $\{y_t\}$ as being generated by a time-invariant linear causal dynamic system driven by another mean zero stationary stochastic process $\{x_t\}$:

$$y_t = \sum_{n=0}^\infty h_n x_{t-n}$$

where $\{h_n\}$ is the impulse response sequence of this system. By causality, h_n is zero for all negative n. The discrete transfer function G(z) is given as the one-sided z-transform

$$G(z) = \sum_0^\infty h_n z^{-n} \text{ or } G(e^{j\omega T}) = \sum_0^\infty h_n e^{-j\omega nT}.$$

Easy calculations show that the covariance matrix for i$\{y_t\}$ is given by

$$\Lambda_{yy}(k) = Ey(t+k)y'(t)$$

$$= E[\{\sum_{m=0}^\infty h_m x_{t+k-m}\}\{\sum_0^\infty h_j x_{t-j}]' \tag{2}$$

$$= \sum_m \sum_j h_m \Lambda_{xx}(k+j-m)h'_j ,$$

where $\Lambda_{xx}(n)$ is the auto-covariance matrix for the x - process and $\Lambda_{yy}(k)$ that for the y - process. By definition the spectrum of the y series equals

$$S_{yy}(\omega) = \sum_{k=-\infty}^{\infty} \Lambda_{yy}(k) e^{-j\omega kT}, \tag{3}$$

and that of the x series is given by

$$S_{xx}(\omega) = \sum_{n=-\infty}^{\infty} \Lambda_{xx}(n)\, e^{-j\omega nT}. \tag{4}$$

Substitute (2) into (3) and use (4) to rewrite the spectral density of y_t in terms of that of x_t as

$$S_{yy}(\omega) = \sum_{\tau=-\infty}^{\infty} \sum_{m=0}^{\infty} \sum_{k}^{\infty} h_m \Lambda_{xx}(\tau+k-m) h_k' e^{-j\omega\tau T}$$

$$= \sum_{m=0}^{\infty} h_m e^{-j\omega mT} \sum_{\tau=-\infty}^{\infty} \Lambda_{xx}(t+k-m) e^{-j\omega(\tau+k-m)T} \sum_{k=0}^{\infty} h_k' e^{j\omega kT}$$

$$= G(e^{j\omega T}) S_{xx}(\omega) G'(e^{-j\omega T}). \tag{5}$$

This important equation relates the spectral densities of the output and input via the transfer function. It is a type of spectral factorization result. Let the variable z correspond with $e^{j\omega T}$. Then we can factor the spectrum thus

$$S_{yy}(z) = G(z) S_{xx}(z) G(z^{-1})^* \tag{6}$$

where * denotes conjugate transpose.

A serially uncorrelated input sequence is called a white noise sequence. Since $\Lambda_{xx}(k) = Ex_{n+k}x_n' = 0$ for $k \neq 0$, its spectrum $U_{xx}(z)$ is a constant independent of z. The spectral density of a dynamic system with a white noise sequence as input can be factored, then, as $G(z)\Delta G(z^{-1})^*$ where Δ is the covariance matrix of the x-process, i.e., $\Lambda_{xx}(0) = \Delta$.

Spectral Factorization

The previous section calculates the spectrum of $\{y_t\}$, given its model dynamics or its transfer function, and a mean zero white noise sequence as its input. The spectral factorization can be thought of as the converse process of generating $\{y_t\}$ as the output of a linear dynamic system driven by white noise, given the spectrum, or equivalently the covariance generating function of the y-process.

From the previous section we know that the covariance, or the correlation coefficient, of a real-valued process $\{y_t\}$ is real and even, $S(z)=S(z^{-1})$, and that $S(z)>0$ on $|z| = 1$ because the Toeplitz matrix $[\Lambda_{|i-j|}]$ is positive semi-definite for any $m, 1 \le i, j \le m$. Since the coefficients are real, the zeros of $S(z)$ are either real or occur in complex conjugate pairs in the complex z-domain. In addition, because $S(z)$ equals $S(z^{-1})$, if $z=z_1$, is a zero, so is z_1^{-1} i.e., The zeros of $S(z)$

therefore occur in pairs of four, with the possible exception of zeros that are exactly on the unit circle. The latter occur in pairs of two (complex conjugate pairs) unless $z=\pm 1$. By collecting appropriate factors, then, $S(z)$ can be factored in a form corresponding to (6), $S(z) = W(z)W'(z^{-1})$, where $W(z)$ collects all zeros lying in $|z| \le 1$. The zero on $|z| = 1$ are equally allocated to $W(z)$ and $W'(z^{-1})$. This is the scalar version of the Spectral Factorization theorem. A basic result for a vector-valued process is:

Theorem. Let $S(z)$ be a real, rational, and full rank covariance generating function. Then it can be factored as $S(z)=W(z)\Delta W^*(z^{-1})$ where $W(z)$ is real, rational, stable, of minimum phase, and $\Delta' = \Delta > 0$.

Consequently, $W^{-1}(z)$ is analytic in $|z| > 1$. We can then write for $|z|>1$, $W^{-1}(z)=\sum_0^\infty c_k z^{-k}$, where the Taylor series expansion is valid in $|z|>1$. The matrix $W^{-1}(z)$ is the z-transform of a stable causal (one-sided) dynamics, hence $W^{-1}(z)$ is a causally stable dynamic system called whitening filter, and $\varepsilon_t = W^{-1}(z)y_t$ is the input white noise. A second way to factor a spectrum is by the Cholesky factorization of the covariance matrix. A third way is to generate innovation sequences by Kalman filters and factor spectrum accordingly.

Let

$$S(q) = \sum_{-m}^{m} \Lambda_h q^{-h}.$$

Then

$$q^m S(q) = \sum_{-2m}^{0} \Lambda_{m+r} q^{-r}$$

$$= \sum_{0}^{2m} \Lambda_{m-s} q^s ,$$

which is a polynominal of degree $2m$.

A stochastic process taking its value in real Euclidean space has a real-valued spectrum $S(e^{j\omega})$, i.e, $S(z)^* = S'(z^{-1})$. So if z_k is a zero of $S(z)$, then so is z_k^{-1}, if z_k is real. If z_k is complex, then z_k^* is also a zero. Let $\{\gamma_h, \gamma_h^*\}$ be a set of complex roots of $z^m S(z) = 0$ with $|\gamma_h| > 1$ and also containing half of those roots $|\gamma_h| = 1$. Let ρ_j be a real root. Then we can factor z^m S(z) as some constant times

$$\left[\prod_{k=1}^{d} (z-\gamma_k)(z-\gamma_k^*) \prod_{j=1}^{r} (z-\gamma_h)\right] \left[\prod_{k=1}^{r} (z-\gamma_k^{-1}) \prod_{j=1}^{s} (z-\rho_j^{-1})\right]$$

where

$$2d + s = m.$$

Let $z\beta(z)=\prod_{k=1}^{d} (z-\gamma_k)(z-\gamma_k^*) \prod_{j=1}^{s} (z-\rho_j)$. Note that $\beta(z)$ has no zero inside the unit disk. If there is no γ_h or ρ_j of modulus 1, then $\beta(z)$ has no zero in

$|z| \leq 1$, i.e., it is the z-transform of a strictly minimum delay, or, in other words, a minimum phase filter.

System theoretic construction provides an alternative to direct spectral factorization of covariance sequences. The spectrum is written as $U(z)=W(z)W^*(z^{-1})$, where $W(z)$ is analytic in $|z| \geq 1$ and of minimum phase, i.e., it has zero inside the unit circle, and rank $W(z)=r$ in $|z| \geq 1$ if rank $U(z)=r$. The matrix $W(z)$, called the spectral factor, is unique up to left multiplication by an orthogonal, real-valued constant matrix.

We next describe an algorithm for calculating the spectral factor due to Anderson et al., [1974]. One of the system theoretic results on positive realness is that J(z) is positive real if there exists a symmetric positive semi -definite matrix P such that

$$\phi(P) = \begin{bmatrix} -APA'+P, & -APC'+M \\ (-APC'+M)', & -CPC'+\Lambda_0 \end{bmatrix} \geq 0, \tag{7}$$

where

$$J(q) = I + C(qI - A)^{-1} M,$$

where rank $(M \ \ AM \ \ \cdots) = n =$ rank $(C' \ \ A'C' \ \ \cdots)$. Suppose such a P exists. Then factor $M(P)$ as

$$\phi(P) = \begin{bmatrix} B \\ I \end{bmatrix} \Delta [B' \ \ I] \tag{8}$$

and construct

$$W(q) = I + C(qI - A)^{-1} B. \tag{9}$$

We can show that $W(q)\Delta W'(q^{-1}) = J(q) + J'(q^{-1})$ by straightforward substitution when $B\Delta$, $\Delta B'$ and $B\Delta B'$ are substituted out by the corresponding expressions from (7).

To recapitulate, let $S(q)$ be a rational spectrum with full rank for almost all z. It is given by $\sum_{-\infty}^{\infty} \Lambda_h q^{-h}$ where $\Lambda_0 = D$ and $\Lambda_h = CA^{h-1}M$, $h>1$. The spectral factorization theorem tells us that $S(q)$ can be uniquely factored as $W(q)\Delta W'(q^{-1})$, $\Delta = \Delta' > 0$, where $W(q)$ has all poles and all zeros inside the unit disc, $|q|<1$, i.e., the poles of $W^{-1}(q)$ are also all inside the unit disc and $\lim_{q \to \infty} W(q) = I$. The spectrum $S(q)$ can also be written as $S(q) = J(q) + J'(q^{-1})$. By construction, the matrix $J(q)$ is readily given by $D/2 + C(qI - A)^{-1}M$ which is realized as the transfer function of the innovation model specified by

$$z_{t+1} = Az_t + M\varepsilon_t,$$

$$y_t = Cz_t + \frac{D}{2}\varepsilon_t .$$

A spectral factor $W(z)$ of $S(z)$ is realizable as the transfer function of a another model

$$z_{t+1} = Az_t + B\varepsilon_t$$

$$y_t = Cz_t + \varepsilon_t .$$

This model has the transfer matrix given by

$$W(q) = I + C(qI - A)^{-1}B ,$$

where

$$B = K\Delta^{-1}, \quad cov\,\varepsilon_t = \Delta,$$

$$M = A\Pi C' + K, \quad \Lambda_0 = C\Pi C' + \Delta,$$

and

$$\Pi = A\Pi A' + K\Delta K'.$$

Spectral factorization naturally arises in filtering problems and in control problems. Consider a Markovian model

$$x_{t+1} = Ax_t + u_t ,$$

$$y_t = Cx_t + v_t ,$$

where $\{u_t, v_t\}$ are jointly serially uncorrelated mean zero processes with covariance

$$cov\begin{bmatrix} u_t \\ v_t \end{bmatrix} = \begin{bmatrix} Q & N \\ N' & R \end{bmatrix} .$$

Its spectrum matrix is given by

$$S(q) = R + C(qI - A)^{-1}N + N'(q^{-1}I - A')C' + C(qI - A)^{-1}Q(q^{-1}I - A')^{-1}C' .$$

This can be factored in terms of a matrix called the return difference matrix

$$T(q) = I + C(qI - A)^{-1}K ,$$

i.e.,

$$S(q) = T(q)(R + C\Pi C')T'(q^{-1}),$$

where

$$K = (A\Pi C' + N)(R + C\Pi C')^{-1}$$

is the optimal Kalman filter gain and where P is the positive definite solution of the algebraic Riccati equation

$$\Pi = A\Pi A' - (A\Pi C' + N)(R + C\Pi C')^{-1}(C\Pi A' + N') + Q.$$

This has been shown by several people. See Shaked [1979] or chapter 10 for example.

The dual problem to the filtering problem discussed above is the optimal regulator problem. Consider minimizing J specified as

$$J = \sum_0^\infty (x'_{t+1}Qx_{t+1} + u_t'Ru_t) .$$

subject to a dynamic constraint

$$x_{t+1} = A_{x_t} + Bu_t .$$

Here the optimal feedback signal is given by

$$u_t = Kx_t$$

where

$$K = R^{-1}B'\Pi.$$

The matrix P is the solution of the same Riccati equation. Here the discrete return difference matrix is given by

$$T(z) = I + K(zI - A)^{-1}B$$

and

$$T'(z^{-1})(R + B'\Pi B)T(z) = R + B'(z^{-1}I - A')^{-1}Q(zI - A)^{-1}B = \psi(z)$$

is a factor of the factorization.

This development shows why optimal regulator problems and the optimal filtering problems are called dual, because the expressions for the regulator gains and filtering gains obey the same equations under suitable one-to-one correspondence.

A.11 Time Series from Intertemporal Optimization

Economic time series are generated as economic agents engage in intertemporal optimization. Although time is an extra complicating factor, problems in dynamic optimization, i.e., optimization over time, have the same basis as static (i.e., linear and nonlinear programming) problems in economics: They arise from the trade-offs that must be made in allocating scarce resources; the only difference being that trade-offs over time, i.e., intertemporal substitution, must also be examined because dynamics constrains choice sets over time. Economic time series are usually nonstationary because circumstances facing optimizing economic agents change with time and do not remain the same. Time series are also nonlinear because the dynamic structures generating data are mostly nonlinear. Faced with nonstationary and nonlinear stochastic processes, various ways of transforming them so that one can approximately deal with weakly stationary and linear problems have been discussed.

Intertemporal optimization of dynamic systems can best be approached using Markovian or state-space representation of dynamic structure. This point of view is inherent in dynamic programming and has been vigorously pursued in the systems literature. Some examples to be introduced presently illustrate how state-space representation may naturally arise in economic intertemporal optimization problems.

It should come as no surprise that a theory of dynamic optimization is best developed for linear dynamic systems. Furthermore, optimization of linear dynamic systems with quadratic performance indices can be developed in an elementary and a self-contained way without elaborate theory. Dynamic programming, when it leads to closed form solutions, is most effective and conceptually straightforward. Linear dynamic systems with quadratic separable cost or performance indices constitute an important class of intertemporal problems, which yield explicit and closed form optimization rules by dynamic programming. For this reason, this section first discusses dynamic optimization with linear dynamic systems with quadratic costs. Optimization of some linear dynamic systems with not-necessarily quadratic performance indices may be iteratively approximated by solving sequences of optimization problems for linear dynamic systems with quadratic costs (Aoki [1962]). This provides further justification for the study of linear dynamic system optimizations with quadratic costs.

When optimization problems with nonquadratic costs or nonlinear dynamics do not yield explicit analytical solutions by dynamic programming, no generally valid analytical tool exists for dealing with them. One must resort to procedures which approximate nonstationary, nonlinear phenomena by locally stationary and locally linear ones. One can proceed in at least two ways. In one approach nonlinear dynamic systems are studied as deviations from some reference paths, i.e., decision or choice variables that are normally chosen to guide nonlinear dynamic systems along some reference paths are assumed known. (In the language of control theory, reference decision variables cause the nonlinear system to "track" or follow the reference time path.) One then focuses on their deviational effects as the decision variables respond to deviations in exogenous variables causing the model to go off the reference paths. In this way, deviation of the actual time path from the reference paths are described by (variational) linear dynamic equations. (See Aoki [1976; pp. 59-62] or Aoki [1981; Chapter 2] for more detailed description of the procedure. Examples in macroeconomics are found in Aoki [1976; pp. 66-68, 239-243] and many places in Aoki [1981].) In econometrics linear (time series) models are often specified for variables that are logarithms of "more basic" variables, yielding so-called log-linear models. These models may be interpreted as arising in the way we described above as deviational or variational models. These models are then converted to state space form to apply a body of well-developed theory for dynamic optimization in state space form.

In the alternative approach one does not explicitly approximate nonlinear problems, but rather directly works with first (and second) order necessary conditions for optimality. Necessary conditions for optimality rarely yield closed-form analytic solutions to optimization problems. Optimization problems are usually too complicated to permit explicit analytic solutions. Necessary conditions are more frequently used to characterize optimal solutions, to narrow a class of possible solutions over which search for optimal solutions are conducted.

Even when explicit closed form solutions are not available, first and second order optimality conditions are often useful in characterizing optimal solutions or reducing the class of solutions from which optimal ones are to be chosen. This is very well understood in the engineering literature. In economics, however, this seems first to have been brought to the attention of the profession by Hall [1978]. Following Hall [1978] a number of recent investigators have employed this approach effectively. Pontryagin's maximum principle is the most systematic way to derive such first and second order conditions. See Canon et al. [1970] for discrete time dynamics. For continuous time version, see Lee and Markus [1967], Flemming and Rischel [1975] or Kamien and Schwarz [1981].

Example: Dynamic Resource Allocation Problem. A simplified model of Long and Plosser [1983] is used to illustrate how economic time series are generated as agents engage in dynamic, i.e., intertemporal optimization. As Long and Plosser mention, this model allows the maximizing consumer sufficient intertemporal and intratemporal substitution opportunities (i.e., among consumption goods and work vs. leisure now and in the future) that he chooses to spread the effects of unanticipated output shocks over time and across commodities. Thus, the output time series of various commodities can show both persistent fluctuations and co-movements. This example captures one way that business cycles may result from such optimizing behavior. This example also illustrates that the concept of state introduced in Chapter 2 naturally arises in formulating the intertemporal optimization problem as a dynamic programming functional equation.

Consider a dynamic allocation decision problem in which an infinitely lived individual allocates his time between leisure and work, and the available output between consumption and input for future production. First we discuss a deterministic version, then a stochastic version. The former is used to introduce and illustrate the dynamic programming procedure for formulating such intertemporal, i.e., sequential decision problems, in particular Bellman's principle of optimality. The latter is used to amplify on the notion of "state" of a dynamic system.

There are two activities producing two goods, each of which may be consumed or used as an input. In its deterministic version the problem is to maximize the present value at time t of the discounted sum of utilities given by

$$U_t = \sum_{\tau=t}^{\infty} \beta^{\tau-t} u(C_\tau, Z_\tau), \qquad 0 < \beta < 1 \tag{1}$$

where

$$u(C_\tau, Z_\tau) = \theta_0 ln Z_\tau + ln C_{1\tau} + ln C_{2\tau} \qquad \theta_0 > 0,\ \theta_i \geq 0,\ i = 1, 2.$$

subject to the next three constraints:

$$C_{jt} + X_{1jt} + X_{2jt} \leq Y_{jt'}, \qquad j = 1, 2, \tag{2}$$

$$Y_{it+1} = L_{it}^{b_i} \prod_{j=1}^{2} X_{ijt}^{a_{ij}}, \tag{3}$$

where

$$b_i + \sum_{j=1}^{2} a_{ij} = 1, \qquad i = 1, 2,$$

and

$$Z_t + L_{1t} + L_{2t} = H. \tag{4}$$

The log-linear utility function is used to yield an analytically closed form solution. Leisure time is denoted by Z_t. In (2), X_{ijt} denotes the amount of good j allocated as input to produce good i. The time devoted to producing good i is denoted by L_{it} in (3). Equation (3) is the Cobb-Douglas production function. The parameters θ, b and a_{ij} express the individual's preferences and production technologies respectively, and do not change with time. They are the structural parameters.

Since H remains constant, the knowledge of $Y_t = (Y_{1t}, Y_{2t})$ at time t completely specifies the maximum attained by U_t. For this reason we call Y_t the state vector of the problem. The constrained maximum of U_t is called the optimal value V_t. Since it depends only on Y_t we write it as

$$V(Y_t) = \max U_t ,$$

subject to (2) through (4).

Note that U_t is maximized with respect to all current and future decision variables. The current allocation decision variables are $L_{it}, C_{it}, X_{ijt}, i, j = 1, 2$. Given the current decision, the immediate or period t return is $u(C_t, Z_t)$. The state is transformed into Y_{t+1} and the problem starts all over again, i.e., the problem of choosing $L_{i\tau}$, $C_{i\tau}$, $X_{ij\tau}$ for $\tau \geq t+1$ has the same structure as the decision problem facing the individual at time t. Given that the optimal sequence of decisions is made from $t+1$ on, the maximum value is $V(Y_{t+1})$. Discounting the value from the future optimal sequence of decisions the decision at t must, therefore, maximize the discounted sum $u(C_t, Z_t) + \beta V(Y_{t+1})$, i.e.,

$$V(Y_t) = \max_{d_t}\{u(C_t, Z_t) + \beta V(Y_{t+1})\} \tag{5}$$

where d_t stands for all current decision variables. Equation (5) thus stands for a sequence of nested decisions

$$V(Y_t) = \max_{d_t}[u(C_t, Z_t) + \max_{d_{t+1}}[u(C_{t+1}, Z_{t+1}) + \max_{d_{t+2}}[u + \ldots].].].$$

If a sequence of decisions $\{d_t, d_{t+1}, d_{t+2}, \cdots\}$ is optimal, then the subsequence of decisions covering decisions from time $t+1$ on $\{d_{t+1}, d_{t+2}, \cdots\}$ must be optimal from time t+1 on. This is an illustration of Bellman's principle of optimality.

Equation (5) is a functional equation which $V(\cdot)$ must satisfy. In general this equation does not admit a closed form solution if a general utility function and a general production technology are employed. Our choice of the log-linear utility function and the Cobb-Douglas production function allows a closed form solution.

This functional equation has a solution of the form given by

$$V(Y_t) = \gamma_1 lnY_{1t} + \gamma_2 lnY_{2t} + v_t. \tag{6}$$

Substituting this into the right hand side of (5) we note

$$\theta_0 lnZ_t + \theta_1 lnC_{1t} + \theta_2 lnC_{2t} + \beta\{\theta_1 lnY_{1t+1} + \theta_2 lnY_{2t+1} + v_{t+1}\}.$$

After (3) is substituted into Y_{it+1}, $i = 1,2$, maximizing the above is a static optimization problem solved by techniques of nonlinear programming. The first order conditions for optimality (these conditions are also sufficient for this problem) are:

$$Z = \theta_0/\lambda, \quad L_{it} = \beta\gamma_i b_i/\lambda,$$

$$C_{it} = \theta_i/\mu_i, \text{ and } X_{ijt} = \beta\gamma_i a_{ij}/\mu_j,$$

where $\gamma_j = \theta_j + \beta_i \Sigma_i \gamma_i a_{ij}$, $j = 1, 2$, where λ and μ_i are the Lagrange multipliers associated with (4) and (2) respectively. (We note that the inequality (2) is always binding for our problem, i.e., the inequality is replaced with the equality.) Determine λ and μ_i from (2) and (4) as

$$\lambda = (\theta_0 + \beta\Sigma\gamma_i b_i)/H,$$

and

$$\mu_j = \gamma_j / Y_{jt}.$$

Hence the optimal decisions are given by

$$C_{it}^* = \theta_i/\gamma_i Y_{it}' \tag{7}$$

$$X_{ijt}^* = (\beta\gamma_i a_{ij}/\gamma_j) Y_{it}'$$

$$L_{it}^* = H\,\beta\gamma_i b_i/(\theta_0 + \beta\Sigma\gamma_i b_i),$$

and

$$Z^* = H\,\theta_0/(\theta_0 + \beta\Sigma\gamma_i b_i).$$

The constant term in (6) evolves with time according to

$$v_t = \beta v_{t+1} + w$$

where

$$w = \theta_0 \ln(\theta_0/\lambda) + \Sigma\theta_i \ln(\theta_i/\gamma_i) + \beta\Sigma_i\gamma_i\ \{b_i \ln(\beta\gamma b_i/\gamma)$$

$$+ \sum_j a_{ij} \ln(\beta\gamma_i a_{ij}/\gamma_i)\}$$

The transversality condition to ensure finite optimal value is

$$\lim_{\tau\to\infty} \beta^\tau v_{t+\tau} = 0,$$

for all $t \geq 0$. Then

$$v_t = w/(1-\beta).$$

Substituting (7) into (3), the optimal outputs are governed by

$$Y_{t+1} = AY_t + n \tag{8}$$

where

$$A = (a_{ij}),\ Y_t = \begin{bmatrix} Y_{1t} \\ Y_{2t} \end{bmatrix},$$

where $Y_{it} = lnY_{it}$, and

$$n_i = b_i \ln(\beta\gamma_i b_i/\lambda) + \sum_j a_{ij} \ln(\beta\gamma_i a_{ij}/\gamma).$$

We next introduce stochastic elements by a random production or technological disturbance

$$Y_{it+1} = \lambda_{it+1}(L_{it}^{b_i} \Pi_j X_{ijt}^{a_{ij}})$$

where $\lambda_{t+1} = (\lambda_{1t+1}, \lambda_{2t+1})$ is assumed to be a Markovian process, i.e., the distribution function $F(\lambda_{t+1} | \lambda_t, \lambda_{t-1}, \ldots)$ equals $F(\lambda_{t+1} | \lambda_t)$. We assume that the value of λ_{t+1} becomes known at time t+1. The notion of state must now be enlarged to include λ_t because Y_t and λ_t now completely determine the future evolution of Y's and λ's. Also we now maximize the expected discounted utility streams.

Equation (5) is replaced with

$$V(S_t) = \max_{d_t} \{u(C_t, Z_t) + (E(V(S_{t+1}) | S_t)\}$$

where

$$S_t = (Y_t, \lambda_t) .$$

Equation (6) changes into

$$V(S_t) = \gamma_1 \ln Y_{1t} + \gamma_2 \ln Y_{2t} + N(\lambda_t) + v_t$$

where

$$N(\lambda_t) = \beta E\{\Sigma_i \gamma_i \ln \lambda_{it+1} + N(\lambda_{t+1}) | S_t\}.$$

With these changes, optimal decisions given by (7) remain valid. The dynamics for Y_t are now stochastic, however, given by

$$Y_{t+1} = AY_t + n_{t+1} + n \tag{9}$$

where

$$n_{it+1} = \ln \lambda_{it+1} .$$

The intertemporal optimization problem of this example has led to the difference equation (8) which generates the sequence of y's. When randomness is introduced into the model, by means of stochastic yields, for example, then the difference equation becomes stochastic difference equation (9) which generates a sequence of random variables, i.e., a time series. This difference equation is in the state space, i.e., in a Markovian model form, because it is a first order difference equation for the state vector. How does it relate to models more familiar to econometricians? Is it an AR, MA or ARMA model? This question is easily answered by applying the Cayley-Hamilton theorem to eliminate the matrix A from the dynamic relations between y's and exogenous noises. (See Aoki [1976; p. 45], for example.) This theorem states that the matrix A being 2 by 2, A^2 can be expressed as a linear combination of A and I, i.e., $A^2 = -\alpha A - \beta I$ for some constant α and β. The dynamic equation is $Y_{t+1} = AY_t$ where A is 2 by 2. Advance t by one to note that $Y_{t+2} = AY_{t+1} = A(AY_t + x) = A^2Y_t$. Multiply Y_{t+1} and Y_t by the constants α and β respectively, and add them to Y_{t+2} to obtain

$$Y_{t+2} + Y_{t+1} + \beta Y_t = (A^2 + \alpha A + \beta I)Y_t + (A + \alpha I)n = (A + \alpha I)n$$

This is an ARMA model involving vector processes $\{Y_t\}$ and $\{V_t\}$.

Note that the elements of the matrix A are the parameters of the production function. The parameters θ_0, θ_1 and θ_2 characterize the utility function. The dynamics exhibit oscillatory behavior if eigenvalues of A are complex, or one peak may exist for a two-dimensional dynamics even when two eigenvalues both have negative real parts.

Will this two sector model exhibit a hump-shaped multiplier profile said to be characteristic of the real output? The dynamic multiplier of (8) is given by $A^k n$. Using the spectral decomposition of A, we can write

$$A^k = \Sigma_i \alpha_i^k u_i v_i',$$

where λ_i is the eigenvalue corresponding to the right eigenvector u_i, and v_i' is its (row) left-hand eigenvector.

For example, the total output multiplier with an exogenous shock to the second sector is equal to $(1\ 1)A^k \begin{pmatrix} 0 \\ 1 \end{pmatrix} = \Sigma_{i=1}^2 \lambda_i^k (u_{i1} + u_{i2})V_{i2},\ k = 0, 1, \cdots$. This is the multiplier time profile of exogenous shocks to the second sector. For the matrix A, the eigenvectors are

$$u_i = \begin{bmatrix} 1 \\ \hline \alpha_i - a_{11}/a_{12} \end{bmatrix}, \qquad i = 1,2,$$

and

$$\begin{pmatrix} v_1' \\ v_2' \end{pmatrix} = \frac{a_{12}}{\lambda_2 - \lambda_1} \begin{pmatrix} (\lambda_2 - a_{11})/a_{12} & -1 \\ (a_{11} - \lambda_1)/a_{12} & 1 \end{pmatrix}.$$

Hence

$$u_{i1} + u_{i2} = (\lambda_i - a_{11} + a_{12})/a_{12}'$$

$$v_1' \begin{pmatrix} 0 \\ 1 \end{pmatrix} = -a_{12}/(\lambda_2 - \lambda_1),$$

and

$$v_2' \begin{pmatrix} 0 \\ 1 \end{pmatrix} = a_{12}/(\lambda_2 - \lambda_1).$$

The multiplier profile is given by

$$m_t = (\lambda_2 - \lambda_1)^{-1}\{-(\lambda_1 - a_{11} + a_{12})\lambda_1^t + (\lambda_2 - a_{11} + a_{12})\lambda_2^t\}.$$

A sufficient condition for the series $\{m_t\}$ to exhibit a peak is to have $m_1 > 1$ because $m_0 = 1$ and $m_\infty = 0$, or $a_{22} + a_{12} > 1$. This condition may be interpreted as good 2 being productive as an input good. Although $a_{i1} + a_{i2} < 1$, $i = 1, 2$, the sum $a_{12} + a_{22}$ may very well be greater than one. For example $a_{11} = 0.3$, $a_{12} = 0.5$, $a_{21} = 0.2$ and $a_{22} = 0.6$ yields $a_{22} + a_{12} = 1.1 > 1$.

Quadratic Regulation Problems

Minimizations of quadratic costs subject to linear dynamic constraints are often called LQ problems, and are basic in many intertemporal optimization formulations. This class of problems is basic partly because the LQ problems are analytically tractable and give us insight into the structure of more general problems, while minimization of nonquadratic costs or inclusion of nonlinear constraints usually leads to analytically intractable problems. Furthermore, optimization problems with nonquadratic criteria and/or nonlinear dynamic constraints can often be iteratively approximated by a sequence of problems with quadratic costs and linear dynamic constraints. See Aoki [1962] for example. This is another reason for studying this class of intertemporal optimization problems.

The LQ problems for continuous-time and discrete-time dynamic systems are extensively discussed in the literature. See Canon et al. [1970] for general statements of the first order necessary conditions for optimality for discrete time problems (discrete maximum principle), for example. Whittle [1982] has a readable treatment of the LQ problems for discrete time dynamics. The maximum principle for continuous time systems is discussed in a number of books, such as Lee and Markus [1967], Fleming and Rishel [1975] and Kamien and Schwarz [1981].

A.12 Time Series from Rational Expectations Models

Anticipated future values of various endogenous and exogenous variables must be incorporated in making rational decisions. Models incorporating expected future values are described by a class of ddifference equations which involve conditionally expected values of future y's as well as current and past y's: a class we have not discussed so far. We follow Gouriéroux et al. [1982] to characterize completely the solutions of the class of first order difference equations in which y_t and a one-step-ahead prediction term $y_{t+1|t}$ appear

$$y_t = ay_{t+1|t} + u_t, \tag{1}$$

where a is a known scalar and $\{u_t\}$ is a mean-zero weakly stationary stochastic process. See also Aoki [1989, Chapter 8] for other, more complicated models. The symbol $y_{t+1|t}$ denotes the conditional expectation of y_{t+1} given an information set I_t where $I_t = \{\varepsilon_t, \varepsilon_{t-1}, \cdots\}$. Here $\{\varepsilon_t\}$ is a sequence of basic, or elementary stochastic processes. Three possibilities are examined: (1) when u_t is related to a basic stochastic process $\{\varepsilon_t\}$ by a MA process, (ii) by an AR process, and (iii) by an ARMA process. See Aoki and Canzoneri [1979] for the solution method in which terms related to $y_{t|t-1}$ rather than $y_{t+1|t}$ appear. As an example leading to dynamics of the type (1), suppose that the money demand function (in a high inflation economy) is specified by $m_t^d - p_t = \alpha\,(p_{t+1|t} - p_t)$ and the money supply is $m_t^s = \mu_t$ where p_t is the logarithm of price level. Then $p_t = ap_{t+1|t} + u_t$ where a = $\alpha/(\alpha - 1)$ and $u_t = \mu_t/(1-\alpha)$. Here $p_{t+1|t} - p_t$ is a proxy for the interest rate because the expected inflation rate completely dominates any other effects in a high inflation economy.

First, recall that the solution of (1) is a sum of a particular solution of the inhomogeneous part and general solutions of the homogeneous part $y_t = ay_{t+1|t}$. The general solutions are martingales. This can be seen by converting $y_t = ay_{t+1|t}$ into $a^t y_{t+1|t}$, , and defining Z_t to be $a^t y_t$. Then this equation is the same as the definition of a martingale $E(Z_{t+1} | I_t) = Z_t$. Denote a particular solution of (1) by y_t^P, and let y_t^h be a general solution of (1). Then $y_t = y_t^P + y_{t_h}$.

This superposition principle also works with respect to the u_t specification. Suppose $u_t = \xi_t + \eta_t$ where ξ 's and η 's are mutually independent mean zero stochastic processes. Then a particular solution for (1) can be made up as the sum of two separate particular solutions of (1), one with $u_t = \xi_t$, and the other with $u_t = \eta_t$ as disturbances. This is because $y_t^\xi = aE(y_{t+1}^\xi | \xi^t) + \xi_t$ and $y_t^\eta = aE(u_{t+1}^\eta | \eta^t) = \eta_t$ can be added together, because the solutions $E(y_{t+1}^\xi | \xi^t) = E(y_{t+1}^\xi | \xi^t, \eta^t)$ and $E(y_{t+1}^\eta | \eta^t) = E(y_{t+1} | \eta^t, \xi^t)$ obtain by the assumed independence of ξ^t and η^t, where the symbol $\{\xi^t\}$ = $\{\xi_t, \xi_{t-1} \cdots\}$ is used, and similarly for η^t.

The method of undetermined coefficients provides a basic procedure for solving (1) if the exogenous noises are independent. First, we illustrate it step by step. After a few practice examples, we can bypass many intermediate steps and proceed more directly to the solutions.

Moving Average Processes

Suppose now that u_t is MA (q), i.e., is related to the elementary stochastic process by $u_t = \theta(L)\varepsilon_t = \varepsilon_t + \theta_1\varepsilon_{t-1} + \cdots + \theta_q\varepsilon_{t-q}$ where ε_t is a mean zero white noise process with unit variance. We assume that all the roots of $\theta(L) = 0$ lie outside the unit circle. Because of linearity of (1) and independence of ε 's, we look for a particular solution of the form

$$y_t^i = ay_{t+1\,|\,t}^i + \varepsilon_{t-i} \qquad i = 0,1,\cdots,q. \tag{2}$$

Then, a particular solution $\Sigma_{i=1}^q \theta_i y_t^i$ satisties (1). Here the conditioning variables are $\varepsilon^t = (\varepsilon_t, \varepsilon_{t-1}, \cdots)$ which are common to all i.

Hypothesize that a solution to (2) is given by

$$y_t^i = \alpha_0\varepsilon_t + \alpha_1\varepsilon_{t-1} + \cdots + \alpha_i\varepsilon_{t-1}$$

where α 's are to be determined by substituting this hypothesized solution form into (2). Then advancing t by one in the above equation and projecting the resulting expression on the subspace spanned by ε^t we obtain

$$\alpha_0\varepsilon_t + \alpha_1\varepsilon_{t-1} + \cdots + \alpha_i\varepsilon_{t-1} = a(\alpha_1\varepsilon_t + \cdots + \alpha_i\varepsilon_{t+1-i}) + \varepsilon_{t-i}.$$

Comparing the coefficient of ε_{t-j} with that on the right hand side, $j = i, i-1, \cdots, 0$, we determine that $\alpha_i = 1$, $\alpha_{i-1} = a, \cdots$, and $\alpha_0 = a^i$, i.e.,

$$y_t^i = T_i(L)\varepsilon_t,$$

where T_i is defined by the above equality and define

$$T(L) = \sum_{i=0}^{q}\theta_i T_i(L).$$

To express y_t in terms of u_t, multiply both sides by $\theta(L)$

$$\theta(L) = T(L)\theta(L)\varepsilon_t = T(L)u_t.$$

By assumption, the zeros of $\theta(L)$ all lie outside the unit circle so $1/\theta(L)$ is a well-defined causal filter. We obtain a particular solution

$$y_t = \{T(L)/\theta(L)\}u_t.$$

This derivation does not reveal how $T(L)$ is related to $\theta(L)$, if at all. An alternative procedure which we discuss later tells us that

$$T(L) = \theta(a) + L\{\theta(L) - \theta(a)\}/(L-a).$$

This can be verified by substitution. It is more convenient for later development to switch from the lag to the z-transform variable, where $L = z^{-1}$. The *MA* polynomial is

$$\theta(z^{-1}) = 1 + \theta_1 z^{-1} + \cdots + \theta_q z^{-q} = z^{-q}(z^q + \theta_1 z^{q-1} + \cdots + \theta_q).$$

So, in terms of the z-variable, all the finite zeros of $\theta(z^{-1})$ lie outside the unit circle.

Now, hypothesize a particular solution of the form

$$y_t = \{\alpha + z^{-1}\gamma(z^{-1})\}\varepsilon_t,$$

for some constant α and polynomial $\gamma(.)$ and see if α and $\gamma(z^{-1})$ exist that satisfy (1). Advance t by one in the above and take its conditional expectation $y_{t+1|t} = \gamma(z^{-1})\varepsilon_t$. Substitute this into (1) to obtain a relation

$$\{\alpha + (z^{-1} - a)\gamma(z^{-1})\}\varepsilon_t = \theta(z^{-1})\varepsilon_t.$$

Setting z^{-1} to a, we see that

$$\alpha = \theta(a).$$

Thus, $\gamma(\bullet)$ must be given by

$$\gamma(z^{-1}) = \{\theta(z^{-1}) - \theta(a)\}/(z^{-1} - a).$$

This right hand side is analytic in z^{-1}. For $\theta(z^{-1})\varepsilon_t$ to be well-defined, $\gamma(z^{-1})$ must be analytic in $|z| \le 1$ and have zeros inside the unit circle.

Autoregressive Processes

Suppose next that $\phi(L)u_t = \varepsilon_t$ where $\phi(L) = 1 + \phi_1 L + \cdots + \phi_p L^p$ with all zeros outside the unit circle. The polynomial $\phi(z^{-1})$ then has all finite zeros inside the unit circle. Try a solution of the form

$$y_t = b(L)u_t \tag{3}$$

where

$$b(L) = b_0 + b_1 L + \cdots + b_{p-1}L^{p-1}.$$

The conditional expectation then becomes,

$$y_{t+1|t} = b_0 u_{t+1|t} + \beta(L)u_t$$

where

$$\beta(L) = b_1 + b_2 L + \cdots + b_{p-1}L^{p-2}.$$

The conditional expectation $u_{t+1|t}$ is calculated analogously as $u_{t+1|t} = -\alpha(L)u_t$, because $u_{t+1} + \alpha(L)u_t = \varepsilon_{t+1}$ where $\alpha(L) = \phi_1 + \phi_2 L + \cdots + \phi_p L^{p-1}$. Hence

$$y_{t+1|t} = \{\beta(L) - b_0\alpha(L)\}u_t. \tag{4}$$

Substituting (3) and (4) into (1), we observe

$$b(L)u_t = a\{\beta(L) - b_0\alpha(L)\}u_t + u_t .$$

If the polynomial $b(\cdot)$ is chosen to satisfy

$$b_0 + \beta(L)(L-a) + ab_0\alpha(L)^{-1} = 0$$

identically in L, then (3) is a particular solution.

Setting L to a in the above, the constant b_0 is equal to

$$b_0 = 1/\{1 + a\,\alpha(a)\} = 1/|\,\phi(a)$$

if $\phi(a)$ is not zero. Assuming this for now, the polynomial $\beta(L)$ is determined then

$$\beta(L) = \{1 - \phi(L)/\phi(a)\}/(L-a),$$

Hence

$$b(L) = b_0 + L(L)$$

$$= \{1/\phi(a)\}\{1 - L\frac{\phi(L)-\phi(a)}{L-a}\}.$$

We can rewrite (3) as

$$\phi(L)y_t = b(L)\varepsilon_t,$$

hence y_t is an $ARMA(p, p-1)$.

When $\phi(a)$ is zero, the trial solution

$$y_t = \{L\beta(L) + t(\gamma_0 + L\Gamma_0(L)\}u_t$$

where $\deg \Gamma_0 = p-2$ works.

If $p = 1$, then $b_0 = 0$ and $\Gamma_0(\cdot)$ is zero. If 1 is a root of $\phi(\,)$ of multiplicity d then y_t is an $ARIMA(p-d, d, p-1)$. See Gouriéroux et al. [1982].

ARMA Models

Consider (1) where

$$\phi(L)u_t = \theta(L)\varepsilon_t$$

where the root of ϕ and θ all lie outside the unit circle.

Multiply the model by $\phi(L)$ to render it as

$$\phi(L)y_t = a\,\phi(L)y_{t+1|t} + \phi(L)u_t \tag{5}$$

$$= a\,\phi(L)y_{t+1|t} + \theta(L)\varepsilon_t .$$

Introduce an auxiliary variable η_t by

$$\eta_t = \phi(L)y_t .$$

Then (5) is a first order difference equation for

$$\eta_t = \phi(L)y_t ,$$

which is the *MA* form discussed above. Its particular solution has been derived as $\eta_t = T(L)\varepsilon_t$ where $T(L) = \theta(a) + L\{\theta(L) - \theta(a)\}/(L-a)$. Hence

$$y_t = \{T(L)/\phi(L)\}\varepsilon_t \tag{6}$$

$$= \{T(L)/\theta(L)\}u_t .$$

Thus a form $\theta(L)y_t = T(L)u_t$ is suggested as a possible solution, where the lag of T is max (q , p-1).

We need $u_{t+1|t}$ in calculating $y_{t+1|t}$. Write u_{t+1} as

$$u_{t+1} = \{\theta(L)/\phi(L)\}\varepsilon_{t+1}$$

$$= (1 + \theta(L)/\phi(L) - 1)\varepsilon_{t+1}$$

$$= \varepsilon_{t+1} + \{\theta(L)/\phi(L) - 1\}\varepsilon_{t+1}.$$

Hence

$$u_{t+1|t} = \frac{1}{L}\left(\frac{\theta(L)}{\phi(L)} - 1\right)\varepsilon_t$$

$$= \frac{1}{L}\left(1 - \frac{\phi(L)}{\theta(L)}\right)u_t .$$

Then advancing t by one in (6), and adding and substracting an undetermined constant τ_0, we express

$$y_{t+1} = \{T(L)/\theta(L)\}u_{t+1}$$

$$= \{\tau_0 + \frac{1}{L}(\frac{T(L)}{\theta(L)} - \tau_0)L\}u_{t+1}$$

$$= \tau_0 u_{t+1} + \frac{1}{L}(\frac{T(L)}{\theta(L)} - \tau_0)u_t,$$

hence

$$y_{t+1|t} = \tau_0 u_{t+1|t} + \frac{1}{L}(\frac{T(L)}{\theta(L)} - \tau_0)u_t.$$

Substituting this into the original equation

$$T(L)/\theta(L) = \frac{a\tau_0}{L}(1 - \frac{\phi(L)}{\theta(L)})u_t + \frac{a}{L}(\frac{T(L)}{\theta(L)} - \tau_0)u_t + u_t$$

or

$$T(L)/\theta(L) = \frac{a\tau_0}{L}(1 - \phi(L)/\theta(L)) + \frac{a}{L}(\frac{T(L)}{\theta(L)} - \tau_0) + 1. \tag{7}$$

Letting $L = a$, τ_0 must satisfy

$$T(a)/\theta(a) = \tau_0(1 - \phi(a)/\theta(a)) + (T/a)/\theta(a) - \tau_0) + 1$$

or if $\phi(a) \neq 0$, then $\tau_0 = \theta(a)/\phi(a)$.

Substituting this back into (7) determines

$$T(L) = \frac{1}{L - a}\{L\theta(L) - a\frac{\theta(a)}{\phi(a)}\phi(L)\}.$$

As pointed out by Gouriéroux et al. [1982] this method is superior to the method of Blanchard [1979] which works only in subcases, as the next example shows. In the example Blanchard's method works only if $|a\phi_1| < 1$.

Examples

Example 1. Suppose

$$u_t + \phi_1 u_{t-1} = \varepsilon_t.$$

Then Blanchard's method solves this equation by successively advancing t, and taking conditional expectations. First take the conditional expectation of $u_{t+1} = -\phi_1 u_t + \varepsilon_{t+1}$ to obtain $u_{t+1|t} = -\phi_1 u_t$. Similarly from $u_{t+2} = -\phi_1 u_{t+1} + \varepsilon_{t+2}$ derive $u_{t+2|t} = -\phi_1 u_{t+1|t} = (-\phi_1)^2 u_t$ and $u_{t+1|t} = (-\phi_1)^i u_t$ in general $i > 0$. Note, however, that

$$y_t = \sum_{i=0}^{\infty} (-a\phi_1)^i u_t$$

converges if and only if $|a\phi_1| < 1$. The procedure due to Gouriéroux et al. shows that $y_t = u_t/(1+\phi_1 a)$ as a particular solution always unless $1+\phi_1 a \neq 0$.

Example 2. Consider a simple model of a closed economy given by

$$y_t = -\sigma\{i_t - (p_{t+1|t} - p_t)\} + \eta_t, \tag{8}$$

$$y_t = a(p_t - p_{t|t-1}) + s_t, \tag{9}$$

$$\mu_t = -ki_t + p_t + y_t. \tag{10}$$

Equation (8) is the aggregate demand equation. The aggregate supply function is given by (9), which is based on specific wage-contracting assumptions. The demand for real balances is given by (10), where the price index term drops out from both sides by assuming unit income elasticity of demand for real balances.

Solve (10) for i_t

$$i_t = \frac{-\mu_t + p_t + y_t}{k}.$$

Equating (8) with (9) we obtain the dynamic equation for the price time series:

$$\pi_1 p_t - \pi_2 p_{t|t-1} - p_{t+1|t} = (1+\sigma/k)s_t - \eta_t - (\sigma/k)\mu_t \tag{11}$$

where

$$\pi_1 = a(1+\sigma/k) + \sigma(1+1/k), \quad \pi_2 = a(1+\sigma/k).$$

Suppose the noises on the right hand side are specified by $\Sigma_{i=1}^{3} f_j(L)\varepsilon_t^j$. Postulate that the solution is of the form $p_t = \Sigma_j(\gamma_j + Lh_j(L))\varepsilon_t^\gamma$. Then $p_{t|t-1} = \Sigma_j Lh_j(L)\varepsilon_t^j$. Because $p_{t+1} = \Sigma_j(\gamma_j + Lhj(L))\varepsilon_{t+1}^j$, we note that $p_{t+1|t} = \Sigma_j h_j(L)\varepsilon_t^j$ (11), we find that γ_j and $h_j(z)$ must satisfy (See Futia [1979]) the following relation

$$\pi_1\gamma_j + (\pi_1 z - \pi_2 z - \sigma)h_j(z) = f_j(z).$$

Let λ be the root of $(\pi_1 - \pi_2)z - \sigma = 0$. Then we solve for γ_j by

$$\gamma_j = \frac{1}{\pi_1} f_j(\lambda)$$

when

$$\lambda = \sigma/(\pi_1 - \pi_2) = (1+1/k)^{-1},$$

and h_j is given by

$$h_j(z) = (f_j(z) - \pi_1\gamma_j)/(\pi_1 z - \pi_2 z - \sigma) = \{f_j(z) - f_j(\lambda)\}/[(\pi_1 - \pi_2)z - \sigma].$$

Note, however, that h_j will not be analytic inside the unit disc $|z| < 1$, unless

$$f_j(z) - f_j(\lambda) = (z - \lambda)^l g_j(z), g_j(\lambda) \neq 0, \quad l \geq 1$$

or

$$f_i(z) = c_j + (z - \lambda)^l g_j(z)$$

for some $c_j \neq 0$.

Some particular cases obtain by specializing $f_j(\cdot)$: Let $f_i(z) = c_j$. Then $\gamma_j = c_j/\pi_1$ and $h_j = 0$; If $f_j(z) = c_j + (z - \lambda)d_j$, then $\gamma_j = c_j/\pi_1$ and $h_j = d_j$.

A.13 Data Sources

Macroeconomic time series from the US, the UK, and Germany are supplied by the Board of Governors of the Federal Reserve System. The real GNP series are supplied by P.Clark, Graduate School of Business, Stanford University.

The German industrial production index (including construction) is seasonally adjusted with 1980 as the base year. The German currency and sight deposits (M1) in billion Deutsch Mark are also seasonally adjusted. The German cost of living index with 1980 as the base year is not seasonally adjusted. Similarly, both the index of industrical production and the currency and private demand deposits (M1 in billion pounds) for the United Kingdom are seasonally adjusted. The retail price index with January 1974 as the base is not seasonally adjusted.

The data sets for these two countries cover the period from January 1974 through September 1984. The series for the industrial production index and the consumer price index for the United States use 1967 as the base year.

The Japanese time series have been taken from Nikkei Data Bank. Five series are used consisting of M2 plus certificates of deposits in .1 billion Yen, call rate (in Tokyo, unconditional average; free rate after April 1979), production index-mining and manufacturing (seasonally adjusted), the whole price index for all commodities, and the current account in million Dollars. The base year for the two indices is 1975. The data set covers the period from January 1975 through January 1982.

References

Achieser, N. I. (1956), *Theory of Approximation*, translated by C. J. Hyman, F. Ungar Pub. Co., New York

Adamjan, V. M., D. Z. Arov and M. G. Krein (1968), "Infinite Hankel Matrices and Generalized Caratheodory-Fejer and Riesz Problems," *Funct. Anal. Appl. 2*, 1-18.

____________ (1971), "Analytic Properties of Schmidt Pairs for a Hankel Operator and the Generalized Schur-Takagi Problem," *Math USSR Sbornik 15*, 31-73.

Ahn, S. K., and G. C. Reinsel (1987), "Estimation for Partially Nonstationary Multivariate Autoregressive Models," presented at the 1987 NSF-NBER Time Series Workshop, Raleigh, NC.

Akaike, H. (1973), "Maximum Likelihood Identification of Gaussian Autoregressive Moving Average Models," *Biometrika, 60*, 255-265.

____________ (1974), "Markovian Representation of Stochastic Processes and Its Application to the Analysis of Autoregressive Moving Average Processes," *Ann. Inst. Statist. Math., 26*, 363-387.

____________ (1975), "Markovian Representation of Stochastic Processes by Canonical Variables," *SIAM Journal on Control, 13*, 162-173.

____________ (1976), "Canonical Correlation Analysis of Time Series and the Use of an Information Criterion," in R. Mehra and K. Lainiotis, eds., *System Identification: Advances and Case Studies*, Academic Press, Inc., New York.

____________ (1980), "Seasonal Adjustment by a Bayesian Modeling," *J. Time Series Analysis, 1*, 1-13.

____________ (1980), "Likelihood and the Bayes Procedure in Bayesian Statistics," in J. N. Bernado et al., eds., *Bayesian Statistics*, Valencia University Press, Valencia, Spain, 141-166.

Akaike, H. et al. (1985), "TIMSAC-84", Computer Science Monographs, No. 22 and 23, *Inst. Stat. Math.*, Tokyo.

Anderson, B. D. O., and S. Vongpanitlerd (1973), *Network Analysis and Synthesis*, Prentice-Hall, Englewood Cliffs, NJ.

Anderson, B. D. O., K. L. Hitz, and N. D. Dieu (1974), ''Recursive Algorithm for Spectral Factorization,'' *IEEE Trans. Circuit and Systems*, CAS-21, 742-750.

Anderson, B. D. O., and J. B. Moore (1975), *Optimal Filtering*, Prentice-Hall, New Jersey.

Anderson, T. W. (1963), ''Determination of the Order of Dependence in Normally Distributed Time Series,'' in *Proc. Symp. Time Series Anal.,* M. Rosenblatt (ed.), J. Wiley.

____________ (1971), *The Statistical Analysis of Time Series*, J. Wiley, New York.

____________ and Kunimoto, N. (1989), ''Asymptotic Robustness In Regression and Autoregression Based on Lindeberg Conditions,'' Tech. Report No. 23, Dept. Statistics, Stanford Univ., June.

Ansley, C. F. (1979), ''An Algorithm for the Exact Likelihood of a Mixed Autoregressive Moving Average Process,'' *Biometrika, 66,* 59-65, 1979.

____________ and P. Newbold (1980), ''Finite Sample Properties of Estimators for Autoregressive Moving Average Models,'' *J. Econometrics 13* 129-183.

Aoki, M. (1962), ''On a Successive Approximation Technique in Solving Some Control System Optimization Problems,'' *J. Math. Anal. Appl., 15*, 418-434.

____________ (1967), *Optimization of Stochastic Systems*, Academic Press, New York.

____________ (1968), ''Control of Large Scale Dynamic Systems by Aggregation,'' *IEEE Trans. Aut. Control*, AC-13, 246-253.

____________ (1968), ''Note on Aggregation and Bounds for Solution of the Matrix Riccati Equation,'' *J. Math. Anal. Appl., 21*, 379-383.

____________ (1971), *Introduction to Optimization Techniques; Fundamentals and Applications of Nonlinear Programming*, Macmillan, New York.

____________ (1971), Aggregation, Chapter 5 in D. A. Wismer, ed., *Optimization Methods for Large-Scale Systems, McGraw-Hill, New York.*

__________ *(1976), Optimal Control and System Theory in Dynamic Economic Analysis*, N. H., Amsterdam.

__________ (1981), *Dynamic Analysis of Open Economies*, Acad. Press, Inc. New York.

__________ (1983), *Note on Economic Time Series Analysis: System Theoretic Approach*, Springer-Verlag, Heidelberg.

__________ (1987), "An Alternative Measure of Random Walk Components in Time Series," *Economics Letters 24*, 227-230.

__________ (1987a), "Studies of Economic Interdependence by State Space Modeling of Time Series: US-Japan Example," *Annal. d'Economie et de Statistique*, No. 617.

__________ (1987b) "Decentralized Monetary Rules in a Three-Country Model: Time Series Evidence of Structural Independence," presented at the 4th International Meeting of Money and Banking, Aix-en-Provence.

__________ (1987c) "Time Series Evidence of Real GNP Interdependence of the USA, West Germany and Japan," presented at the 1987 Far Eastern Meeting of the Econometric Society, Tokyo.

__________ (1988) , "State Space Models for Vector-Valued Time Series with Random Walk Components," invited paper, the 1988 annual meeting of the Am. Stat. Assoc., New Orleans.

__________ (1988) "Alternative State Space Representation of Time Series Models," *J. Econ. Dyn. Control, 12*, 595-607.

__________ (1988) "Cointegration, Error Correction and Aggregation in Dynamic Models: A Comment," *Oxford Bulletin Econ. Stat. 50*, 89-95.

__________ (1989), "Comparisons of Alternative Identification Schemes for the US Real GNP - Unemployment Level Correlation: Sensitivity Analysis," Discussion Paper, Inst. Empirical Macroeconomics, Fed. Res. Bank, Minneapolis.

__________ (1989) *Optimization of Stochastic Systems: Topics in Discrete Time Systems, 2nd Edition*, Academic Press, Cambridge, MA.

____________ (1989) ''Instrumental Variable Estimators for State Space Models'', Discussion Paper 19, Inst. Empirical Macroeconomics, Minneapolis, Nov.

____________ (1990) ''State Space Approach to Modeling Cointegrated Time Series,'' in G. F. Rhodes, Jr., and T. B. Fomby, eds. *Advances in Econometrics, Cointegration, Spurious Regressions and Unit Roots*, JAI Press, Inc., Connecticut.

____________ (1990a), ''A State Space Time Series Modeling Method Without Individual Detrending,'' Forthcoming, *J. Time Series Anal.*

Aoki, M., and M. Canzoneri (1979), ''Reduced Forms of Rational Expectation Models,''*Q. J. E., 93*, 59-71.

Aoki, M., and A. Havenner (1989), ''A Method for Approximate Representation of Vector-Valued Time Series and Its Relation to Two Alternatives,''*J. Econometrics 42*, 181-199.

____________ and A. Havenner (1986), ''Approximate State Space Models of Some Vector-Valued Macroeconomic Time Series for Cross-Country Comparisons,'' *J. Econ. Dyn. Control 10*, 149-155.

Arato, M. (1961), ''On the Sufficient Statistics for Stationary Gaussian Ramdom Processes,'' *Theory. Prob., Its Appl., 6,* 199-201.

Arun, K. S., B. Rao, and S. Y. Kung (1983), ''A New Predictive Efficiency Criterion for Approximate Stochastic Realization,'' 1983 *Conf. on Decision and Control Preprint*, 1353-1355.

Ashenfelter, O., and D. Card (1982), ''Time Series Representations of Economic Variables and Alternative Models of the Labor Market,'' *Rev. Econ. Stud., 49,* 761-782.

Baram, Y. (1981), ''Realization and Reduction of Markovian Models from Non-Stationary Data,'' *IEEE Trans. Aut. Control, AC-26*, 1225-1231.

Bartlett, M. S. (1939), ''A Note on Tests of Significance in Multivariate Analysis,'' *Proc. Cambridge Phil. Soc. 35*, 180-185.

Bassevile, M., Benveniste, A., Moustakido, G., and Roujee, D. (1987), ''Detection and Diagnosis of Changes in the Eigenstructure of Nonstationary Multivariable Systems,'' *Automatica, 23,* 479-89.

Bassevile, M., Benveniste, (eds.) (1989), ''Detection of Abrupt Changes in Signals and Dynamical Systems,'' No. 77, Lecture Notes in *Control and Information Sciences,* Springer-Verlag.

Baumol, W. J. (1970), *Economic Theory and Operations Analysis*, 3rd ed., Prentice-Hall, New Jersey.

Bellman, R. E. (1957), *Dynamic Programming*, Princeton University Press, New Jersey.

__________ (1960), *Introduction to Matrix Analysis*, McGraw-Hill, New York.

Bernauke, B. (1986), ''Alternative Explanations of the Money-Income Correlation,'' Carnegie-Rochester Conf. Public Policy, 25, 49-100.

Beveridge, S., and C. R. Nelson (1981), ''A New Approach to Decomposition of Economic Time Series into Permanent and Transitory Components with Particular Attention to Measurement of the 'Business Cycle,' '' *J. Monetary Economics, 7*, 151-74.

Billingsley, P. (1971), *Weak Convergence of Measures*, Soc. Industrial and Appl. Math., Philadelphia.

Billingsley, P. (1961), "The Lindley-Levy Theorem for Martingales," *Proc. Am. Math Soc., 12,* 788-92.

__________ (1968), *Convergence of Probability Measures*, J. Wiley, New York.

Birkhoff, G., and S. MacLane (1970), *A Survey of Modern Algebra*, Macmillan, New York.

Blanchard, O. J. (1979), ''Backward and Forward Solution for Economics with Rational Expectations,'' *Am. Econ. Rev., 69*, 114-118.

Bosgra, O. H., A. J. J. Van der Weiden (1980), ''Input-Output Invariants for Linear Multivariable Systems,'' *IEEE Trans., AC-25*, 20-36.

Bossaerts, P. (1986) ''Three Essays on Spot and Forward Exchange Rates: Theory of Their Determination, the Estimation of a Markovian Model and an Investigation of the Effect of Government Intervention,'' Ph.D. Thesis, Graduate School of Management, UCLA.

Box, G. E. P., and G. M. Jenkins (1970), *Time Series Analysis, Forecasting and Control*, Holden-Day, San Francisco.

Box, G. E. P., and D. A. Pierce (1970), ''Distribution of Residual Autocorrelations in Autoregressive Integrated Moving Average Time Series Models,'' *J. Am. Stat. Assoc., 65*, 1509-1526.

Box, G. E. P., and G. C. Tiao (1977), "A Canonical Analysis of Multiple Time Series," *Biometrika, 64*, 355-365.

Brissimis, S. (1976), "Multiplier Effects for Higher Than First Order Linear Dynamic Econometric Models," *Econometrica, 44*, 593-595.

Bryson, A. E., and L. J. Henrikson (1968), "Estimation Using Sampled Data Containing Sequentially Correlated Noise," *J. Spacecraft and Rockets, 5*, 662-666.

Byrnes, C. I., and A. L. Lindquist (1982), "The Stability and Instability of Partial Realization," *Systems and Control Letters, 2*, 99-105.

Campbell, J. Y. (1985) "Stock Returns and the Term Structure," NBER Working Paper, No. 1626, Cambridge.

Canon, M. D., C. D. Culum, Jr. and E. Polak (1970), *Theory of Optimal Control and Mathematical Programming*, McGraw-Hill, New York.

Carew, B., and P. R. Belanger (1973), "Identification of Optimum Filter Steady-State Gain for System with Unknown Noise Covariance," *IEEE, AC-18*, 582-587.

Cerchi, M., and A. Havenner (1988) "Cointegration and Stock Prices," *J. Econ. Dyn. Control 12,* p. 333-346.

Chan, N. H. (1988) "The Parameter Inference for Nearly Nonstationary Time Series," *JASA 83*, 857-862.

____________, and C. Z. Wei (1988), "Asymptotic Inference for Nearly Nonstationary AR(1) Processes," *Ann. Stat. 15*, 1050-1063.

Chen, C. T. (1970), *Analysis and Synthesis of Linear Control Systems*, Holt, Rinehart & Winston, New York, Chapters 2, 4, and 8.

Chow, G. C. (1975), *Analysis and Control of Dynamic Economic Systems*, John Wiley, New York.

Clark, D. N. (1968), "On the Spectra of Bounded, Hermitian, Hankel Matrices," *Am. J. Math., 90*, 627-656.

Clarke, T. R. Jr., and W. L. Root (1958), *An Introduction to the Theory of Random Signals and Noise*, McGraw-Hill, New York.

Cochrane, J. H. (1987), "The Application of Unit Root Tests: A Critique," Manuscript (Univ. of Chicago, IL).

Cooper, D. M., and R. Thompson (1977), ''A Note on the Estimation of the Parameters of the Autorepressive-Moving Average Process,'' *Biometrika, 64*, 3, 625-628.

Cryer, J. D., and J. Ledolter (1981), '' Small-Sample Properties of the Maximum Likelihood Estimator in the First-Order Moving Average Model,'' *Biometrika, 68*, 691-694.

Davies, N., and P. Newbold (1979), ''Some Power Studies of a Portmanteau Test of Time Series Model Specification,'' *Biometrika, 66*, 153-155.

de Jong, L. S., (1978),"Numerical Aspects of Recursive Realization Algorithms," *SAIMJ. Control and Optimization, 16, 646-659.*

Deistler, M. (1985), ''General Structure and Parametrization of ARMA and Statespace System and its Relation to Statistical Problems'' in E. J. Hannan, P. R. Krishnaiah and M. M. Rao, eds., *Handbook of Statistics, Vol. 5*, pp. 257-277, Elsever Science Publishers B. V.

Deistler, M., W. Pliberger, and B. M. Rotscher (1982), ''Identifiability and Inference in ARMA Systems,'' in O. D. Anderson ed., *Time Series Analysis*: Theory and Practice 2, North-Holland, Amsterdam.

Denham, M. J. (1984), ''Canonical Forms for the Identification of Multivariable Systems,'' *IEEE Trans., AC-19*, 646-656.

Desai, U. B., and D. Pal (1982), ''A Realization Approach to Stochastic Model Reduction and Balanced Stochastic Realization,'' *Preprint IEEE Conf. on Decision and Control*, pp. 1105-1112.

Dent, W. (1977), "Computation of the Exact Likelihood Function of an ARIMA Process," J. Statisc. Comput. Simul., *5,* 193-206.

Dickinson, B. W. (1983), "Structure of Stationary Finite Observation Record of Discrete-Time Stochastic Linear Systems," *IEEE Trans. Aut. Control., AC-28,* 577-584.

____________ (1981), "Properties and Applications of Gaussian Autoexpressive Processes in Detection Theory," *IEEE Trans. Inf. Theory, IT-27,* 343-47.

Dunsmuir, W. and E. J. Hannan (1976), ''Vector Linear Time Series Models,'' *Adv. Appl. Prob. 8*, 339-364.

Dynkin, E. B., (1961), ''Necessary and Sufficient Statistics for a Family of Probability Distributions," *Selected Transl. in Math. Stat. Prob., 1*, 17-40.

Engle, R. F. (1982), ''Autoregressive Conditional Heteroscedasticity with Estimates of the Variance of United Kingdom Inflation,'' *Econometricka, 50*, 987-1007.

Engle, R. F., and C. W. J. Granger (1987), ''Cointegration and Error Correlation: Representation, Estimation, and Testing,'' *Econometrica, 55*, 251-276.

Evans, G. B. A. and C. Z. Wei (1988), ''Limiting Distributions of Least Squares Estimates of Unstable Autoregressive Processes,'' *Ann. Stat. 16*, 367-401.

____________, and N. E. Savin (1981), ''Testing for Unit Roots: 1,'' *Econometrics 49*, 753-779.

Fan, K. (1951), ''Maximum Properties and Inequalities for the Inequalities for the Eigenvalues of Completely Continuous Operators,'' *Proc. N.A.S., 37*, 760-766.

Faurre, P. L. (1976), ''Stochastic Realization Algorithms,'' in R. Mehra and D. Lainiotis, eds., q.v.

Faurre, P. L., M. Clerget, and F. Germain (1979), *Operateurs Rationels Positifs*, Dunod, Paris.

Findley, D. F. (1985), "Model Selection for Multi-step-ahead Forecasting," *Proc. IFAC Identification and System Diameter,* pp. 1039-1044, Yoh Uk.

Findley, D. F. (1984), "On Some Ambiguities Associated with Fitting of ARMA Models to Time Series," *J. Time Series Anal., 5,* 713-225.

Fleming, W. H., and R. W. Rishel (1975), *Deterministic and Stochastic Optimal Control*, Springer-Verlag, Berlin.

Forsythe, G. E., M. A. Malcolm, and C. B. Moler (1977), *Computer Methods for Mathematical Computations*, Prentice-Hall, Englewood Cliffs, NJ.

Fuller, W. (1976), *Introduction to Statistical Time Series*, Wiley, New York.

Futia, C. A. (1979), ''Stochastic Business Cycles,'' Bell Telephone Labs., Tech. Report, New Jersey.

Gandolfo, G. (1971), *Mathematical Methods and Models in Economic Dynamics*, North-Holland, Amsterdam.

Gantmacher, F. R. (1959), *The Theory of Matrices*, Vol. 1, Chelsea Pub. Co., New York.

Gardner, G., A. C. Harvey, and G. D. A. Phillips (1980), "An Algorithm for Exact Maximum Likelihood Estimation of Autoregressive-Moving Average Models by Means of Kalman Filtering," *Applied Statistics, 29*, 311-322.

Gel'fand, I. M., and A. M. Yaglom (1959), "Calculation of the Amount of Information about a Random Function Contained in Another Such Function," *Am. Math. Soc. Transl. Series 2, 12*, 191-198.

Georgion, T. T., and P. P. Khargoneka (1982), "On the Partial Realization Problem for Covariance Sequences," Proc. Conf. on Inform. Sci., Systems, p. 181, Princeton.

Gersch, W., and G. Kitagawa (1983), "The Prediction of Time Series with Trends and Seasonalities," *J. Business and Econ. Stat., 1*, 253-264.

Geweke, J. (1977), "Dynamic Factor Model," in D. J. Aigner and A. S. Goldberger, eds., *Latent Variables in Socio-Economic Models*, North-Holland, Amsterdam.

Gevers, M. R., and T. Kailath (1973), "An Innovation Approach to Least-Squares Estimation, Part IV: Discrete Time Innovation Representation and Recursive Estimation," *IEEE, AC-18*, 588-600.

Gilbert, E. G. (1963), "Controllability and Observability in Multivariable Control Systems," *SIAM J. Control, 1*, 128-151.

Glover K., (1984), "All Optimal Hankel-norm Approximations of Linear Multivariable Systems and Their L-Error Bounds," *Int'l J. Control, 39*, 1115-1193.

Glover, K., and J. C. Willems (1974), "Parameterizations of Linear Dynamical Systems: Canonical Forms and Identifiability," *IEEE, Trans. Aut. Control, AC-19*, 640-646.

Goldfeld, S. M., and R. Quandt (1976), "Studies in Nonlinear Estimation," Ballinger Publish. Co., Cambridge, MA.

Golub, G. H., and C. Reinsch (1970), "Singular Value Decomposition and Least Squares Solutions," *Numer. Mat., 14*, 403-420.

Golub, G. H. and C. R. van Loan (1983), *Matrix Computations*, Johns Hopkins U. P., Baltimore, MD.

Goodrich, R. L. and P. E. Caines (1979), ''Linear System Identification from Non-stationary Cross-Sectional Data,'' *IEEE Trans. Auto. Control, AC-24*, 403-411.

Goodwin, G. C. and R. L. Payne (1977), *Dynamic System Identification: Experiment Design and Data Analysis*, Academic Press, Inc., New York.

Gourieroux, C., and A. Monfort (1983), *Cours de Series Temporelles*, Economica, Paris.

Gourieroux, C., J. J. Laffont and A. Monfort (1982), ''Rational Expectations Models: Analysis of the Solutions,'' *Econometrica, 50*, 409-425.

Gourieroux, C., and Peancelle, I. (1988), "Detecting a Long Run Relationship," CEPREMAP, N-8902, Dec.

Granger, C. W. J., (1976), ''Some Properties of Time Series Data and Their Use in Econometric Model Specification,'' *J. Econometrics, 16*, 121-130.

____________ , (1983), ''Co-integrated Variables and Error-Correcting Models,'' Univ. Calif. San Diego, Discussion Paper 83-13.

Granger, C. W. J., and M. J. Morris (1976), ''Times Series Modeling and Interpretation,'' *Journal of Royal Statistical Society, Series A, 139*, 246-257.

Granger, C. W. J. and P. Newbold (1974), ''Spurious Regressions In Econometrics,'' *J. of Econometrics, 2*, 111,120.

____________ (1977), *Forecasting Economic Time Series*, Academic Press, New York.

Granger, C. W. J., and A. A. Weiss (1983), ''Time Series Analysis of Error Correction Models,'' in S Karlin, T. Amemiya, and L. A. Goodman, eds., *Studies in Economic Time Series and Multivariate Statistics*, Academic Press, New York.

Grenander, U., and G. Szego (1958), *Toeplitz Forms and Their Applications*, Univ. Calif. Press, Berkeley.

Guidorzi, R. P. (1975), "Canonical Structures in the Identification of Multivariable Systems," *Automatica, 11*, 361-374.

____________ (1981), "Invariants and Canonical Forms for Systems Structural and Parametric Identification," *Automatica, 17*, 117-133.

Hall, R. E. (1978), "Stochastic Implications of the Life Cycle-Permanent Income Hypothesis," *J. Pol. Econ., 6*, 197-988.

Hannan, E. J. and J. Rissanen (1982), "Recursive Estimation of Mixed Autoregressive Moving Average Order," *Biometrika, 69*, 81-94.

Hannan, E. J., and M. Deistler (1988), *Statistical Theory of Linear Systems*, Wiley, New York.

Hansen, L. P. and T. J. Sargent (1980), "Formulating and Estimating Dynamic Rational Expectations Models," *J. Econ. Dynamics and Conntrol, 2*, 7-46.

Hansen, L. P. and K. J. Singleton (1980), "Generalized Instrumental Variables Estimation of Nonlinear Rational Expectations Models," *Econometrica, 50*, 1269-1286.

Harshavardhana, P., E. A. Jonckheere, and L. M. Silverman (1984), "Stochastic Balancing and Approximation - Stability and Minimality," *IEEE Trans. Auto. Control, AD-29*, 744-746.

Harvey, A. C. (1981), *The Econometric Analysis of Time Series*, Philip Allan Pub., Oxford.

____________ (1984), "Trends and Cycles in Macroeconomic Time Series," London School of Econ., Working Paper, London, U. K.

Harvey, A. C., and G. Stein (1978), "Quadratic Weights for Asymptotic Regulator Properties," *IEEE, Trans. Auto. Control, AC-23*, 378-387.

Harvey, A. C., and P. H. J. Todd (1983), "Forecasting Economic Times Series with Structural and Box-Jenkins Models: A Case Study," *J. Business & Economic Statistics, 1*, 299-307.

Harvey, A. C., B. Henry, S. Peters, and S. Wren-Lewis (1985), "Stochastic Trends in Dynamic Regression Models: An Application to the Employment-Output Equation," Discussion Paper No. 83, LSE, London.

Hausman, J. A., and M. Watson, (1985), "Errors in Variables and Seasonal Adjustment Procedures," *JASA, 80*, 531-540.

Hautus, M. L. J. (1969), ''Controllability and Observability Conditions of Linear Autonomous Systems,'' *Ned. Akad. Wetenschappen Proc. Sec. A, 72*, 443-448.

Havenner, A. M., and M. Aoki (1988), ''An Instrumental Variables Interpretation of Linear Systems Theory Estimation,'' *J. Econ. Dyn. Control, 12*, 49-54.

Havenner, A. M., and K. Criddle (1989), ''System Theoretic Time Series: An Application to Inventories and Prices of California Range Cattle,'' *Computers and Mathematics with Applic., 17*, 1177-88.

Helson, Henry (1983), *Harmonic Analysis*, Addision Wesley, Reading, MA.

Hewer, G. A. (1971), ''An Iterative Technique for the Steady State Gains for the Discrete Optimal Regulator,'' *IEEE, Trans. Aut. Control, AC-26*, 382-383.

Hillman, S. C., and Tiao, G. C. (1982), "An ARIMA-Model Based Approach to Seasonal Adjustments," *TASA, 77*, 63-70.

Hodrick, R. J. and E. C. Prescott (1981), ''Post-war U.S. Business Cycle: An Empirical Investigation,'' Discussion Paper No. 451, Carnagie-Mellon University, Pittsburg, PA.

Hosoya, Y. (1977), ''On the Granger Condition for Non-Causality,'' *Econometrica, 45*, 1735-36.

Householder, A. S. (1964), *The Theory of Matrices in Numerical Analysis*, Blaisdell Pub. Co., New York.

Hsiao, C. (1979), ''Autoregressive Modeling of Canadian Money and Income Data,'' *JASA, 74*, 553-560.

Hylleberg, S., R. Engle, F., Granger, G. W. J., and Yoo, B. S. (1988), "Seasonal Integration and Cointegration," mimeo.

Ito, T. (1982), ''A Comparison of Japanese and U.S. Macroeconomic Behavior by a VAR Model,'' Discussion Paper No. 82, University of Minnesota, MN.

Jazwinski, A. H. (1970), *Stochastic Processes and Filtering Theory*, Academic Press, New York.

Johansen, S. (1988), "Statistical Analysis of Cointegrating Vectors," in M. Aoki (ed.), *Economic Time Series with Random Walk and Other Nonstationary Components*, Special Issue, *J. Econ. Dyn. Control 12*, 213-254.

Jonckheere, E. A. and J. W. Helton (1985), "Power Spectrum Reduction by Optimal Hankel Norm Approximation of the Phase of the Outer Spectral Factor," *IEEE Trans. Auto Control, AC-30*, 1192-1207.

Kabamba, P. T. (1985), "Balanced Forms: Canonicity and Parameterization," *IEEE Trans. Aut. Control., AC-30, 1106-1109.*

Kailath, T. (1980), Linear System Theory, Prentice-Hall, New Jersey.

Kalman, R. E. (1960), "A Contribution to the Theory of Optimal Control," *Bol. Socied. Mat. Mexicana, 5*, 102-119.

Kamien, M. F. and N. L. Schwartz (1981), *Dynamic Optimization*, North-Holland, Amsterdam.

Kendall, M. G. (1976), *Time Series*, 2nd ed., Charles Griffin, London, Great Britain.

Kendall, M. G., and A. Stuart (1966), *The Advanced Theory of Statistics*, Charles Griffin, London, England.

Kimura, H. (1982), *Digital Signal Processing and Control* (Japanese), Shokodo, Tokyo.

____________ (1986), "Positive Partial Realization of Covariance Sequences," in C. I. Byrnes and A. Linquist, eds., *Modelling, Identification and Robust Control*, North-Holland, Amsterdam.

Kitagawa, G. (1981), "A Nonstationary Time Series Model and Its Fitting by a Recursive Filter," *J. Time Series Analysis, 2*, 103-116.

Kitagawa, G. (1987), "Non-Gaussian State Space Modeling of Nonstationary Time Series," JASA, *82,* 1032-63.

____________ (1983), "Changing Spectrum Estimation," *J. Sound and Vibration, 89*, 433-445.

Kitagawa, G. and W. Gersch (1984), "A Smoothness Prior-State Space Modeling of Time Series with Trends and Seasonalities," *JASA, 79*, 378-389.

___________ (1985), "A Smoothness Prior Time-Varying AR Coefficient Modeling of Nonstationary Covariance Time Series," *IEEE Trans. Aut. Control, AC-34*, 48-56.

Kshirsagar, A. M. (1972), *Multivariate Analysis*, Marcel Dekker, New York.

Kucera, V. (1972), "The Discrete Riccati Equation of Optimal Control," *Kybernetika (Prague), 8*, 430-447.

Kung, S. (1978), "A New Identification and Model Reduction Algorithm via Singular Value Decompositions," Proc. 12th Annual Asilomar Conf. Circuits and Syst., Asilomar, Ca.

Kung, S. and D. W. Lin (1981), "Optimal Hankel-Norm Model Reductions: Multivariable Systems," *IEEE Trans. Aut. Control, AC-26*, 832-852.

Kwakernaak, H., and R. Sivan (1972), *Linear Optimal Control Systems*, John Wiley, New York.

Kydland, F. and E. C. Prescott (1981), "Time to Build and Aggregate Fluctuations," Carnegie Mellon Working Paper 28-80-81, Pittsburgh, PA.

Lai, T. S., and C. Z. Wei (1985), "Asymptotic Properties of Multivariate Weighted Sums with Applications to Stochastic Regression in Linear Dynamic Systems," *Multivariate Analysis VI*, P. R. Krishnaiah, p. 375-393, Academic Press, New York.

Laning, J. H. Jr. and R. H. Battin (1956), *Random Processes in Automatic Control*, McGraw Hill, New York.

Laub, A. J. (1983), "Numerical Aspects of Solving Algebraic Riccati Equations," *Proc. IEEE Conf. Decision and Control*, 184-186.

Lawley, D. N. (1959), "Test of Significance in Canonical Analysis," *Biometrika, 41*, 59-66.

Lee, E. P. and L. Markus (1967), *Foundations of Optimal Control Theory*, John Wiley, New York.

Li, K-C, (1988), "Sliced Inverse Regression for Dimension Reduction," UCLA Statistics Series 3.

Linquist, A., and G. Picci (1979), "On the Stochastic Realization Problem," *SIAM J. on Control and Optimization, 17*, 365-389.

Litterman, R. B. and T. J. Sargent (1979), ''Detecting Neutral Price Level Changes and Effects of Aggregate Demand with Index Models,'' Research Department Working Paper 125, Fed. Res. Band of Minneapolis, MN.

Ljung, L. (1987), *System Identification*, Prentice Hall, Englewood Cliffs, New Jersey.

Ljung, G. M., and G. E. P. Box (1978), ''On a Measure of Lack of Fit in Time Series Models,'' *Biometrika, 65*, 297-303.

Long, J. B., Jr., and C. I. Plosser (1983), ''Real Business Cycle,'' *J. Pol. Econ., 91*, 39-69.

Lucas, R. E., Jr. (1975), ''An Equilibrium Model of the Business Cycle,'' *J. Pol. Econ., 83*, 1113-1144.

Luenberger, D. G. (1967), ''Canonical Forms for Linear Multivariable Systems,'' *IEEE Trans. Auto. Control, AC-12*, 290-293.

__________ (1979), *Introduction to Dynamic Systems*, John Wiley, New York.

Mandel, J. (1982), ''Use of the Singular Value Decomposition in Regression Analysis, American Statistician, 36, 15-24.

Maravall, A. (1987), "A Note on the Dynamic Structure of a Seasonal Component," Mimeo, Bank of Spain, May.

Mehra, R. K., and D. G. Lainiotis (1976), *System Identification: Advances and Case Studies,* Academic Press, NJ.

Michaletzky, G. (1989), ''Balanced Stochastic Realizations and Canonical Correlations,'' Tech. Report No. 85, UC Santa Barbara, April.

Molinari, B. P. (1975), ''The Stabilizing Solution of the Discrete Algebraic Riccati Equation,'' *IEEE Trans. Aut. Control, AC-20*, 396-399.

Moore, B. C. (1976), ''On the Flexibility Offered by State Feedback in Multivariable Systems Beyond Closed Loop Eigenvalue Assignment,'' *IEEE Trans. Aut. Control, AC-21*, 689-692.

__________ (1978), ''Singular Value Analysis in Linear Systems, Part I and II,'' *Proc. 1978 IEEE Conference on Decision and Control, 66-73*.

____________ *(1981), "Principal Component Analysis in Linear Systems: Controllability, Observability, and Model Reduction," IEEE Trans. Aut. Control, AC-26*, 17-32.

Moore, J. B. and B. D. O. Anderson (1968), "Extensions of Quadratic Minimozation Theory," I: Finite Time Results, II: Infinite Time Results, *International J. Control, 7*, 465-480.

Moustakides, G., and Benveniste, A. (1986), "Detecting Changes in the AR Parameters of a Nonstationary ARMA Process," *Stochastics, 16,* 137-155.

Nehari, Z. (1957), "On Bounded Bilinear Forms," *Ann. Math., 65*, 153-162.

Nelson, Charles R., and Charles I. Plosser (1982), "Trends and Random Walks in Macroeconomic Time Series," *J. of Monetary Economics, 10*, 139-162.

Nerlove, M. D., M. Grether, and J. C. Carvalho (1979), "Analysis of Economic Time Series: A Synthesis," Academic Press, New York.

Newbold, P. (1974), "The Exact Likelihood Function for a Mixed Autoregressive-Moving Average Process," *Bionetiks, 61,* 423-26.

Newton, G. C., L. A. Gould, and J. F. Kaiser (1957), *Analytical Design of Linear Feedback Controls*, John Wiley, New York.

Otter, P. W. (1985), *Dynamic Feature Space Modeling Filtering and Self-Tuning Control of Stochastic Systems*, Springer-Verlag, Heidelberg.

Pagano, M. (1976), "On the Linear Convergence of a Covariance Factorization Algorithm," *J. Asso. Comp. Math., 23*, 310-316.

Parzen, E. (1962), *Stochastic Processes*, Holden-Day, San Francisco, CA.

Pavon, M. (1980), "Stochastic Realization and Invariance Directions of the Matrix Riccati Equations," *J. Control and Optimization, 18*, 155-180.

Pena, D., and G. E. P. Box (1987), "Identifying a Simplifying Structure in Time Series," *JASA, 82*, 836-843.

Pernebo, L., and L. M. Silverman (1982), "Model Reduction via Balanced State Representations," *IEEE, AC-27*, 382-387.

Perron, P. (1986), ''Trends and Random Walks in Macroeconomic Time Series: Further Evidence from a New Approach,'' Working Paper 8650, Department of Economics, University of Montreal, Montreal, Oct, also in Aoki (ed) Special Issue, *J. Econ. Dyn. Control, 12*, Opt. Cir.

Phadke, M. S., and Kedem, G. (1978), "Computation of the Exact Likelihood Function of Multivariate Moving Average Models," *Biometiks, 65,* 511-9.

Phillips, P. C. B. (1987), ''Towards a Unified Asymptotic Theory for Autoregression,'' *Biometrika, 74*, p. 535-547.

____________ (1987), ''Time Series Regression with a Unit Root,'' *Econometrics 55*, p. 277-301.

____________ and S. N. Durlauf (1986), ''Multiple Time Series Regression with Integrated Processes,'' *Rev. Econ. Stud. 54*, p. 439-485.

____________ and B. E. Hansen (1990), ''Statistical Inference in Instrumental Variables Regression with I(1) Processes,'' *Rev. Econ. Stud. 57*, 99-125

____________ and P. Perron (1986), ''Testing for a Unit Root in Time Series Regression,'' Cowles Foundation Discussion Paper No. 795, Yale Univ., July.

Picci, G. (1982), ''Some Numerical Aspects of Multivariable Systems Identification,'' *Math. Progr. Study, 18*, 76-101.

Picci, G., and A. Linquist (1986), ''Dynamic Factor Models,'' Conf. on Optimization and Control, June, Paris.

Plosser, C. I. (1979), "The Analysis of Seasonal Economic Models," *J. Econometrica, 10,* 147-173, 199.

Pollard, D. (1984), *Convergence of Stochastic Processes,*, Springer-Verlag, New York.

Porat, B., and Friedlander, B. (1985), "Asymptotic Accuracy of ARMA Parameter Estimation Methods Based on Sample Covariances," Proc. IFAC Identification and System Parameter Estimation, York, U.K.

Poskitt, D. S. "Precision, Complexity and Bayesian Model Detection," *J. R. Stat. Soc. Sec. B, 45*, 199-208, 1987.

Priel, B., and U. Shaked (1983), ''Cheap Optimal Control of Discrete Single Input Single Output Systems,'' unpublished mimeo, Tel-Aviv University, Tel-Aviv.

Quah, D. (1988) ''The Relative Importance of Permanent and Transitory Components: Identification and Some Theoretical Bounds,'' presented at NSF-NBER Time Series Workshop, U. Chicago.

Rao, C. R. (1979), ''Separation Theorems for Singular Values of Matrices and Their Applications in Multivariate Analysis,'' *J. Multiv. Anal*, *9*, 362-77.

____________ (1965), *Linear Statistical Inference and Its Applications*, J. Wiley, New York.

____________ (1964), ''The Use and Interpretation of Principal Component Analysis in Applied Research,'' *Sankya Series A, 26*, 329-358.

Renyi, A. (1959), ''On Measures of Dependence,'' Acta. Mat. Acad. Sci. Hung., 10, 441-451.

Rissanen, J. (1971), "Recursive Identification of Linear Systems," *SIAM J. Control, 9*, 420-430.

____________ (1974), ''Basis of Invariants and Canonical Forms for Linear Dynamical Systems,'' *Automatica, 10*, 175-182.

____________ (1976), ''Minimax Entropy Estimation of Models for Vector Process,'' in R. Mehra and D. Lainiotis (eds), q.v.

Robinson, E. A., and S. Treitel (1980), *Geophysical Signal Analysis*, Prentice Hall, New Jersey.

Rosenblatt, M. (1978), ''Dependence and Asymptotic Independence for Random Processes,'' in Rosenblatt, ed., *Studies in Probability Theory*, Math Assoc. Am., RI.

Rosenbrock, H. H. (1970), *State Space and Multivariable Theory*, Thomas Nelson and Sons, Ltd., London.

Rothenberg, T. J. (1971), ''Identification in Parametric Models,'' *Econometrica, 39*, 577-591.

Sargan, D. (1964), ''Wages and Prices in the United Kingdom: A Study in Econometric Methodology,'' in Hart et al. (eds), *Econometric Analysis for National Economic Planning*, Butterworth, London.

____________ (1988), *Lectures on Advanced Econometric Theory*, Basil Blackwell, Oxford, U.K.

Sargent, T. J. (1979), *Macroeconomic Theory*, Academic Press, New York.

____________ (1981), ''Interpreting Economic Time Series,'' *J. of Pol. Econ., 89*, 213-248.

Schwarz, G. (1978), ''Estimating the Dimension of a Model,'' *Annals of Statistics, 6*, 461-464.

Schweppe, F. C. (1965), ''Evaluation of Likelihood Functions for Gaussian Signals,'' *IEEE Trans. Inf. Theory, 11*, 61-70.

Schwert, G. W. (1988) ''Test for Unit Roots: A Monte Carlo Investigation,'' *J. Bus. Econ. Stat.* (forthcoming)

Shaked, U. (1979), ''A Transfer Function Approach to the Linear Discrete Stationary Filtering and Steady State Optimal Control Problems,'' *International Journal of Control, 29*, 279-291.

Shiskin, J. and T. J. Plewes (1978), ''Seasonal Adjustment of the U.S. Unemployment Rate,'' *Statistician, 27*, 181-202.

Shumway, R., H., and Stoffer, D. S. (Jan. 1990), "Dynamic Linear Models with Switching," TR191, Div. Statistics, UC Davis.

Silverman, L., and Bettayeb, M. (1980), ''Optimal Approximation of Linear Systems,'' Preprint Joint Automatic Control Conference, FA8-A.

____________ (1982), ''Policy Analysis with Econometric Models,'' *Brookings Papers on Econometric Activity, No. 1*, 107-152.

Sims, C. A. (1980), ''Comparison of Interwar and Postwar Business Cycles: Monetarism Reconsidered,'' *Am. Econ. Rev., 70*, 250-257.

____________ (1986), ''Are Forecasting Models Usable for Policy Analysis?'' *Quarterly Rev. Fed. Res. Bank of Minneapolis*, Winter 2-16.

Solo, Victor (1986), *Topics in Advanced Time Series Analysis*, Lecture Notes, Springer-Verlag, New York.

Spall, J. C. ed. (1988), *Bayesian Analysis of Time Series and Dynamic Models*, Marcel Dekker, New York.

Stock, J. H., and M. W. Watson (1988), ''Variable Trends in Economic Time Series,'' *J. Econ. Perspectives* 2, 147-174.

Swamy, P. A. V. B. and P. A. Tinsley (1980), "Linear Prediction and Estimation Methods for Regression Models with Stationary Stochastic Coefficients," *J. Econometrics, 12*, 103-145.

Takeuchi, K. (1983), "On Statistical Model Selection Based on AIC" (in Japanese), *J. Soc. Instrument & Control Engineers, 22*, 445-453.

Tiao, G. C. and G. E. P. Box (1981), "Modeling Multiple Time Series with Applications," *JASA, 76*, 802-816.

Tong, H. (1983), *Threshold Models in Non-Linear Time Series Analysis*, Lecture Notes in Statistics, No. 21, Springer-Verlag, New York.

Turumi, H. (1988), "Survey of Bayesian and Non-Bayesian Testing of Model Stability in Econometrica," in J.C. Spall (ed.), *Bayesian Analysis of Time Series and Dynamic Models,* Marcel Dekker, New York.

Van Zee, G. A. (1981), *System Identification for Multivariable Control*, Delft University Press, Delft.

Vaughn, D. R. (1970), "A Nonrecursive Algebraic Solution for the Discrete Riccati Equation," *IEEE Trans. Aut. Control, AC-15*, 597-599.

Wertz, V. (1981), "Structure Selection for the Identification of Multivariable Processes," Catholic University of Leuven, Ph. D. Thesis in Applied Science, Leuven.

Wertz, V., M. Gevers, and E. J. Hannan (1982), "The Determination of Optimum Structures for the State Space Representation of Multivariate Stochastic Processes," *IEEE Trans. Aut. Control, AC-27*, 1200-1210.

Whittaker, E. T. (1923), "On a New Method of Graduation," *Proc. Edinburgh Math. Soc., 41*, 57-63.

Whittle, P. (1963), *Prediction and Regulation by Linear Least-Square Methods*, Van Nostrand, Princeton, New Jersey.

——————— (1982), *Optimization Over Time*, Vol. 1, John Wiley, New York.

Wilsky, A. S., and Jones, M. L. (1976), "A Generalized Likelihood Ratio Approach to the Detection and Estimation of Jumps in Linear Systems," IEEE Trans. A. C. AC-21, 108-112.

Yoshikawa, H. (1990), "On the Equilibrium Yen Dollar Rate," *Am. Econ. Rev.*, forthcoming.

Zellner, A. (1984), *Basic Issues in Econometrics*, U. Chicago Press, Chicago.

Subject Index